FOUNDATION OF STATISTICAL ENERGY ANALYSIS IN VIBROACOUSTICS

Foundation of Statistical Energy Analysis in Vibroacoustics

A. Le Bot

CNRS—Tribology and System Dynamics Laboratory,
École centrale de Lyon, Ecully, France

OXFORD
UNIVERSITY PRESS

Great Clarendon Street, Oxford, OX2 6DP,
United Kingdom

Oxford University Press is a department of the University of Oxford.
It furthers the University's objective of excellence in research, scholarship,
and education by publishing worldwide. Oxford is a registered trade mark of
Oxford University Press in the UK and in certain other countries

First Edition published in 2015
Impression: 1

Published in the United States of America by Oxford University Press
198 Madison Avenue, New York, NY 10016, United States of America

British Library Cataloguing in Publication Data
Data available

Library of Congress Control Number: 2015930601

ISBN 978–0–19–872923–5

Printed in Great Britain by
Clays Ltd, St Ives plc

To my children Maé, Milo, Léna, Marilou, and Jules, and my wife Diana

Foreword

Statistical energy analysis was developed using certain results from fluctuation theory, which is applied to linear transducers to calculate their thermal noise spectra. The principal results from that model of interest to us here are:

1. the thermal bath that the system is immersed in acts as a combination of an electrical resistance (a damper) in series with a white noise generator; and
2. each resonance of the transducer gains an amount of energy $k_B T$, where T is the absolute temperature and k_B is Boltzmann's constant.

Condition (1) and our intuition suggest that a system in contact with a thermal bath at temperature T_1 will interact with a second system at temperature T_2 by an energy flow proportional to $\beta(T_1 - T_2)$. Condition (2) indicates that the difference in energies is equivalent to a proportionate difference in temperatures. The requirement that the energy flow from the higher temperature to lower temperature bath requires that $\beta > 0$.

In this carefully crafted and referenced text, Alain Le Bot shows by direct calculation how these two requirements, which are the foundations of statistical energy analysis, are calculated for systems that are modelled as groups or assemblages of resonators excited by white noise sources. The quantity β is positive-definite and proportional to a 'coupling loss factor', which in turn is related to standard acoustics quantities such as transmissibility factors and junction immittances. The analysis and application of statistical energy analysis draw in a number of system descriptions such as modal density and the coupling loss factor that suggest still other descriptors, such as modal spacing, excitation force statistics, and Monte Carlo models. These components of the analytic foundation of statistical energy analysis make it a rich and varied subject for acoustic analysis.

I was once asked if I thought the estimation of energy flow in complex systems by statistical energy analysis was worthy of the Nobel Prize. I responded 'I don't think so', since most of the statistical energy analysis results were intuitive. If the energy were found to flow from lower to higher modal energy, then that might be worthy of the Nobel Prize.

Belmont, Massachusetts
August, 2014

Dick Lyon

Preface

It is more than fifty years since the idea of applying statistical physics methods and results to vibroacoustic problems was established. At the heart of this development by early contributors, and in particular R.H. Lyon, was in essence thermodynamics with modes playing the role of atoms. Since that time, the idea has developed further over the years through publications and debates—sometimes impassioned ones—and a consistent theoretical corpus has been set which is now referred to as *statistical energy analysis*. This, together with reverberation theory in room acoustics, has established statistical vibroacoustics in its own right alongside statistical mechanics, statistical physics, and statistical electromagnetism. The purpose of this textbook is to provide a review of the knowledge in this field and to demonstrate that statistical energy analysis is built on well-founded principles.

Practitioners of statistical energy analysis often encounter difficulties in assessing which cases the theory is applicable to and in appreciating its usefulness for a particular system. The large number of assumptions, often misunderstood, may severely restrict applications. This is why great care has been taken in highlighting the hypothesis throughout the text and in explaining its role and scope in the more important results.

This text is deliberately limited to a strict version of the theory. It is often the case that the employment of more restrictive assumptions ensures that results gain in rigour. Although they are sometimes mentioned, I have also avoided describing the numerous extensions of the theory. Readers interested in the subject may consult the selected references.

This textbook is intended for all people who are interested in statistical energy analysis, including students or teachers who attend or prepare a course in the subject—always more numerous in various universities around the world; engineers looking for the key elements of a discipline they practice; and researchers who have made statistical energy analysis their favourite area. They will find a self-contained and up-to-date textbook that does not require any specific knowledge. My respectful belief is that this book will contribute to disseminating the ideas of statistical energy analysis beyond a narrow set of specialists and help provide recognition for this field, which goes hand in hand with the philosophy of emergentism in physics.

Lyon, France
June, 2014

A. Le Bot

Acknowledgements

I wish to express my thanks to my colleagues F. Sidoroff, N. Totaro, T. Lafont, V. Gouret, and V. Cotoni for the many helpful suggestions patiently given during the preparation of this text. I also wish to acknowledge D. Martin de Argenta and D. Mawer for their help in the writing of the manuscript, with a special thought for the latter.

Contents

1

Random Vibration

In statistical vibroacoustics, sources are random. The random nature of sources with additional assumptions such as stationarity and uncorrelation is essential for apparition of simple features in statistical systems, as will be seen in forthcoming chapters. And for some well-chosen problems, a complete solution using analytical methods becomes possible.

In the simplest case, systems remain deterministic. Consequently, the typical problem that will be outlined in the first chapters is that of predicting the response of a deterministic mechanical system submitted to random forces. Of course, the system response is also a random signal. And since a random signal is characterized by its statistical moments, the fundamental question to resolve is the determination of the statistical moments of linear system output knowing the statistical moments of input.

In this chapter, we introduce the terminology of random processes and definition of their properties—stationarity, correlation function, and power spectral density. Basic results on time-invariant linear systems with stationary random input are also introduced. Many omitted results and proofs may be found in standard texts (Papoulis, 1965; Newland, 1975; Peebles, 1987; Sabot, 2000).

1.1 Random Variable

A random variable is the simplest mathematical object to describe a probabilistic situation. A **random variable** X is a function of the sample space into real numbers. Selecting an element of the sample space can be viewed as a probabilistic experiment consisting, for instance, in throwing a dice or tossing a coin. Each value taken by X is an outcome of the random variable. In the simplest case, this is a finite set, but generally we shall consider random variables for which the set of outcomes is an interval of real numbers or even all real numbers.

For instance, when playing dice, the sample space is the set of the six faces $\{f_A, f_B, f_C, f_D, f_E, f_F\}$ the random variable maps the set of faces to $\{1, 2, 3, 4, 5, 6\}$ (preparing a dice corresponds to writing numbers on faces), and the outcomes are the successive numbers 1, 5, 3, 6, 1, ... that appear during the game. An experiment (throwing the dice) is simply a physical process (rolling the dice) by which a face is selected. The random variable X is the function that maps the faces to their numbers,

Foundation of Statistical Energy Analysis in Vibroacoustics. First Edition. A. Le Bot.

$$X : \{f_A, f_B, f_C, f_D, f_E, f_F\} \rightarrow \{1, 2, 3, 4, 5, 6\} \tag{1.1}$$

Generally, the sample space is never explicitly given. The only useful information is the probabilistic properties of random variables.

Probability density function

A random variable is fully determined by its cumulative distribution function, $x \mapsto \mathrm{pr}(X \leq x)$ which gives the probability that the outcome is less than x. This is the probability of the event $X \leq x$. A random variable is said to be continuous if its cumulative distribution function is derivable. The derivative of the cumulative distribution function is called the **probability density function** (pdf),

$$p(x) = \frac{\mathrm{d}}{\mathrm{d}x}\mathrm{pr}(X \leq x) \tag{1.2}$$

All the probabilistic information on a random variable is contained in its probability density function.

Expectation

The **mean value** or **expectation** $\langle X \rangle$ is the first-order moment. It is defined as

$$\langle X \rangle = \int_{-\infty}^{\infty} x p(x)\,\mathrm{d}x \tag{1.3}$$

A random variable with zero expectation is called zero-mean random variable. The random variable $X - \langle X \rangle$ is always a zero-mean random variable.

Moments

The **moment of order** n is the expectation of the random variable X^n,

$$\langle X^n \rangle = \int_{-\infty}^{\infty} x^n p(x)\,\mathrm{d}x \tag{1.4}$$

The existence of moments of any order is usually not ensured for all random variables, but in practice we shall always assume the existence of moments up to order two.

Variance

The **variance** v_X of a random variable is the second moment of the variable $X - \langle X \rangle$. In terms of probability density function, it reads

$$v_X = \int_{-\infty}^{\infty} x^2 p(x)\,\mathrm{d}x - \langle X \rangle^2 \tag{1.5}$$

The **standard deviation** is defined as $\sigma_X = \sqrt{v_X}$.

Joint probability density function

In the case of two random variables X and Y, the knowledge of individual probability density functions is generally not sufficient to fully characterize their simultaneous probabilistic behaviour. The joint probability distribution function is defined as the map $(x, y) \mapsto \mathrm{pr}(X \leq x \text{ and } Y \leq y)$ that is the probability of the event $X \leq x$ and $Y \leq y$. The **joint probability density function** is the second derivative of the joint distribution function wherever it exists,

$$p(x, y) = \frac{\partial}{\partial x \partial y} \mathrm{pr}(X \leq x \text{ and } Y \leq y) \tag{1.6}$$

Statistical independence

When the events $X \leq x$ and $Y \leq y$ are independent, in the sense that the occurrence of x as outcome of X does not influence the occurrence of any y for Y, the probability of the joint event $X \leq x$ and $Y \leq y$ is simply the product

$$\mathrm{pr}(X \leq x \text{ and } Y \leq y) = \mathrm{pr}(X \leq x)\,\mathrm{pr}(Y \leq y) \tag{1.7}$$

In this case, we shall say that X and Y are **statistically independent**. As an example, we can consider the throwing of two dice, X and Y being the individual results. The result of the first dice has no influence on the result of the second dice.

For statistically independent random variables X and Y, the marginal probability density functions $p_X(x)$ and $p_Y(y)$ are sufficient to predict the joint probabilistic behaviour. By deriving two times eqn (1.7), we get

$$p(x, y) = p_X(x) p_Y(y) \tag{1.8}$$

The joint probability density function is simply the product of the marginal probability density functions.

Joint moments

The **joint moments** are the expectation of the random variable $X^n Y^p$,

$$\langle X^n Y^p \rangle = \int_{-\infty}^{\infty} \int_{-\infty}^{\infty} x^n y^p p(x, y) \,\mathrm{d}x \mathrm{d}y \tag{1.9}$$

We say that the joint moment $\langle X^n Y^p \rangle$ is of order $n + p$. The special cases $p = 0$ and $n = 0$ respectively give the moments of X and Y.

When X and Y are statistically independent, the moments $\langle X^n Y^p \rangle$ can be written as

$$\langle X^n Y^p \rangle = \langle X^n \rangle \langle Y^p \rangle \tag{1.10}$$

This result is obtained by substituting eqn (1.8) into eqn (1.9) and by applying Fubini's theorem (Rudin, 1986, p. 164) to separate the integral into two integrals.

Correlation

The **correlation** of X and Y is the moment $\langle XY \rangle$ and is denoted by R_{XY},

$$R_{XY} = \langle XY \rangle \tag{1.11}$$

If correlation can be written as the product $R_{XY} = \langle X \rangle \langle Y \rangle$, then X and Y are said to be **uncorrelated**.

Statistical independence implies uncorrelation as a special case but the converse is not true.

Statistical moments

Except in some idealized situations, the exact knowledge of a probability density function $p(x)$ cannot be reached by any experiment. The only data available to an experimentalist are the successive outcomes $X_0, X_1, X_2, \ldots$ of the random variable by repeating experiments a large number of times. It is then possible to assess the **statistical moments** from the outcomes of the random variable X with

$$\langle X^n \rangle = \frac{1}{N} \sum_{i=0}^{N} X_i^n \tag{1.12}$$

The larger the N, the more accurate the assessment of moments.

When two random variables are considered, one must observe simultaneously the successive outcomes $X_0, Y_0, X_1, Y_1, X_2, Y_2, \ldots$. The **joint statistical moments** are then

$$\langle X^n Y^p \rangle = \frac{1}{N} \sum_{i=0}^{N} X_i^n Y_i^p \tag{1.13}$$

We must admit that, from a practical point of view, the statistical moments contain all the information available on random variables. The attribution of a probability density function $p(x)$ to a particular probabilistic situation is rather a choice or an assumption. The validity of such a model must be established by observation of experimental results, that is to say the statistical moments.

1.2 Random Process

A **random function**, also called **random process** or **stochastic process**, is a function that maps a subset of real numbers to the set of random variables, $t \mapsto X(t)$. The subset of real numbers is usually an interval, finite or infinite. The variable t is called time.

At a fixed time t, $X(t)$ is a random variable whose outcome is noted $x(t)$. If we do an experiment and consider all outcomes for all times t, we obtain a function $t \mapsto x(t)$

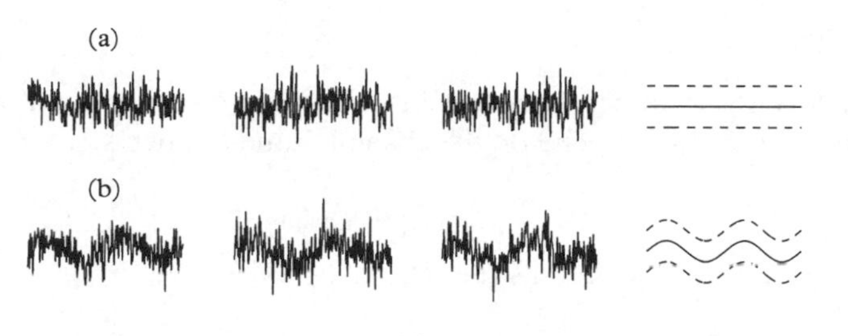

Figure 1.1 *Sample functions of a random process and time evolution of mean (solid line) and standard deviation (broken line). (a) stationary random process; (b) non-stationary random process.*

called **sample function** or **realization** of the random process X (see Fig. 1.1). Thus, a random process can also be viewed as a function of the sample space into the space of real-valued functions. This is just another point of view of what a random process is.

Expectation, variance

At any time t, the random variable $X(t)$ has expectation $\langle X(t)\rangle$ and variance $\langle X^2(t)\rangle - \langle X(t)\rangle^2$. We call expectation and variance of a random process X and we note $\langle X\rangle$ and v_X the deterministic functions,

$$\langle X\rangle : t \mapsto \langle X(t)\rangle \tag{1.14}$$

$$v_X : t \mapsto \langle X^2(t)\rangle - \langle X(t)\rangle^2 \tag{1.15}$$

A zero-mean random process is a random process for which $\langle X(t)\rangle = 0$ at any time. The variance of a zero-mean random process reduces to $t \mapsto \langle X^2(t)\rangle$.

Auto-correlation function

Since $X(t)$ and $X(t')$ are different random variables, it is of interest to study their cross properties. The **auto-correlation function** of a random process X is defined as the expectation of the product $X(t)X(t+\tau)$,

$$R_{XX}(t,\tau) = \langle X(t)X(t+\tau)\rangle \tag{1.16}$$

The auto-correlation function is a deterministic function of two variables, the absolute time t and time delay τ. The auto-correlation at $\tau = 0$ is the second-order moment of $X(t)$,

$$R_{XX}(t,0) = \langle X(t)^2\rangle \tag{1.17}$$

Stationarity

A random process is said to be **wide-sense stationary to order two** if its expectation and auto-correlation do not depend on the absolute time t. That is,

$$\langle X(t) \rangle = \text{constant} \tag{1.18}$$

$$\langle X(t)X(t+\tau) \rangle = R_{XX}(\tau) \tag{1.19}$$

In particular, the auto-correlation function only depends on the time delay τ but not on the absolute time t. We shall say more simply stationary random process for wide-sense stationary to order two. Since the expectation $\langle X \rangle$ is constant, it is convenient to redefine the random process by subtracting its mean value $X(t) - \langle X \rangle$. Thus redefined (and we shall always do it without specifying it), the mean value of a stationary random process X is zero $\langle X(t) \rangle = 0$ at any time, and the variance is the moment of order two $v_{X(t)} = \langle X(t)^2 \rangle$.

The auto-correlation of a stationary random process is an even function,

$$R_{XX}(-\tau) = R_{XX}(\tau) \tag{1.20}$$

This result comes from a direct calculation,

$$\begin{aligned} R_{XX}(-\tau) &= \langle X(t)X(t-\tau) \rangle \\ &= \langle X(t-\tau)X(t) \rangle \\ &= R_{XX}(\tau) \end{aligned}$$

The last equality stems from application of stationarity.

Power spectral density

For any sample function $x(t)$ of a random process X, we can introduce the truncated function $x_T(t)$,

$$x_T(t) = \begin{cases} x(t) & \text{if } t \in [-T, T] \\ 0 & \text{otherwise} \end{cases} \tag{1.21}$$

The support of a truncated function x_T is bounded. If a sample function is locally integrable, continuous for instance, its truncated function x_T has a Fourier transform. We introduce the Fourier pair,

$$\widehat{x_T}(\omega) = \int_{-T}^{T} x(t) \mathrm{e}^{-\imath\omega t}\, \mathrm{d}t \tag{1.22}$$

$$x_T(t) = \frac{1}{2\pi} \int_{-\infty}^{\infty} \widehat{x_T}(\omega) \mathrm{e}^{\imath\omega t}\, \mathrm{d}\omega \tag{1.23}$$

The Fourier transform $\widehat{x_T}(\omega)$ is a complex function in which the square norm is noted $|\widehat{x_T}|^2(\omega)$. We define the random process $|\widehat{X_T}|^2$ as the random process for which the sample functions are $\omega \mapsto |\widehat{x_T}|^2$. Then, the **power spectral density** of a random process X is defined by

$$S_{XX}(\omega) = \lim_{T\to\infty} \frac{\langle|\widehat{X_T}|^2(\omega)\rangle}{2T} \tag{1.24}$$

The Wiener–Khinchin theorem states that for stationary processes, the auto-correlation function R_{XX} and power spectral density S_{XX} form a Fourier pair.

$$S_{XX}(\omega) = \int_{-\infty}^{\infty} R_{XX}(\tau)\mathrm{e}^{-\imath\omega\tau}\,\mathrm{d}\tau \tag{1.25}$$

$$R_{XX}(\tau) = \frac{1}{2\pi}\int_{-\infty}^{\infty} S_{XX}(\omega)\mathrm{e}^{\imath\omega\tau}\,\mathrm{d}\omega \tag{1.26}$$

Since R_{XX} is an even function, S_{XX} is real-valued. This result can be proved with the change of variable $\tau \mapsto -\tau$ in eqn (1.25) and the parity of R_{XX}.

In practice, the Wiener–Khinchin theorem will be more useful than definition (1.24) of the power spectral density.

The second moment of a stationary random process is, according to eqn (1.17), the correlation function at $\tau = 0$. Taken at $\tau = 0$, the second Wiener–Khinchin relationship reads

$$\langle X^2\rangle = \frac{1}{2\pi}\int_{-\infty}^{\infty} S_{XX}(\omega)\,\mathrm{d}\omega \tag{1.27}$$

The above means that the energy of signal X is the integral of its power spectral density divided by 2π. This justifies that a power spectral density is the energy of the signal per unit frequency.

The power spectral density of a stationary random process is an even function,

$$S_{XX}(-\omega) = \overline{S}_{XX}(\omega) = S_{XX}(\omega) \tag{1.28}$$

where the bar denotes the complex conjugate. The first equality stems from eqn (1.25) and the fact that R_{XX} is real-valued and the second from the fact that S_{XX} is also real-valued.

One-sided power spectral density

The power spectral density as introduced in this section is two-sided in the sense that it is defined for positive as well as negative values of ω. But this is an even function. So, in practice, the one-sided power spectral density $s_{XX}(f)$ is used whose variable is the cyclic frequency f ranging from 0 to ∞ instead of the angular frequency ω. The one-sided power spectral density is introduced in such a way that its energy in a frequency band $\mathrm{d}f$ is the same as for the two-sided power spectral density (Fig. 1.2),

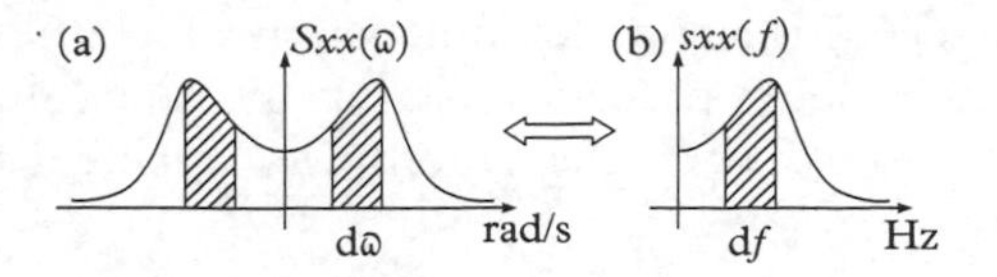

Figure 1.2 *(a) Two-sided and (b) one-sided power spectral densities. Areas under the curves are equal.*

$$s_{XX}(f)\,\mathrm{d}f = [S_{XX}(-\omega) + S_{XX}(\omega)]\,\mathrm{d}\omega \tag{1.29}$$

Since S_{XX} is even, we get

$$s_{XX}(f) = 2S_{XX}(\omega)\frac{\mathrm{d}\omega}{\mathrm{d}f} = 4\pi S_{XX}(\omega) \tag{1.30}$$

Spectrum analysers and acquisition boards often provide the one-sided power spectral density $s_{XX}(f)$.

White noise

A **white noise** is a stationary random process having constant power spectral density $S_{XX}(\omega) = S_0$. The auto-correlation function of a white noise is the delta Dirac function, $R_{XX}(\tau) = S_0\delta(\tau)$ (Fig. 1.3).

Cross-correlation function

If we consider two random processes X and Y, we introduce the **cross-correlation function**

$$R_{XY}(t,\tau) = \langle X(t)Y(t+\tau)\rangle \tag{1.31}$$

When $X = Y$, cross-correlation reduces to auto-correlation.

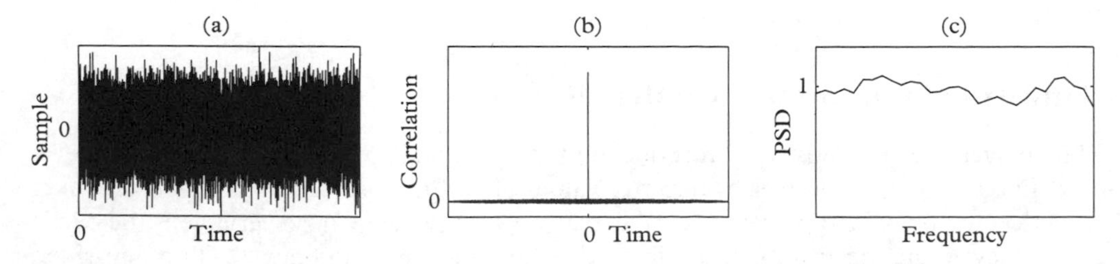

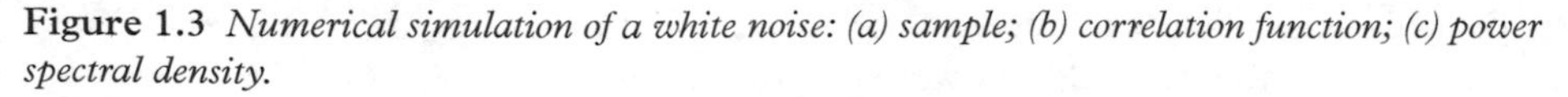

Figure 1.3 *Numerical simulation of a white noise: (a) sample; (b) correlation function; (c) power spectral density.*

The cross-correlation is bilinear. For all linear combinations $\sum_i \alpha_i Y_i$,

$$\begin{aligned} R_{X,\sum_i \alpha_i Y_i}(t,\tau) &= \left\langle X(t) \sum_i \alpha_i Y_i(t+\tau) \right\rangle \\ &= \sum_i \alpha_i \langle X(t) Y_i(t+\tau) \rangle \\ &= \sum_i \alpha_i R_{XY_i}(t,\tau) \end{aligned} \tag{1.32}$$

Similarly,

$$R_{\sum_i \alpha_i X_i, Y}(t,\tau) = \sum_i \alpha_i R_{X_i Y}(t,\tau) \tag{1.33}$$

Joint stationarity

Two processes X and Y are said to be **jointly wide-sense stationary to order two** if they are individually wide-sense stationary to order two and their cross-correlation function does not depend on the absolute time t. We note $R_{XY}(\tau)$.

Let us examine an example in order to illustrate that stationarity does not imply joint stationarity. Let $X_i = \sin(\omega_i t + \Theta)$ and $i = 1, 2$ be two random processes where it is assumed that ω_i is constant and Θ is a random variable with a uniform probability distribution on the interval $[0, \ 2\pi]$. The expectation of X_i is

$$\langle X_i \rangle(t) = \int_0^{2\pi} \sin(\omega_i t + \Theta) \frac{\mathrm{d}\Theta}{2\pi} = 0$$

Auto-correlation and cross-correlation are given by the integral

$$\begin{aligned} \langle X_i(t) X_j(t+\tau) \rangle &= \int_0^{2\pi} \sin(\omega_i t + \Theta) \sin\left[\omega_j(t+\tau) + \Theta\right] \frac{\mathrm{d}\Theta}{2\pi} \\ &= \frac{1}{2} \cos\left[(\omega_i - \omega_j)t + \omega_j \tau\right] \end{aligned}$$

The special case where $\omega_i = \omega_j$ gives $R_{X_i X_i} = {}^1/_2 \cos \omega_i \tau$. It does not depend on t and therefore the random process X_i is stationary. But when $\omega_1 \neq \omega_2$, $R_{X_1 X_2}$ does depend on both t and τ and the random processes are not joint stationary.

The cross-correlation of jointly stationary random processes verifies

$$R_{XY}(-\tau) = R_{YX}(\tau) \tag{1.34}$$

This results from

$$\begin{aligned} R_{XY}(-\tau) &= \langle X(t) Y(t-\tau) \rangle \\ &= \langle Y(t-\tau) X(t) \rangle \\ &= R_{YX}(\tau) \end{aligned}$$

Joint stationarity has been used in the last equality.

Cross-power spectral density

The **cross-power spectral density** of the random processes X and Y is defined as a simple generalization of eqn (1.24),

$$S_{XY}(\omega) = \lim_{T\to\infty} \frac{\left\langle \overline{\widehat{X}_T(\omega)}\, \widehat{Y}_T(\omega) \right\rangle}{2T} \tag{1.35}$$

Indeed, this reduces to power spectral density when $X = Y$.

The Wiener–Khinchin theorem states that

$$S_{XY}(\omega) = \int_{-\infty}^{\infty} R_{XY}(\tau) \mathrm{e}^{-\imath\omega\tau}\, \mathrm{d}\tau \tag{1.36}$$

$$R_{XY}(\tau) = \frac{1}{2\pi} \int_{-\infty}^{\infty} S_{XY}(\omega) \mathrm{e}^{\imath\omega\tau}\, \mathrm{d}\omega \tag{1.37}$$

The cross-correlation function and the cross-power spectral density of jointly stationary random processes form a Fourier pair. In general, S_{XY} is complex-valued.

Taken at time $\tau = 0$, the second Wiener–Khinchin relationship reads

$$\langle XY \rangle = \frac{1}{2\pi} \int_{-\infty}^{\infty} S_{XY}(\omega)\, \mathrm{d}\omega \tag{1.38}$$

Again, S_{XY} is the mutual spectral energy of processes X and Y.

The symmetry relationship of cross-power spectral density is

$$S_{XY}(-\omega) = \overline{S}_{XY}(\omega) = S_{YX}(\omega) \tag{1.39}$$

This is a direct consequence of eqn (1.34).

The cross-power spectral density is also bilinear. For all linear combinations $\sum_i \alpha_i Y_i$ with real coefficients α_i,

$$S_{X,\sum_i \alpha_i Y_i}(\omega) = \sum_i \alpha_i S_{XY_i}(\omega) \tag{1.40}$$

$$S_{\sum_i \alpha_i X_i, Y}(\omega) = \sum_i \alpha_i S_{X_i Y}(\omega) \tag{1.41}$$

This is proved by taking the Fourier transforms of eqns (1.32) and (1.33).

Uncorrelated white noises

Two jointly stationary random processes X and Y are uncorrelated white noises if they are individually white noises $S_{XX}(\omega) = S_1$ and $S_{YY}(\omega) = S_2$ and if their cross-power spectral density is zero $S_{XY}(\omega) = S_{YX}(\omega) = 0$.

Time derivative

The time derivative in the mean square sense of a random process X is defined as the random process $\dot{X}$ for which

$$\lim_{\epsilon\to 0}\left\langle\left|\frac{X(t+\epsilon)-X(t)}{\epsilon}-\dot{X}(t)\right|^2\right\rangle=0 \tag{1.42}$$

for all t. This may be written as

$$\dot{X}(t)=\lim_{\epsilon\to 0}\frac{X(t+\epsilon)-X(t)}{\epsilon} \tag{1.43}$$

in an appropriate topological space. Not all random processes have a derivative but a sufficient condition is the existence of the first two derivatives of auto-correlation at the origin. In the special case where all sample functions have themselves a derivative $\dot{x}(t)$, the time derivative of a random process is also the random process whose sample functions are the time derivatives $\dot{x}(t)$.

The cross-correlation of the processes X and $\dot{Y}$ is

$$\begin{aligned}
R_{X\dot{Y}}(t,\tau) &= \langle X(t)\dot{Y}(t+\tau)\rangle\\
&= \left\langle X(t)\lim_{\epsilon\to 0}\frac{1}{\epsilon}\left[Y(t+\tau+\epsilon)-Y(t+\tau)\right]\right\rangle\\
&= \lim_{\epsilon\to 0}\frac{1}{\epsilon}\left[\langle X(t)Y(t+\tau+\epsilon)\rangle-\langle X(t)Y(t+\tau)\rangle\right]\\
&= \lim_{\epsilon\to 0}\frac{1}{\epsilon}\left[R_{XY}(t,\tau+\epsilon)-R_{XY}(t,\tau)\right]\\
&= \frac{\partial}{\partial\tau}R_{XY}(t,\tau)
\end{aligned} \tag{1.44}$$

where the partial derivative must be taken in the sense of usual functions. In the case of jointly stationary processes, R_{XY} does not depend on t and consequently nor $R_{X\dot{Y}}$. Therefore X and $\dot{Y}$ are also jointly stationary and the above equality reads

$$R_{X\dot{Y}}(\tau)=\frac{\mathrm{d}}{\mathrm{d}\tau}R_{XY}(\tau) \tag{1.45}$$

The similar relationship for the random process X can easily be found by applying the symmetry relationships. For instance,

$$\begin{aligned}
R_{\dot{X}Y}(\tau) &= R_{Y\dot{X}}(-\tau)\\
&= \frac{\mathrm{d}}{\mathrm{d}\tau}R_{YX}(-\tau)
\end{aligned}$$

where the derivative is first applied to the function R_{XY} and the result is taken at the value $-\tau$. But deriving $\tau \mapsto R_{YX}(\tau) = R_{XY}(-\tau)$ and taking the result at $-\tau$ yields

$$R_{\dot{X}Y}(\tau) = -\frac{\mathrm{d}}{\mathrm{d}\tau} R_{XY}(\tau) = -R_{X\dot{Y}}(\tau) \tag{1.46}$$

When applied to the particular value $\tau = 0$, eqns (1.45) and (1.46) reduce to

$$\langle X\dot{Y}\rangle = \frac{\mathrm{d}}{\mathrm{d}\tau} R_{XY}(0) = -\langle \dot{X}Y\rangle \tag{1.47}$$

Thus, moving the dot in a bracket modifies the sign. If furthermore $X = Y$, then

$$\langle X\dot{X}\rangle = 0 \tag{1.48}$$

Since the Fourier transform of a derivative is $\imath\omega$ times the Fourier transform, the Wiener–Khinchin theorem applied to eqn (1.45) gives

$$S_{X\dot{Y}}(\omega) = \imath\omega S_{XY}(\omega) \tag{1.49}$$

Similarly with eqn (1.46),

$$S_{\dot{X}Y}(\omega) = -\imath\omega S_{XY}(\omega) \tag{1.50}$$

Finally, by applying successively (1.49) and (1.50), one obtains

$$S_{\dot{X}\dot{Y}}(\omega) = \omega^2 S_{XY}(\omega) \tag{1.51}$$

1.3 Linear System

A **linear system** is a system for which the response to any linear combination of inputs is the same linear combination of responses to individual inputs. Let $f_i(t)$, $i = 1, \ldots, n$ be a set of input signals where n may be infinite; the response $x(t)$ of a linear system to the input $\sum_i \alpha_i f_i(t)$ where the α_i are arbitrary constants is

$$x(t) = \sum_{i=1}^{n} \alpha_i x_i(t) \tag{1.52}$$

$x_i(t)$ being the response to the input $f_i(t)$. A linear system obeys the linear superposition principle.

The Dirac distribution $\delta(t)$ is defined by

$$f(0) = \int_{-\infty}^{\infty} \delta(\tau) f(\tau)\,\mathrm{d}\tau \tag{1.53}$$

for all test functions f. Mathematically, the Dirac distribution is the continuous linear functional which maps test functions f to $f(0)$ (Rudin, 1991). Similarly, the time-translated distribution $\delta_t = \delta(t-\tau)$ verifies

$$f(t) = \int_{-\infty}^{\infty} \delta(t-\tau)f(\tau)\,\mathrm{d}\tau \tag{1.54}$$

for any signal $f(t)$. Therefore $f(t)$ can be seen as a linear combination of signals $t \mapsto \delta(t-\tau)$ with the coefficients $f(\tau)$. By applying the linearity of system, the response is

$$x(t) = \int_{-\infty}^{\infty} h(t,\tau)f(\tau)\,\mathrm{d}\tau \tag{1.55}$$

where $h(t,\tau)$ is the **impulse response** of the linear system to input $t \mapsto \delta(t-\tau)$.

A linear system is said to be **time-invariant** if a certain input will always give the same output without regard to when the input was applied to the system. In particular, the form of the impulse response is the same whatever the time τ at which the impulse is applied. The impulse response $h(t,\tau)$ of a time-invariant linear system only depends on the difference $t-\tau$. We write $h(t-\tau)$ instead of $h(t,\tau)$.

A linear system is said to be **causal** when its impulse response is identically zero whenever the time t of the response is earlier than the time τ of the impulse. In other words,

$$h(t,\tau) = 0 \quad \text{for} \quad t < \tau \tag{1.56}$$

must hold. For causal linear systems, the convolution product eqn (1.55) reduces to

$$x(t) = \int_{0}^{t} h(t,\tau)f(\tau)\,\mathrm{d}\tau \tag{1.57}$$

whenever $f(\tau) = 0$ if $\tau < 0$. In the subsequent, all linear systems will be time-invariant and causal

An important example of a linear time-invariant causal system is given by linear differential equations which have constant coefficients and are initially at rest. The condition of initial rest is important since an initial movement would introduce a preferential time that would break down time-invariance.

Response to deterministic signal

A linear time-invariant system is fully determined by its impulse response $h(t)$ (Fig. 1.4). The response to any input $f(t)$ is

$$x(t) = \int_{-\infty}^{\infty} h(t-\tau)f(\tau)\,\mathrm{d}\tau \tag{1.58}$$

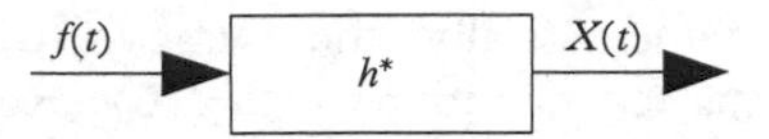

Figure 1.4 *Linear system. The output $x(t)$ is the convolution product of the impulse response $h(t)$ and the input $f(t)$.*

The above integral is a convolution product, which is also written $x = h * f$. The convolution product is commutative, i.e. $h*f = f*h$. This follows from the change of variable $\tau' = t - \tau$,

$$x(t) = \int_{-\infty}^{\infty} h(\tau')f(t-\tau')\,\mathrm{d}\tau' \tag{1.59}$$

the minus sign appearing in $\mathrm{d}\tau' = -\mathrm{d}\tau$ being neutralized by the inversion of integration limits.

Frequency response function

An impulse response is a temporal characterization of a linear system. The equivalent characterization in the frequency domain is the so-called **frequency response function**. The frequency response function $H(\omega)$ is the Fourier transform of the impulse response $h(t)$. The impulse response and frequency response function form a Fourier pair,

$$H(\omega) = \int_{-\infty}^{\infty} h(t)\mathrm{e}^{-\imath\omega t}\,\mathrm{d}t \tag{1.60}$$

$$h(t) = \frac{1}{2\pi}\int_{-\infty}^{\infty} H(\omega)\mathrm{e}^{\imath\omega t}\,\mathrm{d}\omega \tag{1.61}$$

Since the Fourier transform of a convolution product is the product of Fourier transforms, the Fourier transforms of input $\widehat{f}(\omega)$ and output $\widehat{x}(\omega)$ are related by

$$\widehat{x}(\omega) = H(\omega)\widehat{f}(\omega) \tag{1.62}$$

This relationship is the equivalent of eqn (1.59) in the frequency domain.

Response to random signal

So far, the signals $f(t)$ and $x(t)$ have been tacitly assumed to be deterministic. But since eqn (1.59) applies to any deterministic function $f(t)$, it also applies to any outcome of a random process F. We shall then write

$$X(t) = \int_{-\infty}^{\infty} h(\tau)F(t-\tau)\,\mathrm{d}\tau \tag{1.63}$$

where X and F are two random processes. The above equation simply means that eqn (1.59) holds for all outcomes $f(t)$ and $x(t)$.

When an input signal F is a random process, the response X is also a random process. The problem is then to determine the statistical characteristics of the response X assuming the statistical characteristics of the input F to be known. From now, we assume that F is a stationary random process with zero-mean.

The expectation of the response $\langle X(t)\rangle$ is

$$\langle X(t)\rangle = \int_{-\infty}^{\infty} h(\tau)\langle F(t-\tau)\rangle\,\mathrm{d}\tau = 0 \tag{1.64}$$

since $\langle F\rangle = 0$. The response X is therefore also a zero-mean random process

Correlation function of response

Let us now consider two linear time-invariant systems whose impulse responses are $h_1(t)$ and $h_2(t)$. The frequency response functions are $H_1(\omega)$ and $H_2(\omega)$. The inputs are F_1 and F_2 and the responses are X_1 and X_2. The inputs F_1 and F_2 are assumed to be jointly stationary processes with zero-mean. The cross-correlation of the responses is

$$\begin{aligned}
R_{X_1X_2}(t,\tau) &= \langle X_1(t)X_2(t+\tau)\rangle \\
&= \left\langle \int_{-\infty}^{\infty} h_1(\theta_1)F_1(t-\theta_1)\,\mathrm{d}\theta_1 \times \int_{-\infty}^{\infty} h_2(\theta_2)F_2(t+\tau-\theta_2)\,\mathrm{d}\theta_2 \right\rangle \\
&= \int_{-\infty}^{\infty}\int_{-\infty}^{\infty} h_1(\theta_1)h_2(\theta_2)\langle F_1(t-\theta_1)F_2(t+\tau-\theta_2)\rangle\,\mathrm{d}\theta_1\mathrm{d}\theta_2
\end{aligned}$$

where the sum signs and the expectation have been interchanged. In this equation, the cross-correlation $R_{F_1F_2}(\tau+\theta_1+\theta_2)$ of forces can be recognized. It leads to

$$R_{X_1X_2}(\tau) = \int_{-\infty}^{\infty}\int_{-\infty}^{\infty} h_1(\theta_1)h_2(\theta_2)R_{F_1F_2}(\tau+\theta_1-\theta_2)\,\mathrm{d}\theta_1\mathrm{d}\theta_2 \tag{1.65}$$

The above equation gives the cross-correlation function of responses in terms of the cross-correlation of forces. But it also gives another result. The right-hand side of this equation does not depend on the absolute time t. We have also seen in eqn (1.64) that the expectations $\langle X_1\rangle$ and $\langle X_1\rangle$ do not depend on the absolute time t. Therefore, the responses X_1 and X_2 are also jointly stationary random processes. This justifies that we have omitted the absolute time t in the arguments of $R_{X_1X_2}$ in eqn (1.65).

Power spectral density of response

We can now turn to the calculation of the cross-power spectral density of responses. Using the Wiener–Khinchin theorem (1.36), $S_{X_1X_2}$ is the Fourier transform of the cross-correlation function $R_{X_1X_2}$,

$$\begin{aligned} S_{X_1X_2} &= \int_{-\infty}^{\infty} R_{X_1X_2}(\tau)e^{-\imath\omega\tau}\,d\tau \\ &= \int_{-\infty}^{\infty}\int_{-\infty}^{\infty}\int_{-\infty}^{\infty} h_1(\theta_1)h_2(\theta_2)R_{F_1F_2}(\tau+\theta_1-\theta_2)e^{-\imath\omega\tau}\,d\theta_1 d\theta_2 d\tau \\ &= \int_{-\infty}^{\infty}\int_{-\infty}^{\infty} h_1(\theta_1)h_2(\theta_2)\left(\int_{-\infty}^{\infty} R_{F_1F_2}(\tau+\theta_1-\theta_2)e^{-\imath\omega\tau}\,d\tau\right) d\theta_1 d\theta_2 \end{aligned}$$

With the change of variable $\tau' = \tau + \theta_1 - \theta_2$, it yields

$$S_{X_1X_2} = \int_{-\infty}^{\infty}\int_{-\infty}^{\infty} h_1(\theta_1)h_2(\theta_2)\left(\int_{-\infty}^{\infty} R_{F_1F_2}(\tau')e^{-\imath\omega(\tau'-\theta_1+\theta_2)}\,d\tau'\right) d\theta_1 d\theta_2$$

By splitting the exponential, the power spectral density of force appears,

$$S_{X_1X_2} = \int_{-\infty}^{\infty} h_1(\theta_1)e^{\imath\omega\theta_1}\,d\theta_1 \times \int_{-\infty}^{\infty} h_2(\theta_2)e^{-\imath\omega\theta_2}\,d\theta_2 \times S_{F_1F_2}(\omega)$$

Introducing the frequency response functions $H_i(\omega)$ of systems yields

$$S_{X_1X_2}(\omega) = \overline{H}_1(\omega)H_2(\omega)S_{F_1F_2}(\omega) \tag{1.66}$$

This relationship gives the cross-power spectral density of responses in terms of the cross-power spectral density of inputs.

The special case of similar linear systems is of interest. Since eqn (1.66) applies for two different systems, it also applies for the same system with $H_1 = H_2 = H$, $X_1 = X_2 = X$, and $F_1 = F_2 = F$,

$$S_{XX}(\omega) = |H|^2(\omega)S_{FF}(\omega) \tag{1.67}$$

The power spectral density of the response X of a linear system is the square norm of frequency response function $|H|^2$ times the power spectral density of input S_{FF}.

Another special case is when a linear system, say the first one, is trivial $H_1 = 1$. In this case, the response X_1 equals the input F_1. Applying eqn (1.66) to a single linear system with $F_1 = F_2 = F$, $X_2 = X$ and $H_2 = H$, we get

$$S_{FX}(\omega) = H(\omega)S_{FF}(\omega) \tag{1.68}$$

This is the cross-power spectral density of input–output for a linear system. The dual relationship for the cross-power spectral density of output–input is obtained by applying eqn (1.39),

$$S_{XF}(\omega) = \overline{H}(\omega) S_{FF}(\omega) \tag{1.69}$$

since the power spectral density S_{FF} is real-valued, i.e. $\overline{S}_{FF} = S_{FF}$.

Power spectral density of time-derived response

We have found the statistical moments, auto-correlation of output, cross-correlations input–output, and the related power spectral densities of any linear time-invariant system. But we could have made another choice for the response. We could have admitted that the response was $\dot{X}$ instead of X. In that case, all the calculations would remain unchanged, except that the impulse response of the system would be $\dot{h}$ instead of h. As a result, the cross-correlation $R_{\dot{X}_1\dot{X}_2}$ is

$$R_{\dot{X}_1\dot{X}_2}(\tau) = \int_{-\infty}^{\infty}\int_{-\infty}^{\infty} \dot{h}_1(\theta_1)\dot{h}_2(\theta_2) R_{F_1F_2}(\tau + \theta_1 - \theta_2)\, \mathrm{d}\theta_1 \mathrm{d}\theta_2 \tag{1.70}$$

In a similar way, the cross-power spectral density of eqn (1.66) is still valid, but with $\imath\omega H_i(\omega)$, the Fourier transform of $\dot{h}_i$ instead of $H_i(\omega)$.

$$S_{\dot{X}_1\dot{X}_2}(\omega) = \omega^2 \overline{H}_1(\omega) H_2(\omega) S_{F_1F_2}(\omega) \tag{1.71}$$

Another way to get this result is by applying successively eqn (1.51) and eqn (1.66).

The special case of a single system gives

$$S_{\dot{X}\dot{X}}(\omega) = \omega^2 |H|^2(\omega) S_{FF}(\omega) \tag{1.72}$$

for the power spectral density of the time derivative of response,

$$S_{F\dot{X}}(\omega) = \imath\omega H(\omega) S_{FF}(\omega) \tag{1.73}$$

for the cross-power spectral density input derivative of output, and

$$S_{\dot{X}F}(\omega) = -\imath\omega \overline{H}(\omega) S_{FF}(\omega) \tag{1.74}$$

for the cross-power spectral density derivative of output–input.

Multiple input–multiple output system

Up to now, we have only considered the case of **single input–single output** (SISO) linear systems. However, most of the results remain unchanged if we consider **multiple input–multiple output** (MIMO) systems. In a MIMO linear system, the inputs are noted f_i where $i = 1, \ldots, P$ and the outputs x_j where $j = 1, \ldots, N$ (Fig. 1.5). The numbers N and P can be different. Formally, all the results of this section hold provided that $\mathbf{f}$ and $\mathbf{x}$ are now interpreted as column vectors of dimension respectively P and N.

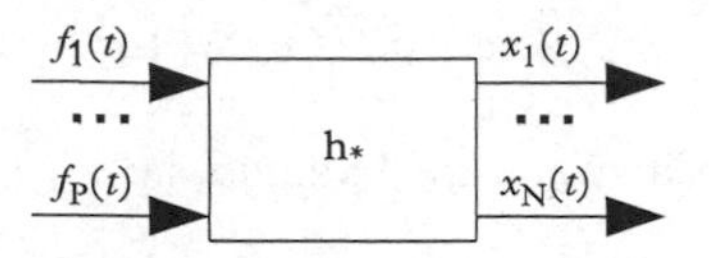

Figure 1.5 *Multiple input–multiple output linear filter.*

For instance, the response of a MIMO linear time-invariant system to inputs f is

$$\mathbf{x}(t) = \int_{-\infty}^{\infty} \mathbf{h}(\tau)\mathbf{f}(t-\tau)\,\mathrm{d}\tau \tag{1.75}$$

where $\mathbf{h}$ is the impulse response matrix of dimension $N \times P$. The component $h_{ij}(t)$ is the output i when the system is submitted to a unit impulse $f_j(t) = \delta(t)$ on input j while $f_k(t) = 0$ for $k \neq j$.

In the Fourier plane, the matrix $\mathbf{H}$ of frequency response functions is simply the term by term Fourier transform of the impulse response matrix. The response of the system is

$$\widehat{\mathbf{x}}(\omega) = \mathbf{H}(\omega)\widehat{\mathbf{f}}(\omega) \tag{1.76}$$

When the inputs are random, we form the random vectors $\mathbf{F}$ and $\mathbf{X}$ as the column vectors whose components are the stochastic processes $F_i(t)$ and $X_j(t)$. Their correlations are matrices $\mathbf{R_{FF}}$ and $\mathbf{R_{XX}}$ of dimension respectively $P \times P$ and $N \times N$ constituted by the cross-correlations $R_{F_iF_j}$ and $R_{X_iX_j}$. Similar definitions hold for power spectral densities.

The correlation matrix of outputs of a MIMO time-invariant linear systems is

$$\mathbf{R_{XX}}(\tau) = \int_{-\infty}^{\infty}\int_{-\infty}^{\infty} \mathbf{h}(\theta_1)\mathbf{R_{FF}}(t+\theta_1-\theta_2)\mathbf{h}^T(\theta_2)\,\mathrm{d}\theta_1\,\mathrm{d}\theta_2 \tag{1.77}$$

where the upperscript T denotes the transpose (not conjugate!). Similarly, the power spectral density matrix of outputs is

$$\mathbf{S_{XX}}(\omega) = \overline{\mathbf{H}}(\omega)\mathbf{S_{FF}}(\omega)\mathbf{H}^T(\omega) \tag{1.78}$$

where the bar denotes the conjugate matrix (not transpose!). For input–output power spectral densities, we have

$$\mathbf{S_{FX}}(\omega) = \mathbf{S_{FF}}(\omega)\mathbf{H}^T(\omega) \tag{1.79}$$

and

$$\mathbf{S_{XF}}(\omega) = \overline{\mathbf{H}}(\omega)\mathbf{S_{FF}}(\omega) \tag{1.80}$$

Similar equations involving $\dot{\mathbf{X}}$ are derived by substituting $\mathbf{H}(\omega) \leftrightarrow \imath\omega\mathbf{H}(\omega)$ into the above equations.

It is sometimes more convenient to manage explicitly the components F_i and X_j instead of the more compact matrix and vector products. In a scalar form, the above equation becomes

$$x_i(t) = \sum_{j=1}^{P} \int_{-\infty}^{\infty} h_{ij}(\tau) f_j(t-\tau)\, d\tau, \qquad i = 1, \ldots, N \tag{1.81}$$

In the Fourier plane, the response is

$$\widehat{x_i}(\omega) = \sum_{j=1}^{P} H_{ij}(\omega)\widehat{f_j}(\omega), \qquad i = 1, \ldots, N \tag{1.82}$$

The cross-power spectral densities become

$$S_{X_iX_j}(\omega) = \sum_{k=1}^{P}\sum_{l=1}^{P} \overline{H}_{ik}(\omega) S_{F_kF_l}(\omega) H_{jl}(\omega) \tag{1.83}$$

For input–output power spectral densities

$$S_{F_iX_j}(\omega) = \sum_{k=1}^{P} S_{F_iF_k}(\omega) H_{jk}(\omega) \tag{1.84}$$

and

$$S_{X_iF_j}(\omega) = \sum_{k=1}^{P} \overline{H}_{ik}(\omega) S_{F_kF_j}(\omega) \tag{1.85}$$

When we consider the time derivative of responses $\dot{X}_i$ the rule is always to substitute $\imath\omega H_{ik}(\omega)$ for $H_{ik}(\omega)$.

2

Mechanical Resonators

This chapter investigates the behaviour of mechanical resonators excited by uncorrelated random forces, with special emphasis on energy and power.

2.1 Single Resonator

Let us consider the single resonator shown in Fig. 2.1. The resonator is composed of a mass m attached to a rigid foundation by a spring of stiffness k and a dashpot of viscous damping coefficient c. The moving mass is submitted to an external force $f(t)$ and its position is noted $x(t)$. The equation of motion is

$$m\ddot{x} + c\dot{x} + kx = f \tag{2.1}$$

The impulse response is the response of the oscillator initially at rest and submitted to a mechanical shock at time $t = 0$. More exactly, this is the solution noted $h(t)$ of eqn (2.1) with initial conditions $h(t) = \dot{h}(t) = 0$ when $t < 0$ and a Dirac delta right-hand side $f(t) = \delta(t)$. Introducing the natural frequency of the undamped oscillator $\omega_0 = (k/m)^{1/2}$ and the damping ratio $\zeta = c/2m\omega_0$, the impulse response is

$$h(t) = \begin{cases} Y(t)\frac{1}{m\omega_0\sqrt{1-\zeta^2}}\mathrm{e}^{-\zeta\omega_0 t}\sin\left(\omega_0\sqrt{1-\zeta^2}t\right) & \text{for } \zeta < 1 \\ Y(t)\frac{t}{m}\mathrm{e}^{-\omega_0 t} & \text{for } \zeta = 1 \\ Y(t)\frac{1}{m\omega_0\sqrt{\zeta^2-1}}\mathrm{e}^{-\zeta\omega_0 t}\sinh\left(\omega_0\sqrt{\zeta^2-1}t\right) & \text{for } \zeta > 1 \end{cases} \tag{2.2}$$

In these expressions $Y(t)$ denotes the Heaviside function,

$$Y(t) = \begin{cases} 1 & \text{if } t \geq 0 \\ 0 & \text{otherwise} \end{cases} \tag{2.3}$$

The impulse response is plotted in Fig. 2.2 for the three cases $\zeta < 1$, $\zeta = 1$, and $\zeta > 1$.

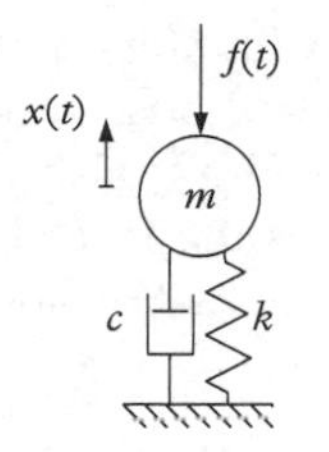

Figure 2.1 *Single oscillator.*

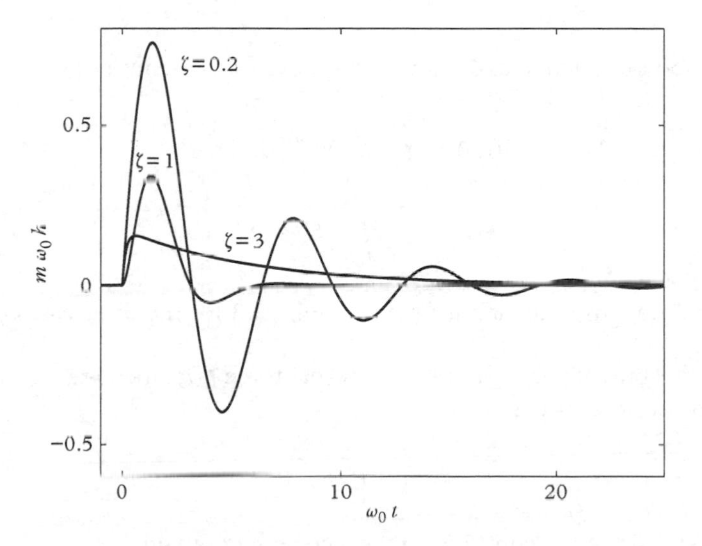

Figure 2.2 *Impulse response of a single oscillator.*

The initial-value problem given by eqn (2.1) and $x(0) = x_0$ and $\dot{x}(0) = v_0$ admits the unique solution

$$x(t) = \int_0^t f(\tau)h(t-\tau)\,\mathrm{d}\tau + m(v_0 + 2\zeta\omega_0 x_0)h(t) + mx_0\dot{h}(t) \tag{2.4}$$

The convolution product of the right-hand side is the so-called particular integral or **forced response** $x_f(t)$. It verifies $x_f(0) = \dot{x}_f(0) = 0$. The second and third terms constitute the **transient response** $x_t(t)$. For a damped oscillator $\zeta > 0$, $x_t \to 0$ when $t \to \infty$. We shall always neglect the transient response.

For any external force $f(t)$ verifying $f(t) = 0$ when $t < 0$, the solution to eqn (2.1) with vanishing initial conditions $x(0) = \dot{x}(0) = 0$ is given by the convolution product

$$x(t) = \int_0^t f(\tau)h(t-\tau)\,\mathrm{d}\tau = \int_{-\infty}^{\infty} f(\tau)h(t-\tau)\,\mathrm{d}\tau \tag{2.5}$$

This is also the solution to eqn (2.1) with non-vanishing initial conditions after a certain time. The resonator is therefore a linear, causal, and time-invariant system.

The frequency response function $H(\omega)$, also called **receptance** (Appendix D), is defined in such a way that $H(\omega)e^{\iota\omega t}$ is the forced response of the oscillator submitted to a pure harmonic excitation $f(t) = e^{\iota\omega t}$. For any value of ζ, the frequency response function of eqn (2.1) is

$$H(\omega) = \frac{1}{m\omega_0^2\left[1 + 2\iota\zeta\frac{\omega}{\omega_0} - \left(\frac{\omega}{\omega_0}\right)^2\right]} \tag{2.6}$$

Frequency response function and impulse response form a Fourier pair,

$$H(\omega) = \int_{-\infty}^{\infty} h(t)e^{-\iota\omega t}\,dt \tag{2.7}$$

$$h(t) = \frac{1}{2\pi}\int_{-\infty}^{\infty} H(\omega)e^{\iota\omega t}\,d\omega \tag{2.8}$$

The frequency response function is plotted in Fig. 2.3 for the three cases $\zeta < 1$, $\zeta = 1$, and $\zeta > 1$.

If $\widehat{f}(\omega)$ is the Fourier transform of the external force $f(t)$, then the Fourier transform $\widehat{x}(\omega)$ of the forced response is

$$\widehat{x}(\omega) = H(\omega)\widehat{f}(\omega) \tag{2.9}$$

Equation (2.9) is similar to eqn (2.5) in the frequency domain.

Power balance

Let us multiply eqn (2.1) by $\dot{x}$,

$$m\ddot{x}\dot{x} + c\dot{x}\dot{x} + kx\dot{x} = f\dot{x}$$

The first and third term in the left-hand side may be integrated,

$$\frac{d}{dt}\left(\frac{1}{2}m\dot{x}^2 + \frac{1}{2}kx^2\right) + c\dot{x}^2 = f\dot{x}$$

We recognize the expressions of the kinetic energy,

$$K = \frac{1}{2}m\dot{x}^2 \tag{2.10}$$

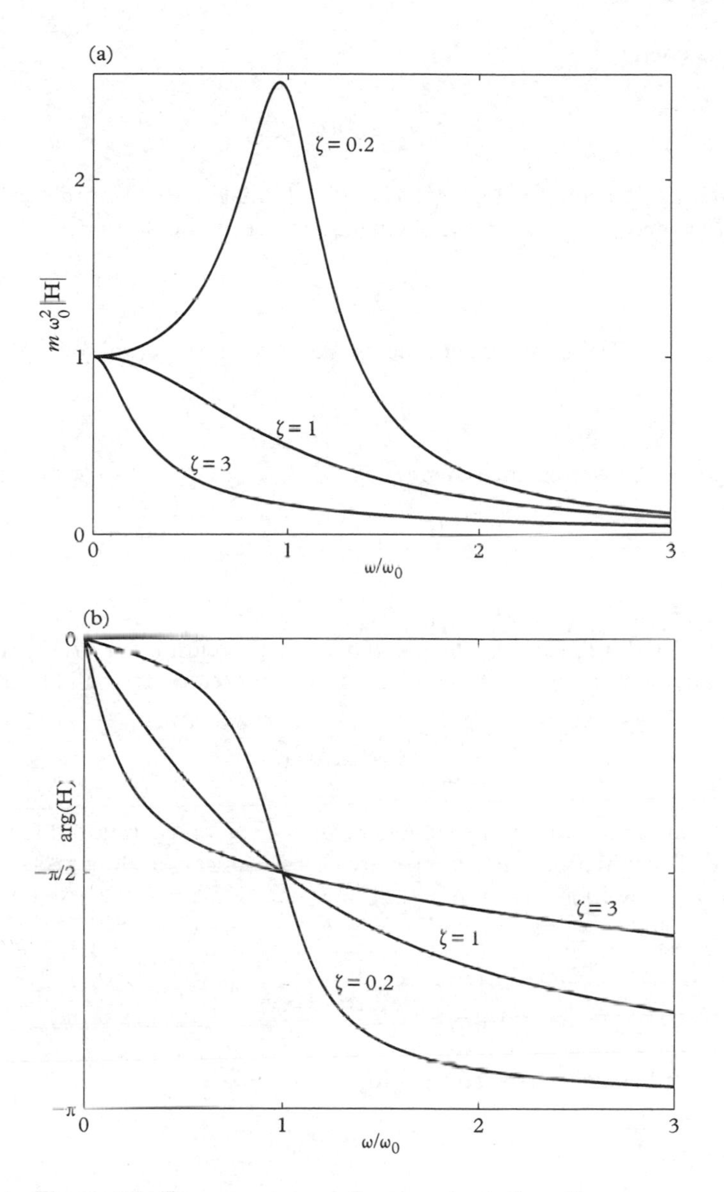

Figure 2.3 *Frequency response function of a single oscillator. (a) magnitude; (b) phase.*

and the elastic energy,

$$V = \frac{1}{2}kx^2 \tag{2.11}$$

We may also introduce the total energy $E = K + V$ that we shall simply call vibrational energy. Furthermore, $c\dot{x}^2$ is the power being dissipated in the dashpot,

$$P_{\text{diss}} = c\dot{x}^2 \tag{2.12}$$

while $f\dot{x}$ is the power being injected in the oscillator by the external force $f(t)$,

$$P_{\text{inj}} = f\dot{x} \tag{2.13}$$

The instantaneous power balance then reads

$$\frac{\mathrm{d}E}{\mathrm{d}t} + P_{\text{diss}} = P_{\text{inj}} \tag{2.14}$$

From now on, we shall assume that $f(t)$ is a random process. The position $x(t)$ is therefore also a random process. The instantaneous power balance (2.14) is valid at any time and for any sample functions $f(t)$ and $x(t)$. It is therefore also valid for mean values,

$$\left\langle \frac{\mathrm{d}E}{\mathrm{d}t} \right\rangle + \langle P_{\text{diss}} \rangle = \langle P_{\text{inj}} \rangle$$

This power balance is valid for any stationary or not stationary random force f.

In the case of stationary force f, x and $\dot{x}$ are also stationary random processes by virtue of Section 1.3. But we have seen in eqn (1.48) that $\langle X\dot{X} \rangle = 0$ for a stationary process X. Applied to x and $\dot{x}$, we get $\langle x\dot{x} \rangle = \langle \dot{x}\ddot{x} \rangle = 0$ and therefore

$$\left\langle \frac{\mathrm{d}E}{\mathrm{d}t} \right\rangle = m\langle \dot{x}\ddot{x} \rangle + k\langle x\dot{x} \rangle = 0 \tag{2.15}$$

The mean power balance (2.15) reduces to

$$\langle P_{\text{diss}} \rangle = \langle P_{\text{inj}} \rangle \tag{2.16}$$

What is injected in the system by external force is dissipated in the dashpot. There is no variation of the mean stored energy $\langle E \rangle$.

Injected power

The power being injected by an external force f is the product $f\dot{x}$. But when f is a white noise of power spectral density S_0, it is possible to find a relationship between the mean injected power $\langle P_{\text{inj}} \rangle$ and S_0.

By virtue of eqn (1.38), the expectation of $f\dot{x}$ is

$$\langle f\dot{x}\rangle = \frac{1}{2\pi}\int_{-\infty}^{\infty} S_{f\dot{x}}(\omega)\, d\omega \tag{2.17}$$

Since, $\dot{x}$ is the time-derived output of the linear system $h(t)$ to the input f, eqn (1.73) applies,

$$S_{f\dot{x}}(\omega) = \imath\omega H(\omega) S_0 \tag{2.18}$$

The following integral is calculated in Appendix C (see eqn (C.34)) by means of the residue technique,

$$\int_{-\infty}^{\infty} \imath\omega H(\omega)\, d\omega = \frac{\pi}{m} \tag{2.19}$$

where $H(\omega)$ is given by eqn (2.6). Combining (2.13), (2.17), (2.18), and (2.19) gives

$$\langle P_{\text{inj}}\rangle = \frac{S_0}{2m} \tag{2.20}$$

This relationship is useful for practical purpose since the measurement of S_0 only requires a single force sensor, whereas the measurement of $\langle f\dot{x}\rangle$ requires the synchronized acquisition of force and velocity signals.

Equality of kinetic and elastic energies

We now introduce a fundamental result that will turn out to be useful throughout this text: the equality of kinetic and elastic energies. Indeed, this equality is generally not true at a particular time since, when oscillating, a resonator permanently transforms kinetic energy into elastic energy (when decelerating) and conversely (when accelerating). However, the equality holds in a certain sense under well-chosen assumptions. The exact statement that we shall prove is the following. *When a mechanical resonator is submitted to a stationary white noise force, the expectations of kinetic and elastic energies are equal.*

The expected value of the elastic energy is

$$\langle V\rangle = \frac{1}{2}k\langle x^2\rangle = \frac{k}{4\pi}\int_{-\infty}^{\infty} S_{xx}(\omega)\, d\omega$$

where (1.27) has been substituted. Now, by introducing eqn (1.67) which gives the power spectral density of response to a linear time-invariant system,

$$\langle V\rangle = \frac{k}{4\pi}\int_{-\infty}^{\infty} |H|^2(\omega) S_0\, d\omega \tag{2.21}$$

Since the force f is assumed to be a white noise, its power spectral density S_0 is constant and may leave the integral. We get

$$\langle V\rangle = \frac{S_0}{4\pi} k \int_{-\infty}^{\infty} |H|^2(\omega)\,\mathrm{d}\omega \tag{2.22}$$

The expected value of elastic energy therefore depends on the integral of the square norm of the frequency response function $H(\omega)$. But $H(\omega)$ is given by eqn (2.6) and an integration of rational fractions by the residue technique gives (see Appendix C, eqn (C.24))

$$\int_{-\infty}^{\infty} |H|^2(\omega)\,\mathrm{d}\omega = \frac{\pi}{2\zeta m^2 \omega_0^3} \tag{2.23}$$

We finally obtain

$$\langle V\rangle = \frac{S_0}{8\zeta m\omega_0} \tag{2.24}$$

where the relation $\omega_0^2 = k/m$ has been used.

The same reasoning can be done for kinetic energy:

$$\langle K\rangle = \frac{1}{2} m\langle \dot{x}^2\rangle = \frac{m}{4\pi}\int_{-\infty}^{\infty} S_{\dot{x}\dot{x}}(\omega)\,\mathrm{d}\omega$$

The power spectral density of $\dot{x}$ is given by eqn (1.72) in terms of the power spectral density of f,

$$\langle K\rangle = \frac{m}{4\pi}\int_{-\infty}^{\infty} \omega^2 |H|^2(\omega) S_0\,\mathrm{d}\omega$$

Again, the power spectral density S_0 is constant therefore

$$\langle K\rangle = \frac{S_0}{4\pi} m \int_{-\infty}^{\infty} \omega^2 |H|^2(\omega)\,\mathrm{d}\omega \tag{2.25}$$

The integral of the right-hand side is calculated in Appendix C, eqn (C.29)

$$\int_{-\infty}^{\infty} \omega^2 |H|^2(\omega)\,\mathrm{d}\omega = \frac{\pi}{2\zeta m^2 \omega_0} \tag{2.26}$$

Substituting this quadrature formula into eqn (2.25) gives

$$\langle K\rangle = \frac{S_0}{8\zeta m\omega_0} \tag{2.27}$$

Thus, by comparing eqns (2.24) and (2.27), we obtain the following fundamental result,

$$\langle V \rangle = \langle K \rangle \tag{2.28}$$

A single resonator excited by a white noise force satisfies the equality of mean kinetic and elastic energies.

Dissipated power

Dissipation is imposed by the dashpot attached to the mass. We have found that the mean dissipated power is proportional to $\langle \dot{x}^2 \rangle$. We naturally guess to obtain a proportionality with the mean kinetic energy. Comparison of eqn (2.10) and (2.12) shows that

$$\langle P_{\text{diss}} \rangle = \frac{2c}{m} \langle K \rangle \tag{2.29}$$

The power dissipated by a viscous force is proportional to the kinetic energy. But since we have established the equality of kinetic and elastic energies in eqn (2.28), the vibrational energy is simply twice the kinetic energy. Thus, we may re-formulate the above law as

$$\langle P_{\text{diss}} \rangle = \frac{c}{m} \langle E \rangle \tag{2.30}$$

The mean dissipated power is proportional to the mean vibrational energy $\langle E \rangle$.

Statistical energy analysis of a single oscillator

Statistical energy analysis is somewhat trivial in the special case of a single resonator. Nevertheless, we shall apply it to introduce the method on a simple example in order to highlight the main steps and results that will be generalized to more complex systems.

The principle of statistical energy analysis consists in writing a power balance in steady-state conditions. In this case, it reduces to the equality of input and dissipated power,

$$\langle P_{\text{diss}} \rangle = \langle P_{\text{inj}} \rangle \tag{2.31}$$

The injected power $\langle P_{\text{inj}} \rangle$ is the source term exactly as the force f is the source term in the governing equation. Both force and injected power play the same role, but for different frameworks of analysis; the force-displacement analysis for the former, the energetic analysis for the latter. The link between the two has already been established, in eqn (2.20),

$$\langle P_{\text{inj}} \rangle = \frac{S_0}{2m} \tag{2.32}$$

where S_0 is the power spectral density of the external force. We can therefore choose to assume either $\langle P_{inj} \rangle$ or S_0 to be known.

Concerning the dissipated power, we have found a proportionality with the mean vibrational energy,

$$\langle P_{diss} \rangle = \frac{c}{m} \langle E \rangle \tag{2.33}$$

Substituting the above equality into the steady-state power balance gives

$$\langle P_{inj} \rangle = \frac{c}{m} \langle E \rangle \tag{2.34}$$

This is a simple linear equation between the vibrational energy $\langle E \rangle$ and the injected power $\langle P_{inj} \rangle$. Thus, if the injected power is assumed to be known or is assessed by eqn (2.31), the mean vibrational energy $\langle E \rangle$ is easily computed by the above equality. That was the main goal of statistical energy analysis in computing vibrational energies without solving the governing equation (2.1).

In Section 2.2, we generalize these results to the case of a set of coupled resonators. In particular, we shall establish the proportionality of injected power vector to vibrational energy vector, as in eqn (2.34).

2.2 Coupled Resonators

Let us now consider a set of N resonators as shown in Fig. 2.4. Each resonator consists of a moving mass m_i, a spring k_i attached to a fixed point, and a dashpot c_i. The values of m_i, c_i, and k_i can be different for each resonator. The uncoupled natural frequency of resonators is $\omega_i = (k_i/m_i)^{1/2}$ and the damping ratio $\zeta_i = c_i/2m_i\omega_i$. The position of resonator i is noted x_i. Resonators are linked to each other by couplings that will be described later on.

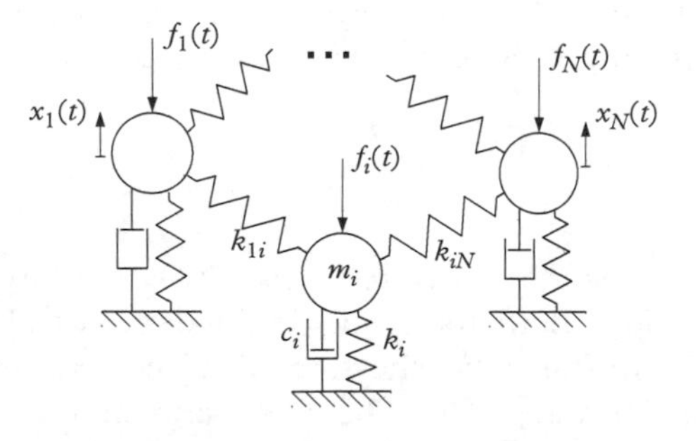

Figure 2.4 *Multiple oscillators coupled by conservative couplings.*

Lagrange's equation

We start by introducing the Lagrangian point of view of mechanics. A system is characterized by a Lagrangian $L(x_i, \dot{x}_i)$ which depends on positions x_i and velocities $\dot{x}_i$ of moving points of mass m_i. We exclude the case where the Lagrangian explicitly depends on time $L(x_i, \dot{x}_i, t)$. If the system is dissipative, the so-called Rayleigh dissipation function $R(\dot{x}_i)$ is also required. The Lagrange equation is

$$\frac{\mathrm{d}}{\mathrm{d}t}\left(\frac{\partial L}{\partial \dot{x}_i}\right) + \frac{\partial R}{\partial \dot{x}_i} - \frac{\partial L}{\partial x_i} = f_i \tag{2.35}$$

where f_i are the generalized external forces. They are defined by the expression of the differential form giving the infinitesimal work $\delta W = \sum_i f_i \,\mathrm{d}x_i$. The Lagrange equation (2.35) gives the governing equation of any mechanical system.

The energy of a system is defined by

$$E = \left(\sum_{i=1}^{N} \dot{x}_i \frac{\partial L}{\partial \dot{x}_i}\right) - L \tag{2.36}$$

In the presence of both dissipation and external forces, the power balance reads

$$\frac{\mathrm{d}E}{\mathrm{d}t} + \sum_{i=1}^{N} \dot{x}_i \frac{\partial R}{\partial \dot{x}_i} = \sum_{i=1}^{N} f_i \dot{x}_i \tag{2.37}$$

In the above, the second term of the left-hand side is the power being dissipated and the right-hand side is the power being injected in the system by external forces.

To prove eqn (2.37), we start by deriving the energy,

$$\frac{\mathrm{d}E}{\mathrm{d}t} = \sum_{i=1}^{N} \ddot{x}_i \frac{\partial L}{\partial \dot{x}_i} + \dot{x}_i \frac{\mathrm{d}}{\mathrm{d}t}\left(\frac{\partial L}{\partial \dot{x}_i}\right) - \frac{\mathrm{d}L}{\mathrm{d}t}$$

The chain rule applied to $L(x_i, \dot{x}_i)$ gives

$$\frac{\mathrm{d}L}{\mathrm{d}t} = \sum_{i=1}^{N} \frac{\mathrm{d}x_i}{\mathrm{d}t}\frac{\partial L}{\partial x_i} + \frac{\mathrm{d}\dot{x}_i}{\mathrm{d}t}\frac{\partial L}{\partial \dot{x}_i}$$

Combining the two above equations gives

$$\frac{\mathrm{d}E}{\mathrm{d}t} = \sum_{i=1}^{N} \dot{x}_i \left[\frac{\mathrm{d}}{\mathrm{d}t}\left(\frac{\partial L}{\partial \dot{x}_i}\right) - \frac{\partial L}{\partial x_i}\right]$$

The conclusion is obtained by applying the Lagrange equation (2.35).

Conservative coupling

We are now going to introduce couplings between oscillators. Several types of coupling will be considered. But, whatever the type of coupling, we shall never place a dashpot between resonators and more generally, no dissipative process is allowed between resonators in our model. This constitutes the first important assumption of statistical energy analysis.

Assumption 1 *Couplings are conservative.*

We now describe three types of conservative couplings.

Elastic coupling

Resonators are coupled by springs of stiffness k_{ij}. But the coefficients k_{ij} cannot take any value. They are constrained by the condition of conservative coupling. The force acting on resonator i and imposed by resonator j is $k_{ij}(x_j - x_i)$ and the work done by infinitesimal displacements dx_i and dx_j is therefore

$$\delta W = k_{ij}(x_j - x_i)\,dx_i + k_{ji}(x_i - x_j)\,dx_j \tag{2.38}$$

The force is conservative if the above differential form is closed. This imposes the condition

$$\frac{\partial}{\partial x_j} k_{ij}(x_j - x_i) = \frac{\partial}{\partial x_i} k_{ji}(x_i - x_j) \tag{2.39}$$

that is

$$k_{ij} = k_{ji} \tag{2.40}$$

Another argument to justify the above symmetry condition is to invoke the third Newton law which states that the sum of action $k_{ij}(x_j - x_i)$ and reaction $k_{ji}(x_i - x_j)$ is zero.

Since the differential form (2.38) is closed, δW is the differential of a function that is easily found to be $V_{ij}(x_i, x_j) = 1/2 \times k_{ij}(x_i - x_j)^2$. This is simply the elastic energy stored in the stiff coupling. So we can construct the total elastic energy stored in springs by simply summing all individual energies,

$$\begin{aligned} V &= \sum_{i=1}^{N} \left[\frac{1}{2} k_i x_i^2 + \sum_{j>i} \frac{1}{2} k_{ij} (x_i - x_j)^2 \right] \\ &= \sum_{i=1}^{N} \frac{1}{2} \left[\left(k_i + \sum_{j \neq i} k_{ij} \right) x_i^2 - \sum_{j \neq i} k_{ij} x_i x_j \right] \end{aligned} \tag{2.41}$$

where the second sum of the first equation runs over $j > i$ not to take into account two times the same coupling energy. The second equality requires a proof.

The sum $\sum_{i,j\neq i}$ is separated in two groups $\sum_{i,j>i}$ and $\sum_{i,j<i}$.

$$\sum_{i,j\neq i} k_{ij}\left(x_i^2 - x_i x_j\right) = \sum_{i=1}^{N}\sum_{j>i} k_{ij}\left(x_i^2 - x_i x_j\right) + \sum_{i=1}^{N}\sum_{j<i} k_{ij}\left(x_i^2 - x_i x_j\right)$$

In the second term of the right-hand side, the subscripts i and j are dummy indices and may be permuted,

$$\sum_{i,j\neq i} k_{ij}\left(x_i^2 - x_i x_j\right) = \sum_{i=1}^{N}\sum_{j>i} k_{ij}\left(x_i^2 - x_i x_j\right) + \sum_{j=1}^{N}\sum_{i<j} k_{ji}\left(x_j^2 - x_j x_i\right)$$

But from Fig. 2.5, it can be seen that the set of subscripts $j,\ i < j$ is the same as $i,\ j > i$. One obtains

$$\sum_{i,j\neq i} k_{ij}\left(x_i^2 - x_i x_j\right) = \sum_{i=1}^{N}\sum_{j>i} k_{ij}\left(x_i^2 - x_i x_j\right) + \sum_{i=1}^{N}\sum_{j>i} k_{ji}\left(x_j^2 - x_j x_i\right)$$

Since $k_{ji} = k_{ij}$, the two double sums of the right-hand side may collapse,

$$\sum_{i,j\neq i} k_{ij}\left(x_i^2 - x_i x_j\right) = \sum_{i=1}^{N}\sum_{j>i} k_{ij}\left(x_i^2 - 2x_i x_j + x_j^2\right) = \sum_{i=1}^{N}\sum_{j>i} k_{ij}\left(x_i - x_j\right)^2$$

This completes the proof.

We introduce the stiffness matrix $\mathbf{K}$,

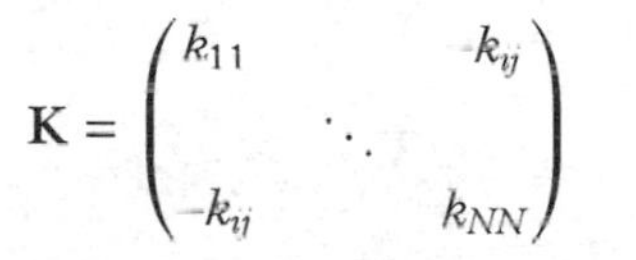

(2.42)

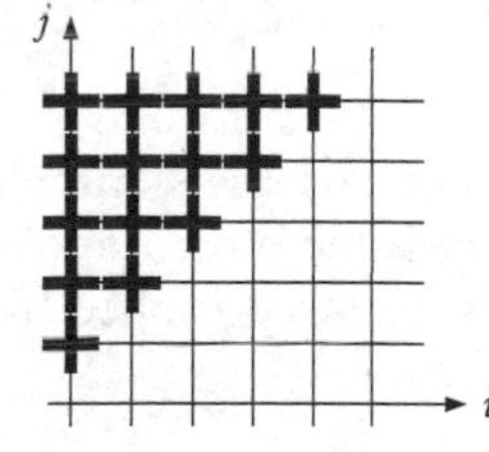

Figure 2.5 *The set of subscripts $j = 1$ to N and $i < j$ is the same as $i = 1$ to N and $j > i$.*

where diagonal terms are $k_{ii} = k_i + \sum_{j \neq i} k_{ij} > 0$ and off-diagonal terms are $-k_{ij} \leq 0$. By virtue of eqn (2.40) $\mathbf{K}$ is symmetric,

$$\mathbf{K}^T = \mathbf{K} \tag{2.43}$$

The stiffness matrix is useful for many purposes and will appear frequently throughout this chapter. However, the first important fact is that the total elastic energy introduced in eqn (2.41) may be written as

$$V = \frac{1}{2}\mathbf{x}^T \mathbf{K} \mathbf{x} \tag{2.44}$$

where $\mathbf{x} = (x_1, \ldots, x_N)^T$ is a column vector. Since $V > 0$ whenever $\mathbf{x} \neq 0$, $\mathbf{K}$ is positive definite,

$$\mathbf{x}^T \mathbf{K} \mathbf{x} > 0 \tag{2.45}$$

for all $\mathbf{x} \neq 0$.

From the analytic mechanics point of view, the elastic forces may be derived from the Lagrangian $L_{\text{stiff}} = -V$,

$$L_{\text{stiff}} = -\frac{1}{2}\mathbf{x}^T \mathbf{K} \mathbf{x} \tag{2.46}$$

The force acting on resonator i imposed by all springs is

$$\begin{aligned}
\frac{\partial L_{\text{stiff}}}{\partial x_i} &= -\frac{\partial}{\partial x_i}\left[\sum_l \frac{1}{2} k_{ll} x_l^2 - \sum_{l,m \neq l} \frac{1}{2} k_{lm} x_l x_m\right] \\
&= -k_{ii} x_i - \sum_{j \neq i} \frac{1}{2}\left(k_{ij} x_j + k_{ji} x_j\right) \\
&= -k_i x_i + \sum_{j \neq i} k_{ij}\left(x_j - x_i\right)
\end{aligned} \tag{2.47}$$

where the term k_{ii} has been expanded and the symmetry $k_{ij} = k_{ji}$ has been substituted into the last equation. The first term is the force imposed by the spring k_i attached to a fixed point while the second term is the force applied by all other oscillators through the coupling springs k_{ij}.

Gyroscopic coupling

Gyroscopic coupling is less common and therefore less intuitive than elastic coupling. But it will turn out to be useful when studying a vibroacoustic system in Chapter 4. For a gyroscopic coupling, the interaction force acting on resonator i imposed by resonator j is $-g_{ij}\dot{x}_j$. The work done by infinitesimal displacements $\mathrm{d}x_i$ and $\mathrm{d}x_j$ is therefore

$$\delta W = -g_{ij}\dot{x}_j\,\mathrm{d}x_i - g_{ji}\dot{x}_i\,\mathrm{d}x_j \tag{2.48}$$

This differential form cannot be closed. But along a trajectory $x_i(t)$, $x_j(t)$, the rate of work is

$$\frac{\mathrm{d}W}{\mathrm{d}t} = -g_{ij}\dot{x}_j\frac{\mathrm{d}x_i}{\mathrm{d}t} - g_{ji}\dot{x}_i\frac{\mathrm{d}x_j}{\mathrm{d}t} = -\left(g_{ij} + g_{ji}\right)\dot{x}_i\dot{x}_j \tag{2.49}$$

Therefore if the antisymmetry condition

$$g_{ij} = -g_{ji} \tag{2.50}$$

is verified then $\dot{W} = 0$ and the coupling is conservative. This is what will be called a gyroscopic coupling.

The gyroscopic matrix is

$$\mathbf{G} = \begin{pmatrix} 0 & & g_{ij} \\ & \ddots & \\ g_{ij} & & 0 \end{pmatrix} \tag{2.51}$$

The condition (2.50) implies that $\mathbf{G}$ is antisymmetric,

$$\mathbf{G}^T = -\mathbf{G} \tag{2.52}$$

The Lagrangian associated with a gyroscopic coupling is

$$L_{\text{gyr}} = \sum_{i=1}^{N}\sum_{j\neq i}\frac{1}{2}g_{ij}\dot{x}_i x_j = \frac{1}{2}\dot{\mathbf{x}}^T\mathbf{G}\mathbf{x} \tag{2.53}$$

The force acting on resonator i and imposed by all other resonators is

$$\begin{aligned} -\frac{\mathrm{d}}{\mathrm{d}t}\left(\frac{\partial L_{\text{gyr}}}{\partial \dot{x}_i}\right) + \frac{\partial L_{\text{gyr}}}{\partial x_i} &= -\frac{\mathrm{d}}{\mathrm{d}t}\sum_{j\neq i}\frac{1}{2}g_{ij}x_j + \frac{\partial}{\partial x_i}\sum_{l,m\neq l}\frac{1}{2}g_{lm}\dot{x}_l x_m \\ &= -\sum_{j\neq i}\frac{1}{2}\left(g_{ij} - g_{ji}\right)\dot{x}_j \\ &= -\sum_{j\neq i} g_{ij}\dot{x}_j \end{aligned} \tag{2.54}$$

where the antisymmetry (2.50) has been substituted.

Inertial coupling

An inertial coupling may arrive in a mechanical problem in various circumstances. The simplest situation to reveal the meaning of inertial coupling consists in introducing a system without inertial coupling in a rectangular coordinate frame and rewriting the problem in a generalized coordinate frame. So, consider a set of points of rectangular coordinate x'_i and mass m'_i. Their kinetic energy is of the form

$$K = \sum_{l=1}^{N} \frac{1}{2} m'_l \dot{x}'^2_l \tag{2.55}$$

Now, if we consider a change of variable $x'_1, \ldots, x'_N \mapsto x_1, \ldots, x_N$, the chain rule applied to the functions $x'_l(x_1, \ldots, x_N)$ gives

$$\dot{x}'_l = \sum_{i=1}^{N} \frac{\partial x'_l}{\partial x_i} \dot{x}_i \tag{2.56}$$

Hence the kinetic energy,

$$K = \sum_{i=1}^{N} \sum_{j=1}^{N} \frac{1}{2} m_{ij} \dot{x}_i \dot{x}_j \tag{2.57}$$

where

$$m_{ij} = \sum_{l=1}^{N} m'_l \frac{\partial x'_l}{\partial x_i} \frac{\partial x'_l}{\partial x_j} \tag{2.58}$$

m_{ij} is called the generalized mass. From eqn (2.58) it is clear that

$$m_{ij} = m_{ji} \tag{2.59}$$

Generally, the generalized mass m_{ij} may depend on generalized coordinates x_j but we shall always consider the linear case where the generalized masses are constant.

The mass matrix is

$$\mathbf{M} = \begin{pmatrix} m_1 & & m_{ij} \\ & \ddots & \\ m_{ij} & & m_N \end{pmatrix} \tag{2.60}$$

where we have noted m_i instead of m_{ii}. By virtue of eqn (2.59), $\mathbf{M}$ is symmetric,

$$\mathbf{M}^T = \mathbf{M} \tag{2.61}$$

The kinetic energy reads

$$K = \frac{1}{2}\dot{\mathbf{x}}^T \mathbf{M}\dot{\mathbf{x}} \tag{2.62}$$

And since K is always a positive quantity for non-zero masses, $\mathbf{M}$ is positive-definite,

$$\mathbf{x}^T \mathbf{M}\mathbf{x} > 0 \tag{2.63}$$

for all $\mathbf{x} \neq 0$. Let us remark that positive-definiteness would be violated for systems having an isolated resonator with a null mass $m_i = 0$. Indeed, we shall exclude such pathological cases and we shall always assume positive-definiteness of the mass matrix.

In the Lagrange equation, an inertial coupling is taken into account through a term $L_{\text{mass}} = K$,

$$L_{\text{mass}} = \sum_{i=1}^{N}\sum_{j=1}^{N} \frac{1}{2} m_{ij}\dot{x}_i\dot{x}_j = \frac{1}{2}\dot{\mathbf{x}}^T \mathbf{M}\dot{\mathbf{x}} \tag{2.64}$$

The inertial force acting on resonator i is

$$\begin{aligned} \frac{\mathrm{d}}{\mathrm{d}t}\left(\frac{\partial L_{\text{mass}}}{\partial \dot{x}_i}\right) &= \frac{\mathrm{d}}{\mathrm{d}t}\sum_{j=1}^{N}\frac{1}{2}m_{ij}\dot{x}_j + \frac{\mathrm{d}}{\mathrm{d}t}\sum_{j=1}^{N}\frac{1}{2}m_{ji}\dot{x}_j \\ &= \sum_{j=1}^{N} m_{ij}\ddot{x}_j \end{aligned} \tag{2.65}$$

where eqn (2.59) has been used. Of course, the self-inertia m_{ii} of resonator i is included in such an expression. But the action of resonator j on resonator i is a fictitious force $m_{ij}\ddot{x}_j$. Thus, an inertial coupling does not mean that a mass is added between the resonators and consequently that an additional degree of freedom must be introduced in the coupling. An inertial coupling results in a coupling of the differential equations through inertial terms without modifying the number of degrees of freedom.

Dissipation

To take into account dissipation by dashpots attached to resonators as shown in Fig. 2.4, we consider the Rayleigh function,

$$R = \sum_{i=1}^{N} \frac{1}{2} c_i \dot{x}_i^2 = \frac{1}{2}\dot{\mathbf{x}}^T \mathbf{C}\dot{\mathbf{x}} \tag{2.66}$$

In the second relationship, we have introduced the damping matrix,

$$\mathbf{C} = \begin{pmatrix} c_1 & & \bigcirc \\ & \ddots & \\ \bigcirc & & c_N \end{pmatrix} \tag{2.67}$$

where $c_i \geq 0$ for all i. The damping matrix $\mathbf{C}$ is diagonal and therefore trivially symmetric,

$$\mathbf{C}^T = \mathbf{C} \tag{2.68}$$

The fact that $\mathbf{C}$ is diagonal stems from Assumption 1, following which dashpots cannot be placed between resonators.

In the Lagrange equation, the Rayleigh function leads to the viscous force,

$$\frac{\partial R}{\partial \dot{x}_i} = c_i \dot{x}_i \tag{2.69}$$

acting on resonator i.

Differential equations

Let us go back to the set of oscillators drawn in Fig. 2.4. The most general system in the presence of elastic, gyroscopic, and inertial couplings has Lagrangian $L = L_{\text{mass}} + L_{\text{gyr}} + L_{\text{stiff}}$,

$$L = \frac{1}{2}\dot{\mathbf{x}}^T\mathbf{M}\dot{\mathbf{x}} + \frac{1}{2}\dot{\mathbf{x}}^T\mathbf{G}\mathbf{x} - \frac{1}{2}\mathbf{x}^T\mathbf{K}\mathbf{x} \tag{2.70}$$

The first term is the kinetic energy of resonators, the second term is for gyroscopic interaction, and the third term is the elastic energy. By expanding the products, it yields

$$L = \frac{1}{2}\sum_{i=1}^{N}\left[\sum_{j=1}^{N} m_{ij}\dot{x}_i\dot{x}_j + \sum_{j\neq i} g_{ij}\dot{x}_i x_j - k_{ii}x_i^2 + \sum_{j\neq i} k_{ij}x_i x_j\right] \tag{2.71}$$

Applied to the above Lagrangian and the Rayleigh function (2.66), the Lagrange equation (2.35) leads to

$$\sum_j m_{ij}\ddot{x}_j + c_i\dot{x}_i + \sum_{j\neq i} g_{ij}\dot{x}_j + k_i x_i + \sum_{i\neq j} k_{ij}(x_i - x_j) = f_i \tag{2.72}$$

where f_i is the generalized force applied to resonator i. This is the set of differential equations that we study in this chapter.

Let us rewrite eqn (2.72) in a matrix form. Let $\mathbf{x}$ and $\mathbf{f}$ be the column vectors whose elements are respectively x_i and f_i with $i = 1, \ldots, N$. Then, eqn (2.72) becomes

$$\mathbf{M}\ddot{\mathbf{x}} + (\mathbf{C} + \mathbf{G})\,\dot{\mathbf{x}} + \mathbf{K}\mathbf{x} = \mathbf{f} \tag{2.73}$$

where $\mathbf{M}$, $\mathbf{C}$, $\mathbf{G}$, and $\mathbf{K}$ are respectively the mass, damping, gyroscopic, and stiffness matrices.

Equation (2.72) is linear and the general solution with null initial conditions is

$$x_i(t) = \sum_{j=1}^{N} \int_{-\infty}^{\infty} h_{ij}(\tau) f_j(t-\tau)\, d\tau, \qquad i = 1, \ldots, N \tag{2.74}$$

where $h_{ij}(\tau)$ denotes the impulse response of resonator i when resonator j is excited. The system composed of N coupled resonators is therefore a linear causal and time-invariant system with N inputs and N outputs.

In a matrix notation, we introduce the matrix of impulse responses $\mathbf{h}(t)$ whose entries are $h_{ij}(t)$. The solution to eqn (2.73) with null initial conditions is

$$\mathbf{x}(t) = \int_{-\infty}^{\infty} \mathbf{h}(\tau)\mathbf{f}(t-\tau)\, d\tau \tag{2.75}$$

Frequency response function

The frequency response functions are found by taking the Fourier transform of eqn (2.73). It yields

$$\left(-\omega^2\mathbf{M} + \imath\omega\,(\mathbf{C}+\mathbf{G}) + \mathbf{K}\right)\widehat{\mathbf{x}} = \widehat{\mathbf{f}} \tag{2.76}$$

This is a set of linear equations on unknowns $\widehat{x}_i$. We introduce the dynamic stiffness matrix (Appendix D),

$$\mathbf{D}(\omega) = -\omega^2\mathbf{M} + \imath\omega\,(\mathbf{C}+\mathbf{G}) + \mathbf{K} \tag{2.77}$$

and the above equation becomes $\mathbf{D}\widehat{\mathbf{x}} = \widehat{\mathbf{f}}$. The solution can be written in a matrix form as

$$\widehat{\mathbf{x}} = \mathbf{H}\widehat{\mathbf{f}} \tag{2.78}$$

where

$$\mathbf{H}(\omega) = \mathbf{D}^{-1} = \left(-\omega^2\mathbf{M} + \imath\omega\,(\mathbf{C}+\mathbf{G}) + \mathbf{K}\right)^{-1} \tag{2.79}$$

The matrix $\mathbf{H}$ is called the frequency response function. Its entries are noted H_{ij}.

Since transpose and inverse commute, the symmetries of $\mathbf{M}$, $\mathbf{C}$, $\mathbf{K}$, and the antisymmetry of $\mathbf{G}$ imply

$$\mathbf{H}^T(\omega) = \left(-\omega^2\mathbf{M} + \imath\omega\,(\mathbf{C}-\mathbf{G}) + \mathbf{K}\right)^{-1} \tag{2.80}$$

Therefore $\mathbf{H}$ is generally not symmetric except if $\mathbf{G} = 0$.

Frequency response function and impulse response form a Fourier pair,

$$\mathbf{H}(\omega) = \int_{-\infty}^{\infty} \mathbf{h}(t) \mathrm{e}^{-\imath\omega t} \, \mathrm{d}t \tag{2.81}$$

$$\mathbf{h}(t) = \frac{1}{2\pi} \int_{-\infty}^{\infty} \mathbf{H}(\omega) \mathrm{e}^{\imath\omega t} \, \mathrm{d}\omega \tag{2.82}$$

where the integrals apply separately on each entry of matrices.

Global energy

The general expression of the energy of a system is given by eqn (2.36). Applied to the Lagrangian introduced in eqn (2.70), we get

$$E = \frac{1}{2} \dot{\mathbf{x}}^T \mathbf{M} \dot{\mathbf{x}} + \frac{1}{2} \mathbf{x}^T \mathbf{K} \mathbf{x} \tag{2.83}$$

By expanding all terms,

$$E = \frac{1}{2} \sum_{i=1}^{N} \left[\sum_{j} m_{ij} \dot{x}_i \dot{x}_j + k_{ii} x_i^2 - \sum_{j \neq i} k_{ij} x_i x_j \right] \tag{2.84}$$

In this expression, one can recognize the terms of isolated resonators and the terms added by elastic and inertial couplings. It may be pointed out that the gyroscopic forces do not contribute to the total energy. This statement may be verified directly by calculating

$$E_{\text{gyr}} = \sum_{i=1}^{N} \dot{x}_i \frac{\partial L_{\text{gyr}}}{\partial \dot{x}_i} - L_{\text{gyr}} = \sum_{i=1}^{N} \dot{x}_i \left(\sum_{j \neq i} \frac{1}{2} g_{ij} x_j \right) - \left(\sum_{i=1}^{N} \sum_{j \neq i} \frac{1}{2} g_{ij} \dot{x}_i x_j \right) = 0$$

The gyroscopic term does not modify the expression of energy although it appears in the Lagrangian. The main consequence will be to modify the equilibrium between kinetic and elastic energy as will be seen later. For usual mechanical systems, the Lagrangian is the difference of kinetic and potential energy $L = K - V$ while the total energy is the sum $E = K + V$. Consequently, the kinetic energy is half the sum $E + L$ and the potential energy is half the difference $E - L$. But in the presence of a gyroscopic term,

$$E + L = \dot{\mathbf{x}}^T \mathbf{M} \dot{\mathbf{x}} + \frac{1}{2} \dot{\mathbf{x}}^T \mathbf{G} \mathbf{x} \tag{2.85}$$

and

$$E - L = \mathbf{x}^T \mathbf{K} \mathbf{x} - \frac{1}{2} \dot{\mathbf{x}}^T \mathbf{G} \mathbf{x} \tag{2.86}$$

These relationships will be useful later on.

Local energy

In the previous subsections we have found the energy and Lagrangian of the whole set of resonators. But in statistical energy analysis the spirit of the method is to divide the structure into subsystems considered as energy tanks able to exchange their energy. In the present context, the structure is the whole set of resonators while the subsystems are single resonators. Thus the underlying question to the definition of subsystems is how to affect the different terms appearing in the expressions of energy (2.84) and Langrangian (2.71) to the individual resonators? Of course, the answer is not unique and the path we shall pursue in this subsection must rather be considered as a definition of subsystems. A second question that will be discussed in the next subsection is: what is the expression of the exchanged power between resonators? We shall see that this is a technical point that admits an unambiguous response once we know the energy of resonators.

In constructing the Lagrangian (2.71) of the whole set of resonators, we have introduced five terms. Two terms are clearly associated with the kinetic and elastic energies of uncoupled oscillators. These are

$$\begin{array}{ll} \text{kinetic energy} & 1/2 \times m_i\dot{x}_i^2 \\ \text{elastic energy} & 1/2 \times k_i x_i^2 \end{array} \tag{2.87}$$

These two terms must clearly be affected to the Lagrangian of resonator i. But the three types of conservative couplings lead us to introduce three additional terms in the Lagrangian. These are

$$\begin{array}{ll} \text{inertial coupling} & 1/2 \times m_{ij}\dot{x}_i\dot{x}_j \\ \text{gyroscopic coupling} & 1/2 \times g_{ij}\dot{x}_i x_j \\ \text{elastic coupling} & 1/2 \times k_{ij}(x_i - x_j)^2 \end{array} \tag{2.88}$$

If the kinetic $1/2 \times m_i\dot{x}_i^2$ and elastic $1/2 \times k_i x_i^2$ terms do not raise any questions, the status of the interaction terms $1/2 \times m_{ij}\dot{x}_i\dot{x}_j$, $1/2 \times g_{ij}\dot{x}_i x_j$, and $1/2 \times k_{ij}(x_i - x_j)^2$ must be clarified.

When carrying out a partition of a system into subsystems, the first property that is expected is that the whole is the sum of the parts. Thus, either we decide to separate the energy of oscillators $1/2 \times m_i\dot{x}_i^2$ and $1/2 \times k_i x_i^2$ from the coupling terms $1/2 \times m_{ij}\dot{x}_i\dot{x}_j$, $1/2 \times g_{ij}\dot{x}_i x_j$, and $1/2 \times k_{ij}(x_i - x_j)^2$, or we decide to share the terms $1/2 \times m_{ij}\dot{x}_i\dot{x}_j$, $1/2 \times g_{ij}\dot{x}_i x_j$, and $1/2 \times k_{ij}(x_i - x_j)^2$ in two parts and to affect them to oscillators i and j. In the first case, we must admit that couplings are true subsystems since they contain a part of vibrational energy. This method is not natural in statistical energy analysis for at least two reasons. The first is that couplings, as defined in this canonical system of N oscillators, are conservative. Therefore they exhibit particular features that are not representative of a regular subsystem. The second reason is that, usually, couplings do not contain a significant part of the total energy. So, in practice, it is not useful to separate couplings from other subsystems. It is more reasonable and meaningful to consider that a coupling is just a boundary between adjacent subsystems by which the vibrational energy can flow.

We therefore adopt the principle that the terms $1/2 \times m_{ij}\dot{x}_i\dot{x}_j$, $1/2 \times g_{ij}\dot{x}_i x_j$, and $1/2 \times k_{ij}(x_i - x_j)^2$ must be split and affected to oscillators i and j. Let us first examine the third term. The most symmetrical way to proceed is to affect the term $1/2 \times k_{ij}x_i^2$ to oscillator i, $1/2 \times k_{ij}x_j^2$ to oscillator j, and to share the cross-product $-1/2 \times 2k_{ij}x_ix_j$ in two equal parts.

$$\frac{1}{2}k_{ij}\left(x_i^2 - x_ix_j\right) \mapsto \text{resonator } i \quad \text{and} \quad \frac{1}{2}k_{ij}\left(x_j^2 - x_ix_j\right) \mapsto \text{resonator } j$$

Similarly for an inertial coupling, the subscripts i and j appear in the two terms $1/2 \times m_{ij}\dot{x}_i\dot{x}_j$ and $1/2 \times m_{ji}\dot{x}_i\dot{x}_j$. So, if we affect half the sum to each resonator,

$$\frac{1}{4}\left(m_{ij} + m_{ji}\right)\dot{x}_i\dot{x}_j = \frac{1}{2}m_{ij}\dot{x}_i\dot{x}_j \mapsto \text{resonators } i,\ j$$

since $m_{ij} = m_{ji}$. For a gyroscopic coupling, the half sum of the two involved terms is

$$\frac{1}{4}\left(g_{ij}\dot{x}_ix_j + g_{ji}x_i\dot{x}_j\right) = \frac{1}{4}g_{ij}\left(\dot{x}_ix_j - x_i\dot{x}_j\right) \mapsto \text{resonators } i,\ j$$

since $g_{ij} = -g_{ji}$. One finally obtains

$$L_i = \frac{1}{2}\left[\sum_j m_{ij}\dot{x}_i\dot{x}_j + \frac{1}{2}\sum_{j\neq i} g_{ij}\left(\dot{x}_ix_j - x_i\dot{x}_j\right) - k_{ii}x_i^2 + \sum_{j\neq i} k_{ij}x_ix_j\right] \tag{2.89}$$

$$E_i = \frac{1}{2}\left[\sum_j m_{ij}\dot{x}_i\dot{x}_j + k_{ii}x_i^2 - \sum_{j\neq i} k_{ij}x_ix_j\right] \tag{2.90}$$

for respectively the Lagrangian and energy of resonator i. These expressions ensure that the total Lagrangian and energy are simply the sum of Lagrangians and energies of subsystems,

$$L = \sum_{i=1}^{N} L_i \tag{2.91}$$

$$E = \sum_{i=1}^{N} E_i \tag{2.92}$$

Power balance

The expression of local energy E_i has been found in the previous subsection. The problem is now to find the expression of the exchanged power between two resonators i and j.

The power balance equation is obtained by multiplying eqn (2.72) by $\dot{x}_i$,

$$\sum_j m_{ij}\ddot{x}_j\dot{x}_i + c_i\dot{x}_i^2 + \sum_{j\neq i} g_{ij}\dot{x}_j\dot{x}_i + k_{ii}x_i\dot{x}_i - \sum_{j\neq i} k_{ij}x_j\dot{x}_i = f_i\dot{x}_i$$

where we have substituted $k_{ii} = k_i + \sum_{j \neq i} k_{ij}$. After adding and subtracting $\dot{x}_j \ddot{x}_i$ and $x_i \dot{x}_j$ in the left-hand side,

$$\frac{1}{2}\sum_j m_{ij}(\ddot{x}_j \dot{x}_i + \dot{x}_j \ddot{x}_i) + \frac{1}{2}\sum_{j \neq i} m_{ij}(\ddot{x}_j \dot{x}_i - \dot{x}_j \ddot{x}_i) + c_i \dot{x}_i^2 + \sum_{j \neq i} g_{ij} \dot{x}_j \dot{x}_i + k_{ii} x_i \dot{x}_i$$
$$+ \frac{1}{2}\sum_{j \neq i} k_{ij}\left(-\dot{x}_i x_j - x_i \dot{x}_j + x_i \dot{x}_j - \dot{x}_i x_j\right) = f_i \dot{x}_i$$

The terms $\ddot{x}_j \dot{x}_i + \dot{x}_j \ddot{x}_i$, $x_i \dot{x}_i$, and $-\dot{x}_i x_j - x_i \dot{x}_j$ are the time derivatives of respectively $\dot{x}_j \dot{x}_i$, $x_i^2/2$, and $-x_i x_j$. It yields

$$\frac{1}{2}\frac{\mathrm{d}}{\mathrm{d}t}\left(\sum_j m_{ij} \dot{x}_i \dot{x}_j + k_{ii} x_i^2 - \sum_{j \neq i} k_{ij} x_i x_j\right) + c_i \dot{x}_i^2$$
$$+ \sum_{j \neq i}\left[\frac{1}{2} m_{ij}(\ddot{x}_j \dot{x}_i - \dot{x}_j \ddot{x}_i) + g_{ij} \dot{x}_i \dot{x}_j + \frac{1}{2} k_{ij}(x_i \dot{x}_j - \dot{x}_i x_j)\right] = f_i \dot{x}_i$$

The vibrational energy E_i introduced in eqn (2.90) can be recognized. We get

$$\frac{\mathrm{d}E_i}{\mathrm{d}t} + P_{\mathrm{diss},i} + \sum_{j \neq i} P_{ij} = P_i \tag{2.93}$$

This equation is the instantaneous power balance of subsystem i. The term

$$P_{\mathrm{diss},i} = c_i \dot{x}_i^2 \tag{2.94}$$

is the power being dissipated in the dashpot. In the right-hand side, the term

$$P_i = f_i \dot{x}_i \tag{2.95}$$

is the power being injected in subsystem i by the external force f_i. Finally, the term

$$P_{ij} = \frac{1}{2} m_{ij}(\dot{x}_i \ddot{x}_j - \ddot{x}_i \dot{x}_j) + g_{ij} \dot{x}_i \dot{x}_j + \frac{1}{2} k_{ij}\left(x_i \dot{x}_j - \dot{x}_i x_j\right) \tag{2.96}$$

is the net power transmitted from resonator i to resonator j. Since this expression is antisymmetric under the permutation $i \leftrightarrow j$,

$$P_{ji} = -P_{ij} \tag{2.97}$$

The net power transmitted from subsystem j to i is the opposite of the net power transmitted from subsystem i to j.

Mean energies and powers

Let us now consider the case of stationary random forces. The mean values of L_i, E_i, and P_{ij} are obtained by simply adding brackets $\langle\rangle$ to their instantaneous expressions. But in a bracket, moving a dot changes the sign by virtue of eqn (1.47). Therefore $\langle \dot{x}_i x_j \rangle = -\langle x_i \dot{x}_j \rangle$, $\langle \ddot{x}_i \dot{x}_j \rangle = -\langle \dot{x}_i \ddot{x}_j \rangle$, and eqns (2.89), (2.90), and (2.96) successively reduce to

$$\langle L_i \rangle = \frac{1}{2}\left[\sum_j m_{ij}\langle \dot{x}_i \dot{x}_j \rangle + \sum_{j \neq i} g_{ij}\langle \dot{x}_i x_j \rangle - k_{ii}\langle x_i^2 \rangle + \sum_{j \neq i} k_{ij}\langle x_i x_j \rangle \right] \tag{2.98}$$

for the mean Lagrangian,

$$\langle E_i \rangle = \frac{1}{2}\left[\sum_j m_{ij}\langle \dot{x}_i \dot{x}_j \rangle + k_{ii}\langle x_i^2 \rangle - \sum_{j \neq i} k_{ij}\langle x_i x_j \rangle \right] \tag{2.99}$$

for the mean energy, and

$$\langle P_{ij} \rangle = m_{ij}\langle \dot{x}_i \ddot{x}_j \rangle + g_{ij}\langle \dot{x}_i \dot{x}_j \rangle + k_{ij}\langle x_i \dot{x}_j \rangle \tag{2.100}$$

for the mean exchanged power.

Power balance (2.93) is valid at any time and for any outcome of the random processes f_i, $i =, 1 \ldots, N$. It is therefore also valid for mean values,

$$\left\langle \frac{\mathrm{d}E_i}{\mathrm{d}t} \right\rangle + \langle P_{\mathrm{diss},i} \rangle + \sum_{j \neq i} \langle P_{ij} \rangle = \langle P_i \rangle \tag{2.101}$$

But since x_i, x_j, $\dot{x}_i$, and $\dot{x}_j$ are also stationary, eqn (1.47) gives $\langle \dot{x}_i x_j \rangle = -\langle x_i \dot{x}_j \rangle$, $\langle \ddot{x}_i \dot{x}_j \rangle = -\langle \dot{x}_i \ddot{x}_j \rangle$, $\langle x_i \dot{x}_i \rangle = 0$, and

$$\left\langle \frac{\mathrm{d}E_i}{\mathrm{d}t} \right\rangle = \frac{1}{2}\sum_j m_{ij}(\langle \ddot{x}_i \dot{x}_j \rangle + \langle \dot{x}_i \ddot{x}_j \rangle) + k_{ii}\langle x_i \dot{x}_i \rangle - \frac{1}{2}\sum_{j \neq i} k_{ij}(\langle \dot{x}_i x_j \rangle + \langle x_i \dot{x}_j \rangle) = 0$$

Therefore, the time-derivative term vanishes in the mean power balance. The mean power balance equation of subsystem i in steady-state conditions then reads

$$\langle P_{\mathrm{diss},i} \rangle + \sum_{j \neq i} \langle P_{ij} \rangle = \langle P_i \rangle \tag{2.102}$$

One might also raise the question of the global power balance, that is for the entire set of oscillators. It is simply obtained by summing the local power balance over all subsystems. Starting from the general case of non-stationary forces (2.101),

$$\sum_i \left\langle \frac{\mathrm{d}E_i}{\mathrm{d}t} \right\rangle + \sum_i \langle P_{\mathrm{diss},\,i} \rangle + \sum_i \sum_{j \neq i} \langle P_{ij} \rangle = \sum_i \langle P_i \rangle$$

Re-ordering the third sum of the left-hand side,

$$\sum_i \sum_{j \neq i} \langle P_{ij} \rangle = \sum_i \sum_{j > i} \left(\langle P_{ij} \rangle + \langle P_{ji} \rangle \right) = 0$$

Since $\langle P_{ji} \rangle = -\langle P_{ij} \rangle$, this sum vanishes. Furthermore, by virtue of eqn (2.92), the energy of the complete system E is the sum of partial energies E_i. The first sum simplifies,

$$\left\langle \frac{\mathrm{d}E}{\mathrm{d}t} \right\rangle + \sum_i \langle P_{\mathrm{diss},\,i} \rangle = \sum_i \langle P_i \rangle \tag{2.103}$$

This is the global power for the entire system.

Indeed, in the case of stationary forces, the power balance reduces to the last two terms.

$$\sum_i \langle P_{\mathrm{diss},\,i} \rangle = \sum_i \langle P_i \rangle \tag{2.104}$$

In steady-state condition, the sum of dissipated powers exactly balances the sum of injected powers.

Random forces

From now on, the external forces applied to resonators will be mutually uncorrelated white noises. This is the second important assumption of statistical energy analysis.

Assumption 2 *External forces are uncorrelated white noises.*

This assumption contains several results. First of all, the force f_i is a stationary random process whose power spectral density is constant. We shall denote it by S_i. In the time domain, it is equivalent to say that the auto-correlation is a delta Dirac function, $R_{f_i f_i}(\tau) = S_i \delta(\tau)$. Secondly, the forces f_i and f_j are jointly stationary with a zero cross-power spectral density $S_{ij} = 0$ when $i \neq j$. In time domain, f_i are δ-correlated, $R_{f_i f_j}(\tau) = S_i \delta_{ij} \delta(\tau)$ where δ_{ij} is Kronecker's symbol.

Mean injected power

According to eqn (2.95), the power being injected by forces in the mechanical system is $\langle P_i \rangle = \langle f_i \dot{x}_i \rangle$ where x_i and f_i are two random processes. The expectation of the cross-product of two random processes is given by eqn (1.38),

$$\langle f_i \dot{x}_i \rangle = \frac{1}{2\pi} \int_{-\infty}^{\infty} S_{f_i \dot{x}_i}(\omega) \, \mathrm{d}\omega \tag{2.105}$$

Since $\dot{x}_i$ is the time-derived output of the MIMO linear system whose impulse responses are $h_{ij}(t)$, eqns (1.49) and (1.84) apply,

$$S_{f_i\dot{x}_i}(\omega) = \imath\omega \sum_{j=1}^{N} H_{ij}(\omega) S_{f_i f_j}(\omega) \tag{2.106}$$

But we have assumed that the forces f_i are uncorrelated white noises and therefore $S_{f_i f_j} = S_i \delta_{ij}$. One obtains

$$\langle f_i \dot{x}_i \rangle = \frac{S_i}{2\pi} \int_{-\infty}^{\infty} \imath\omega H_{ii}(\omega)\, \mathrm{d}\omega \tag{2.107}$$

But $\int \imath\omega H_{ii}\, \mathrm{d}\omega / 2\pi$ is the inverse Fourier transform of $\dot{h}_{ii}$ at $t = 0$. However, $t \mapsto \dot{h}_{ii}(t)$ is discontinuous at $t = 0$ and great attention must be paid in calculating the inverse Fourier transform. In the special case of a single resonator, the impulse response given in eqn (2.2) shows that the limit of $\dot{h}$ is 0 when $t \to 0^-$ but $1/m$ when $t \to 0^+$. In general, the inverse Fourier transform of an integrable and piecewise continuous function f at most finitely discontinuous is equal to $1/2 \times \left[f(t^+) + f(x^-) \right]$ where $f(t^+)$ and $f(t^-)$ respectively denote the right and left limits of f at t. In the present case, $\dot{h}_{ii}(0^-) = 0$ since the system is initially at rest but $\dot{h}_{ii}(0^+) \neq 0$. Therefore,

$$\frac{1}{2\pi} \int_{-\infty}^{\infty} \imath\omega H_{ii}(\omega)\, \mathrm{d}\omega = \frac{1}{2} \dot{h}_{ii}(0^+) \tag{2.108}$$

Substituting into eqn (2.107) gives

$$\langle f_i \dot{x}_i \rangle = \frac{S_i}{2} \dot{h}_{ii}(0^+) \tag{2.109}$$

To calculate the right limit $\dot{h}_{ii}(0^+)$, let us remark that the impulse response is zero for $t < 0$ by causality and verifies the homogeneous version ($f = 0$) of eqn (2.73) when $t > 0$. Let i be the system which is subjected to the impulse force $f_i(t) = \delta(t)$ and let us form the column vector $\mathbf{x} = (h_{1i}, \ldots, h_{Ni})^T$. We construct $\mathbf{x}$ as a product of a twice continuously differentiable vector χ and the Heaviside function $Y(t)$. By deriving

$$\begin{aligned}
\mathbf{x}(t) &= Y(t)\chi(t) \\
\dot{\mathbf{x}}(t) &= Y(t)\dot{\chi}(t) + \delta(t)\chi(0) \\
\ddot{\mathbf{x}}(t) &= Y(t)\ddot{\chi}(t) + \delta(t)\dot{\chi}(0) + \delta'(t)\chi(0)
\end{aligned}$$

and combining

$$\begin{aligned}
\mathbf{M}\ddot{\mathbf{x}} + (\mathbf{C} + \mathbf{G})\,\dot{\mathbf{x}} + \mathbf{K}\mathbf{x} = Y(t)\,[\mathbf{M}\ddot{\chi} + (\mathbf{C} + \mathbf{G})\,\dot{\chi} + \mathbf{K}\chi] + \\
\delta(t)\,[\mathbf{M}\dot{\chi}(0) + (\mathbf{C} + \mathbf{G})\,\chi(0)] + \delta'(t)\mathbf{M}\chi(0)
\end{aligned} \tag{2.110}$$

Since the functions $Y(t)$, $\delta(t)$, and $\delta'(t)$ are linearly independent, it is clear that $\mathbf{x}$ is the impulse response if and only if $\mathbf{M}\ddot{\chi} + (\mathbf{C}+\mathbf{G})\,\dot{\chi} + \mathbf{K}\chi = 0$, $\chi(0) = (0,\ldots,0)^T$, and $\mathbf{M}\dot{\chi}(0) = (0,\ldots,1,0,\ldots,0)^T$ where the one is the i-th component. The impulse response vector $\mathbf{x}$ is therefore the homogeneous solution to eqn (2.73) with initial conditions $\mathbf{x}(0) = (0,\ldots,0)^T$ and $\dot{\mathbf{x}}(0) = \mathbf{M}^{-1}(0,\ldots,1,0,\ldots,0)^T$. Since $\dot{h}_{ii}(0^+)$ is the entry in the i-th row of $\dot{x}(0)$,

$$\dot{h}_{ii}(0^+) = \left[\mathbf{M}^{-1}\right]_{ii} \tag{2.111}$$

where $\left[\mathbf{M}^{-1}\right]_{ii}$ is the entry of the i-th row i and i-th column of the inverse of the matrix $\mathbf{M}$.

Finally, the mean power $\langle P_i \rangle = \langle f_i \dot{x}_i \rangle$ supplied to resonator i is

$$\langle P_i \rangle = \frac{S_i}{2}\left[\mathbf{M}^{-1}\right]_{ii} \tag{2.112}$$

This result generalizes eqn (2.20) to an arbitrary number of resonators conservatively coupled. It appears (Gersh, 1969) in the special case of elastic couplings.

Mean energy sharing

We have seen at the beginning of this chapter that for a single resonator the equality $\langle V \rangle = \langle K \rangle$ is verified under the condition that the external force f is a white noise random process. We are now in a position to generalize this result for a set of oscillators. But two new facts lead us to modify this statement. The first one is that in the presence of a gyroscopic coupling, the kinetic and elastic energies must rather be replaced by respectively $E + L$ and $E - L$. The expected equality is therefore $\langle E + L \rangle = \langle E - L \rangle$ or equivalently,

$$\langle L \rangle = 0 \tag{2.113}$$

The second point is that in statistical energy analysis, we guess to prove a stronger result. We aim to state that the equality of $E + L$ and $E - L$ is valid *locally*, that is for each individual oscillator. Nevertheless, this result will turn out to be valid only for asymptotically stable systems under the assumption of random forces. The key result of this subsection is the following.

Let a set of oscillators be governed by eqn (2.72) where the forces f_i are uncorrelated white noises. Let L_i be the Lagrangian of individual resonators defined by eqn (2.89). If the system is asymptotically stable then

$$\langle L_i \rangle = 0 \tag{2.114}$$

for all i.

Before giving the proof, let us comment on this result. It is clear that for a unique resonator, eqn (2.114) reduces to eqn (2.28). The case of a single damped resonator is therefore embodied in the above result. Let us recall that for a single resonator, a null

Lagrangian means that the mean kinetic energy equals the mean elastic energy, a result that has been proved in Section 2.1.

The definition of asymptotic stability is given in Appendix A. But in the particular case of our set of resonators, asymptotic stability means that if all excitations are switched off, the system will stop in the limit of infinite time. In practice, this is of course the case for all structures of interest since perpetual motion does not exist! But from the mathematical point of view, asymptotic stability is ensured under the condition that either all resonators are damped, or all undamped resonators are connected to damped resonators (Appendix A). The special case of an isolated undamped resonator must therefore be avoided. It is easy to observe why the result fails for a single undamped resonator. In the proof of equality $\langle V \rangle = \langle K \rangle$ in Section 2.1, we have calculated both $\langle V \rangle$ and $\langle K \rangle$ and found that they are equal but inversely proportional to the damping ratio which, of course, requires that the damping ratio is not zero.

For a set of resonators we have defined L_i such that $\sum L_i = L$. The argument was that the whole must be the sum of the parts. Among several advantages of this definition, eqn (2.113) becomes a mere consequence of eqn (2.114). So, we need only prove eqn (2.114).

In the absence of gyroscopic coupling, the Lagrangian L_i defined by eqn (2.89) may clearly be separated into two terms that may be attributed to the kinetic and elastic energy of the resonator. Therefore, if we introduce

$$K_i = \frac{1}{2}\sum_j m_{ij}\dot{x}_i\dot{x}_j, \quad V_i = \frac{1}{2}\left[\sum_j k_{ii}x_i^2 - \sum_{j\neq i} k_{ij}x_i x_j\right]$$

as the natural definitions of local kinetic and elastic energies, the equality $\langle L_i \rangle = 0$ reads $\langle K_i \rangle = \langle V_i \rangle$. Thus, $\langle K_i \rangle = \langle V_i \rangle$ holds for all resonators or, in other words, the equilibrium of kinetic and elastic energies is verified locally. This result is fundamental for several reasons. In practice, when we want to measure the vibrational energy of a subsystem, it is sufficient to measure its kinetic energy and to multiply the result by two. The kinetic energy is most often easier to measure than the elastic energy, especially for thin structures for which measuring the elastic energy requires assessing the second-order spatial derivatives of deflection, a task generally difficult to perform. The equality of kinetic and elastic energies is invoked in many circumstances, sometimes without giving the exact conditions under which it is valid.

We now turn to the proof of eqn (2.114). To establish eqn (2.114) where $\langle L_i \rangle$ is given by eqn (2.98), we must examine $\langle x_i x_j \rangle$, $\langle \dot{x}_i x_j \rangle$, and $\langle \dot{x}_i \dot{x}_j \rangle$. Let us start with $\langle x_i x_j \rangle$. The expectation of a product $x_i x_j$ is, following eqn (1.38),

$$\langle x_i x_j \rangle = \frac{1}{2\pi}\int_{-\infty}^{\infty} S_{x_i x_j}(\omega)\,\mathrm{d}\omega \tag{2.115}$$

But x_i, x_j are the outputs of the MIMO systems h_{ik}, h_{jl} to the inputs f_k, f_l,

$$S_{x_i x_j}(\omega) = \sum_{k=1}^{N}\sum_{l=1}^{N} S_{kl}\overline{H}_{ik}(\omega)H_{jl}(\omega) \tag{2.116}$$

where $H_{ik}(\omega)$ is the entry in the i-th row and the k-th column of the matrix $\mathbf{H}$. By virtue of Assumption 2, all terms $k \neq l$ vanish,

$$S_{x_i x_j}(\omega) = \sum_{k=1}^{N} S_k \overline{H}_{ik}(\omega) H_{jk}(\omega) \tag{2.117}$$

Hence,

$$\langle x_i x_j \rangle = \sum_{k=1}^{N} \frac{S_k}{2\pi} \int_{-\infty}^{\infty} \overline{H}_{ik}(\omega) H_{jk}(\omega)\, \mathrm{d}\omega \tag{2.118}$$

The product $\langle \dot{x}_i x_j \rangle$ is calculated in exactly the same manner, except that $\dot{x}_i$ is the time-derived output of the system h_{ij} to the inputs f_k. A factor $-\imath\omega$ is therefore introduced. The result is

$$\langle \dot{x}_i x_j \rangle = \sum_{k=1}^{N} \frac{S_k}{2\pi} \int_{-\infty}^{\infty} -\imath\omega \overline{H}_{ik}(\omega) H_{jk}(\omega)\, \mathrm{d}\omega \tag{2.119}$$

By a similar reasoning,

$$\langle \dot{x}_i \dot{x}_j \rangle = \sum_{k=1}^{N} \frac{S_k}{2\pi} \int_{-\infty}^{\infty} \omega^2 \overline{H}_{ik}(\omega) H_{jk}(\omega)\, \mathrm{d}\omega \tag{2.120}$$

Introducing the above equations in eqn (2.98) gives

$$\begin{aligned}\langle L_i \rangle = \sum_{k=1}^{N} \frac{S_k}{4\pi} \int_{-\infty}^{\infty} \Bigg[& \omega^2 \sum_j m_{ij} \overline{H}_{ik} H_{jk} \\ & - \sum_{j\neq i} g_{ij} \imath\omega \overline{H}_{ih} H_{jh} - k_{ii} \,|\, H_{ik} \,|^2 + \sum_{j\neq i} k_{ij} \overline{H}_{ik} H_{jk} \Bigg] \mathrm{d}\omega\end{aligned} \tag{2.121}$$

Thus $\langle L_i \rangle$ is found to be a linear combination of S_k. Since the equality $\langle L_i \rangle = 0$ must be valid for any S_k, all coefficients of the linear combinations must be zero. It yields

$$\int_{-\infty}^{\infty} \overline{H}_{ik} \left[-\omega^2 \sum_j m_{ij} H_{jk} + \sum_{j\neq i} g_{ij} \imath\omega H_{jk} + k_{ii} H_{ik} - \sum_{j\neq i} k_{ij} H_{jk} \right] \mathrm{d}\omega = 0 \tag{2.122}$$

where we have factorized $\overline{H}_{ik}$. This equality is established in what follows.

To prove the above equality, we first rewrite it by using matrix notations introduced in the previous subsection. The sum $\sum_j m_{ij} H_{jk}$ may be viewed as the entry in the i-th row and the k-th column of the matrix $\mathbf{MH}$. Let us denote it by $[\mathbf{MH}]_{ik}$. Similarly, the sum $\sum_j g_{ij} H_{jk}$ is $[\mathbf{GH}]_{ik}$. Furthermore, the last two terms of eqn (2.122) may be written as

$$k_{ii}H_{ik} - \sum_{j \neq i} k_{ij}H_{jk}$$

and may be viewed as the entry in the i-th row and the k-th column of the matrix **KH**. We shall denote it by $[\mathbf{KH}]_{ik}$. With these considerations, equality (2.122) becomes

$$\int_{-\infty}^{\infty} \overline{H}_{ik} \left[-\omega^2 \mathbf{MH} + \imath\omega \mathbf{GH} + \mathbf{KH} \right]_{ik} \, \mathrm{d}\omega = 0 \tag{2.123}$$

But by factorizing **H** the matrix inside the brackets becomes $\left(-\omega^2\mathbf{M} + \imath\omega\mathbf{G} + \mathbf{K}\right)\mathbf{H}$. Adding and subtracting the matrix $\imath\omega\mathbf{C}$ and remarking that **H** is the inverse of $-\omega^2\mathbf{M} + \imath\omega(\mathbf{C} + \mathbf{G}) + \mathbf{K}$, eqn (2.123) becomes

$$\int_{-\infty}^{\infty} \overline{H}_{ik} \, [\mathbf{I}]_{ik} \, \mathrm{d}\omega - \int_{-\infty}^{\infty} \imath\omega\overline{H}_{ik} \, [\mathbf{CH}]_{ik} \, \mathrm{d}\omega = 0 \tag{2.124}$$

where **I** is the identity matrix whose i, k-component is $[\mathbf{I}]_{ik} = \delta_{ik}$. The matrix **C** is also diagonal and the i, k-component of the product **CH** is $c_i H_{ik}$. Finally, the equality to prove reduces to

$$\delta_{ik} \int_{-\infty}^{\infty} \overline{H}_{ik}(\omega) \, \mathrm{d}\omega - \imath c_i \int_{-\infty}^{\infty} \omega \, |\, H_{ik} \,|^2(\omega) \, \mathrm{d}\omega = 0 \tag{2.125}$$

We are now going to prove that both integrals are zero.

Let us start by the second integral of eqn (2.125). The integrand factor is the function $\omega \mapsto \omega \, |\, H_{ik} \,|^2(\omega)$. Since **M**, **C**, **G**, **K** are real-valued, the conjugate of the dynamic stiffness matrix is

$$\overline{\mathbf{D}}(\omega) = -\omega^2\mathbf{M} - \imath\omega(\mathbf{C} + \mathbf{G}) + \mathbf{K} = \mathbf{D}(-\omega) \tag{2.126}$$

Remember that $\overline{\mathbf{D}}$ is defined as the matrix whose entries are the conjugate entries of **D**. After eqn (2.79), **H** is the inverse of the dynamic stiffness matrix and since matrix inversion and complex conjugation commute,

$$\overline{\mathbf{H}}(\omega) = \overline{\mathbf{D}^{-1}} = \left(\overline{\mathbf{D}}\right)^{-1} = \mathbf{H}(-\omega) \tag{2.127}$$

In particular, for the entry i, k,

$$H_{ik}(-\omega) = \overline{H}_{ik}(\omega) \tag{2.128}$$

By using this property two times, we obtain that $\omega \mapsto \omega \, |\, H_{ik} \,|^2(\omega)$ is an odd function. The integral from $-\infty$ to ∞ is therefore zero,

$$\int_{-\infty}^{\infty} \omega \, |\, H_{ik} \,|^2(\omega) \, \mathrm{d}\omega = 0 \tag{2.129}$$

The first integral $\int H_{ik}\, d\omega$ may be calculated by the residue theorem. Since $\mathbf{H} = \mathbf{D}^{-1}$, we first form the determinant,

$$\Delta(\omega) = \det\left(-\omega^2\mathbf{M} + \imath\omega(\mathbf{C} + \mathbf{G}) + \mathbf{K}\right) \tag{2.130}$$

We also form the cofactors $(-1)^{i+k}\Delta_{ik}$ with

$$\Delta_{ik}(\omega) = \det\left(-\omega^2\mathbf{M}^{ik} + \imath\omega(\mathbf{C}^{ik} + \mathbf{G}^{ik}) + \mathbf{K}^{ik}\right) \tag{2.131}$$

where $\mathbf{A}^{ik}$ designates the matrix extracted from $\mathbf{A}$ by cancelling row i and column k. Frequency response functions are then given by

$$H_{ik}(\omega) = (-1)^{i+k}\frac{\Delta_{ki}(\omega)}{\Delta(\omega)} \tag{2.132}$$

Both $\Delta_{ki}(\omega)$ and $\Delta(\omega)$ are polynomials and consequently $\omega \mapsto H_{ik}(\omega)$ is a rational function.

The degree of the polynomial $\Delta(\omega)$ is determined in the following manner. The determinant (2.130) is

$$\Delta(\omega) = \begin{vmatrix} -m_1\omega^2 + \imath c_1\omega + k_{11} & & -m_{ij}\omega^2 + \imath\omega g_{ij} - k_{ij} \\ & \ddots & \\ -m_{ij}\omega^2 + \imath\omega g_{ij} - k_{ij} & & -m_N\omega^2 + \imath c_N\omega + k_{NN} \end{vmatrix} \tag{2.133}$$

where diagonal entries are $D_{ii} = -\omega^2 m_i + \imath\omega c_i + k_{ii}$ and off-diagonal entries are $D_{ij} = -m_{ij}\omega^2 + \imath\omega g_{ij} - k_{ij}$. These are polynomials of degree ≤ 2. By using the definition of a determinant (Lang, 2002), we get

$$\Delta(\omega) = \sum_{\sigma\in S_N} \epsilon_\sigma D_{\sigma(1)1} D_{\sigma(2)2} \dots D_{\sigma(N)N} \tag{2.134}$$

where σ is a permutation of the symmetric group S_N and ϵ_σ its sign. All products $\epsilon_\sigma D_{\sigma(1)1} D_{\sigma(2)2} \dots D_{\sigma(N)N}$ are polynomials of degree $\leq 2N$ in which the coefficient of ω^{2N} is $(-1)^N \epsilon_\sigma m_{\sigma(1)1} m_{\sigma(2)2} \dots m_{\sigma(N)N}$. It is then clear that the coefficient of ω^{2N} in $\Delta(\omega)$ is

$$(-1)^N \sum_{\sigma\in S_N} \epsilon_\sigma m_{\sigma(1)1} m_{\sigma(2)2} \dots m_{\sigma(N)N} = (-1)^N \det(\mathbf{M}) \tag{2.135}$$

Since $\det(\mathbf{M}) \neq 0$ ($\mathbf{M}$ is definite positive), the polynomial $\Delta(\omega)$ has degree $2N$.

For Δ_{ki}, the reasoning is similar. Recall that Δ_{ki} is the determinant extracted from Δ by cancelling row k and column i. We conclude that $\Delta_{ki}(\omega)$ is a polynomial of degree $\leq 2(N-1)$ and that the coefficient of $\omega^{2(N-1)}$ is $\det(\mathbf{M}^{ki})$.

Consequently $\omega \mapsto H_{ik}(\omega)$ is a rational function with $2(N-1)$ zeros (at most) and $2N$ poles. In the complex plane, $z \mapsto H_{ik}(z)$ is holomorphic except at its poles. We choose the closed path shown in Fig. 2.6 which consists of the segment $[-R, R]$ and the lower semicircle C_R of radius R oriented clockwise. The residue theorem applied to H_{ik} with this path gives

$$\int_{-R}^{R} H_{ik}(z)\,\mathrm{d}z + \int_{C_R} H_{ik}(z)\,\mathrm{d}z = -2\imath\pi \sum_j \mathrm{Res}(z_j, H_{ik}) \tag{2.136}$$

where the sum of the right-hand side runs over all poles enclosed by the path. But H_{ik} is a rational function whose difference in degrees of denominator and numerator is greater than 2 so that $zH_{ik}(z) \to 0$ when $|z| \to \infty$. By Jordan's lemma (see Appendix B),

$$\lim_{R\to\infty} \int_{C_R} H_{ik}(z)\,\mathrm{d}z = 0 \tag{2.137}$$

Taking the limit $R \to \infty$ in both sides of eqn (2.136) gives

$$\int_{-\infty}^{\infty} H_{ik}(z)\,\mathrm{d}z = -2\imath\pi \sum_j \mathrm{Res}(z_j, H_{ik}) \tag{2.138}$$

where the sum now runs over all poles located in the half-plane $\mathrm{Im}(z) < 0$.

The only remaining question is therefore the localization of poles of $H_{ik}(\omega)$. To solve this question, we must examine the roots of $\Delta(\omega)$. Equivalently, we may introduce the polynomial,

$$\tilde{\Delta}(p) = \det(p^2\mathbf{M} + p(\mathbf{C} + \mathbf{G}) + \mathbf{K}) \tag{2.139}$$

which implies $\Delta(\omega) = \tilde{\Delta}(\imath\omega)$. Thus, the roots of Δ are the roots of $\tilde{\Delta}$ rotated by $\pi/2$ clockwise. But, the position of the zeros of $\tilde{\Delta}$ in the complex plane is related to the stability of the system $\mathbf{M}\ddot{\mathbf{X}}+(\mathbf{C}+\mathbf{G})\dot{\mathbf{X}}+\mathbf{K}\mathbf{X} = \mathbf{F}$. More precisely, if the system is asymptotically

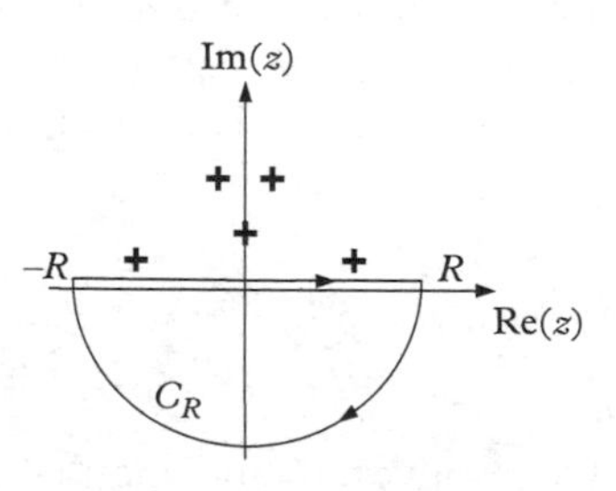

Figure 2.6 *Path of integration in complex plane and position of poles of H_{ik}.*

stable (see Appendix A for definitions and conditions implying asymptotic stability), that is if,

$$\lim_{t\to\infty} \left\| \mathbf{X}(t) \right\| = 0 \tag{2.140}$$

for any $\mathbf{X}(0)$, then all roots of $\tilde{\Delta}$ are positioned in the complex half-plane $\mathrm{Re}(z) < 0$. Hence, all roots of Δ are positioned in the complex half-plane $\mathrm{Im}(z) > 0$. Therefore the sum in the right-hand side of eqn (2.138) is empty and

$$\int_{-\infty}^{\infty} H_{ik}(z)\, \mathrm{d}z = 0 \tag{2.141}$$

This completes the proof.

2.3 Weakly Coupled Resonators

In this section, we make the assumption that the coupling of resonators is light. This is the third assumption of statistical energy analysis.

Assumption 3 *Couplings are light.*

This means that $k_{ij} \ll k_i$ for all j, $g_{ij} \ll c_i$, and $m_{ij} \ll m_i$. This enables us to introduce a small parameter $\epsilon \ll 1$ such that

$$k_{ij} = \epsilon\alpha_{ij}, \quad g_{ij} = \epsilon\alpha'_{ij}, \quad \text{and} \quad m_{ij} = \epsilon\alpha''_{ij} \tag{2.142}$$

The perturbation technique consists in seeking the solution x_i as a development in powers of ϵ,

$$x_i(t) = x_{i0}(t) + \epsilon x_{i1}(t) + \epsilon^2 x_{i2}(t) + o(\epsilon^2) \tag{2.143}$$

By deriving term by term, we get similar developments for $\dot{x}_i$ and $\ddot{x}_i$. Substituting eqns (2.142) and (2.143) into (2.72) and identifying all terms of same order ϵ^n gives

$$\begin{cases} m_i\ddot{x}_{i0} + c_i\dot{x}_{i0} + k_i x_{i0} = f_i(t) \\ m_i\ddot{x}_{i1} + c_i\dot{x}_{i1} + k_i x_{i1} = \sum_{j\neq i} \alpha_{ij}\left(x_{j0} - x_{i0}\right) - \alpha'_{ij}\dot{x}_{j0} - \alpha''_{ij}\ddot{x}_{j0} \\ m_i\ddot{x}_{i2} + c_i\dot{x}_{i2} + k_i x_{i2} = \sum_{j\neq i} \alpha_{ij}\left(x_{j1} - x_{i1}\right) - \alpha'_{ij}\dot{x}_{j1} - \alpha''_{ij}\ddot{x}_{j1} \\ \ldots \end{cases} \tag{2.144}$$

This is a set of linear differential equations of second order with constant coefficients where the unknowns are the functions x_{in}. The right-hand side is the source term. The unknown x_{in} of order n thus depends on all unknowns $x_{j,n-1}$ of order $n-1$. It is therefore possible to solve recursively this set of equations. Let us introduce the impulse response $h_i(t)$ of the resonator. The solution is

$$\begin{cases} x_{i0} = h_i * f_i \\ x_{i1} = h_i * \sum_{j\neq i} \alpha_{ij}\left(x_{j0} - x_{i0}\right) - \alpha'_{ij}\dot{x}_{j0} - \alpha''_{ij}\ddot{x}_{j0} \\ x_{i2} = h_i * \sum_{j\neq i} \alpha_{ij}\left(x_{j1} - x_{i1}\right) - \alpha'_{ij}\dot{x}_{j1} - \alpha''_{ij}\ddot{x}_{j1} \\ \dots \end{cases} \tag{2.145}$$

where * denotes a convolution product. We have thus obtained the solution to the governing equation (2.72) as a sequence of linear time-invariant systems, the output of one being the input of the next.

Mean exchanged power

The mean power exchanged between i and j is, following eqn (2.100),

$$\langle P_{ij}\rangle = \epsilon\left[\alpha''_{ij}\langle\dot{x}_i\ddot{x}_j\rangle + \alpha'_{ij}\langle\dot{x}_i\dot{x}_j\rangle + \alpha_{ij}\langle x_i\dot{x}_j\rangle\right]$$

Introducing the development (2.143) of x_j and x_i gives

$$\begin{aligned}\langle P_{ij}\rangle = \epsilon\Big[&\alpha''_{ij}\big\langle(\dot{x}_{i0} + \epsilon\dot{x}_{i1})(\ddot{x}_{j0} + \epsilon\ddot{x}_{j1})\big\rangle + \alpha'_{ij}\big\langle(\dot{x}_{i0} + \epsilon\dot{x}_{i1})(\dot{x}_{j0} + \epsilon\dot{x}_{j1})\big\rangle \\ &+ \alpha_{ij}\big\langle(x_{i0} + \epsilon x_{i1})(\dot{x}_{j0} + \epsilon\dot{x}_{j1})\big\rangle\Big] + o(\epsilon^2)\end{aligned}$$

Developing and keeping all terms up to order 2,

$$\begin{aligned}\langle P_{ij}\rangle = &\ \epsilon\left[\alpha''_{ij}\langle\dot{x}_{i0}\ddot{x}_{j0}\rangle + \alpha'_{ij}\langle\dot{x}_{i0}\dot{x}_{j0}\rangle + \alpha_{ij}\langle x_{i0}\dot{x}_{j0}\rangle\right] \\ &+ \epsilon^2\Big[\alpha''_{ij}\left(\langle\dot{x}_{i0}\ddot{x}_{j1}\rangle + \langle\dot{x}_{i1}\ddot{x}_{j0}\rangle\right) + \alpha'_{ij}\left(\langle\dot{x}_{i0}\dot{x}_{j1}\rangle + \langle\dot{x}_{i1}\dot{x}_{j0}\rangle\right) \\ &+ \alpha_{ij}\left(\langle x_{i0}\dot{x}_{j1}\rangle + \langle x_{i1}\dot{x}_{j0}\rangle\right)\Big] + o(\epsilon^2)\end{aligned} \tag{2.146}$$

The first cross-correlation to be evaluated is $\langle x_{i0}\dot{x}_{j0}\rangle$. Following eqn (1.38),

$$\langle x_{i0}\dot{x}_{j0}\rangle = \frac{1}{2\pi}\int_{-\infty}^{\infty} S_{x_{i0}\dot{x}_{j0}}(\omega)\,\mathrm{d}\omega$$

Since x_{i0} is the output of the linear system h_i to the input f_i and $\dot{x}_{j0}$ the time-derived output of the linear system h_j to the input f_j,

$$S_{x_{i0}\dot{x}_{j0}} = \overline{H}_i(\omega) \times \imath\omega H_j(\omega) S_{f_if_j} = 0 \tag{2.147}$$

where the last equality stems from $S_{f_if_j} = 0$ by Assumption 2. So, $\langle x_{i0}\dot{x}_{j0}\rangle = 0$. In a similar fashion,

$$\langle \dot{x}_{i0}\ddot{x}_{j0}\rangle = \langle \dot{x}_{i0}\dot{x}_{j0}\rangle = \langle x_{i0}\dot{x}_{j0}\rangle = 0 \tag{2.148}$$

All first-order terms in eqn (2.146) vanish.

For the second-order terms, we start by

$$\langle x_{i0}\dot{x}_{j1}\rangle = \frac{1}{2\pi}\int_{-\infty}^{\infty} S_{x_{i0}\dot{x}_{j1}}(\omega)\,\mathrm{d}\omega \tag{2.149}$$

Since $\dot{x}_{j1}$ is the time-derived output of the linear system h_j to the input $\sum_{l\neq j}\alpha_{jl}(x_{l0}-x_{j0})-\alpha'_{jl}\dot{x}_{l0}-\alpha''_{jl}\ddot{x}_{l0}$ and x_{i0} is the output of the linear system h_i to the input f_i,

$$S_{x_{i0}\dot{x}_{j1}} = \sum_{l\neq j}\overline{H}_i(\omega) \times \imath\omega H_j(\omega)\left(\alpha_{jl}S_{f_ix_{l0}} - \alpha_{jl}S_{f_ix_{j0}} - \alpha'_{jl}S_{f_i\dot{x}_{l0}} - \alpha''_{jl}S_{f_i\ddot{x}_{l0}}\right) \tag{2.150}$$

Again x_{l0} is the output of the linear system h_l to the input f_l,

$$S_{x_{i0}\dot{x}_{j1}} = \sum_{l\neq i}\imath\omega\overline{H}_iH_j\left(\alpha_{jl}H_lS_{f_if_l} - \alpha_{jl}H_jS_{f_if_j} - \imath\omega\alpha'_{jl}H_lS_{f_if_l} + \omega^2\alpha''_{jl}H_lS_{f_if_l}\right)$$

As usual $S_{f_if_j} = 0$ and among all terms $l\neq j$ the index $l = i$ is the only one for which $S_{f_if_l}\neq 0$. Therefore,

$$S_{x_{i0}\dot{x}_{j1}} = \imath\omega\left(\alpha_{ji} - \imath\omega\alpha'_{ji} + \omega^2\alpha''_{ji}\right)H_j(\omega)\,|\,H_i\,|^2(\omega)S_{f_if_i}$$

And finally,

$$\langle x_{i0}\dot{x}_{j1}\rangle = \frac{S_i}{2\pi}\int_{-\infty}^{\infty}\imath\omega\left(\alpha_{ji} - \imath\omega\alpha'_{ji} + \omega^2\alpha''_{ji}\right)H_j(\omega)\,|\,H_i\,|^2(\omega)\,\mathrm{d}\omega \tag{2.151}$$

Interchanging i and j in eqn (2.151) and moving the dot by eqn (1.47) leads to

$$\langle x_{i1}\dot{x}_{j0}\rangle = -\frac{S_j}{2\pi}\int_{-\infty}^{\infty}\imath\omega\left(\alpha_{ij} - \imath\omega\alpha'_{ij} + \omega^2\alpha''_{ij}\right)H_i(\omega)\,|\,H_j\,|^2(\omega)\,\mathrm{d}\omega \tag{2.152}$$

The calculation of $\langle \dot{x}_{i0}\dot{x}_{j1}\rangle$ follows the proof of eqn (2.151) except that $-\iota\omega\overline{H}_i$ must replace $\overline{H}_i$ at the first step. The result is

$$\langle \dot{x}_{i0}\dot{x}_{j1}\rangle = \frac{S_i}{2\pi}\int_{-\infty}^{\infty} \omega^2 \left(\alpha_{ji} - \iota\omega\alpha'_{ji} + \omega^2\alpha''_{ji}\right) H_j(\omega)\,|\,H_i\,|^2(\omega)\,\mathrm{d}\omega \tag{2.153}$$

By permuting i and j,

$$\langle \dot{x}_{i1}\dot{x}_{j0}\rangle = \frac{S_j}{2\pi}\int_{-\infty}^{\infty} \omega^2 \left(\alpha_{ij} - \iota\omega\alpha'_{ij} + \omega^2\alpha''_{ij}\right) H_i(\omega)\,|\,H_j\,|^2(\omega)\,\mathrm{d}\omega \tag{2.154}$$

Again, $\langle \dot{x}_{i0}\ddot{x}_{j1}\rangle$ follows the proof of eqn (2.151) but with an additional factor ω^2,

$$\langle \dot{x}_{i0}\ddot{x}_{j1}\rangle = \frac{S_i}{2\pi}\int_{-\infty}^{\infty} \iota\omega^3 \left(\alpha_{ji} - \iota\omega\alpha'_{ji} + \omega^2\alpha''_{ji}\right) H_j(\omega)\,|\,H_i\,|^2(\omega)\,\mathrm{d}\omega \tag{2.155}$$

By moving one dot and permuting i and j,

$$\langle \dot{x}_{i1}\ddot{x}_{j0}\rangle = -\frac{S_j}{2\pi}\int_{-\infty}^{\infty} \iota\omega^3 \left(\alpha_{ij} - \iota\omega\alpha'_{ij} + \omega^2\alpha''_{ij}\right) H_i(\omega)\,|\,H_j\,|^2(\omega)\,\mathrm{d}\omega \tag{2.156}$$

Applying the symmetries $\alpha_{ji} = \alpha_{ij}$, $\alpha''_{ji} = \alpha''_{ij}$ and antisymmetry $\alpha'_{ji} = -\alpha'_{ij}$ and substituting the above six relationships into eqn (2.146) gives

$$\begin{aligned}\langle P_{ij}\rangle = \epsilon^2 \Bigg[& \frac{S_i}{2\pi}\int_{-\infty}^{\infty} \iota\omega\left[(\alpha_{ij} + \omega^2\alpha''_{ij})^2 + \omega^2\alpha'^2_{ij}\right] H_j\,|\,H_i\,|^2\,\mathrm{d}\omega \\ & -\frac{S_j}{2\pi}\int_{-\infty}^{\infty} \iota\omega\left[(\alpha_{ij} + \omega^2\alpha''_{ij})^2 + \omega^2\alpha'^2_{ij}\right] H_i\,|\,H_j\,|^2\,\mathrm{d}\omega\Bigg] + o(\epsilon^2)\end{aligned} \tag{2.157}$$

This equation shows that the exchanged power is a linear combination of power spectral densities of forces. When only one resonator is excited, the power flows from the excited resonator to the passive resonator. But when both resonators are excited, the direction of flux depends not only on the power spectral densities S_i but also on the coefficients $\int \iota\omega[(\alpha_{ij} + \omega^2\alpha''_{ij})^2 + \omega^2\alpha'^2_{ij}]H_j\,|\,H_i\,|^2\,\mathrm{d}\omega$.

By the same method that we obtained the exchanged power in terms of spectral densities, we can derive the vibrational energy E_i in terms of spectral densities S_i. The mean vibrational energy is given by eqn (2.90). By introducing the development (2.142), (2.143) we get at order zero

$$\langle E_i\rangle = \frac{1}{2}\left[m_i\langle \dot{x}_{i0}^2\rangle + k_i\langle x_{i0}^2\rangle\right] + o(1) \tag{2.158}$$

But,

$$\langle x_{i0}^2 \rangle = \frac{S_i}{2\pi} \int_{-\infty}^{\infty} |H_i|^2(\omega)\, \mathrm{d}\omega \tag{2.159}$$

and

$$\langle \dot{x}_{i0}^2 \rangle = \frac{S_i}{2\pi} \int_{-\infty}^{\infty} \omega^2 |H_i|^2(\omega)\, \mathrm{d}\omega \tag{2.160}$$

This is the same problem as the determination of the vibrational energy of a single resonator solved in Section 2.1 and for which we have shown that the mean kinetic and elastic energies are equal. Hence,

$$\langle E_i \rangle = \frac{S_i}{2\pi} \int_{-\infty}^{\infty} m_i \omega^2 |H_i|^2(\omega)\, \mathrm{d}\omega + o(1) \tag{2.161}$$

The order of this development is zero although the development in eqn (2.157) is of order two. However, since (2.157) does not contain any term of order zero or one, combining eqns (2.157) and (2.161) gives

$$\langle P_{ij} \rangle = \epsilon^2 \left[\frac{\int_{-\infty}^{\infty} \imath\omega \left[(\alpha_{ij} + \omega^2 \alpha''_{ij})^2 + \omega^2 \alpha'^2_{ij} \right] H_j |H_i|^2\, \mathrm{d}\omega}{\int_{-\infty}^{\infty} m_i \omega^2 |H_i|^2\, \mathrm{d}\omega} \langle E_i \rangle \right.$$

$$\left. - \frac{\int_{-\infty}^{\infty} \imath\omega \left[(\alpha_{ij} + \omega^2 \alpha''_{ij})^2 + \omega^2 \alpha'^2_{ij} \right] |H_j|^2 H_i\, \mathrm{d}\omega}{\int_{-\infty}^{\infty} m_j \omega^2 |H_j|^2\, \mathrm{d}\omega} \langle E_j \rangle \right] + o(\epsilon^2)$$

The exchanged power is therefore also a linear combination of vibrational energies. This may be written as

$$\langle P_{ij} \rangle = \beta_{ij} \langle E_i \rangle - \beta_{ji} \langle E_j \rangle + o(\epsilon^2) \tag{2.162}$$

where

$$\beta_{ij} = \epsilon^2 \frac{\int_{-\infty}^{\infty} \imath\omega \left[(\alpha_{ij} + \omega^2 \alpha''_{ij})^2 + \omega^2 \alpha'^2_{ij} \right] H_j |H_i|^2\, \mathrm{d}\omega}{\int_{-\infty}^{\infty} m_i \omega^2 |H_i|^2\, \mathrm{d}\omega} \tag{2.163}$$

And if we restore the original notation $k_{ij} = \epsilon\alpha_{ij}$, $g_{ij} = \epsilon\alpha'_{ij}$, and $m_{ij} = \epsilon\alpha''_{ij}$,

$$\beta_{ij} = \frac{\int_{-\infty}^{\infty} \imath\omega \left[(k_{ij} + \omega^2 m_{ij})^2 + \omega^2 g_{ij}^2 \right] H_j |H_i|^2\, \mathrm{d}\omega}{\int_{-\infty}^{\infty} m_i \omega^2 |H_i|^2\, \mathrm{d}\omega} \tag{2.164}$$

The integral (2.164) is quite general and also applies for inertial, gyroscopic, and elastic couplings and even in the presence of the three types of couplings. Furthermore, no assumption has been made on the damping level and therefore (2.164) applies for light and strong damping.

In order to get explicit expressions of β_{ij} in terms of resonator characteristics m_i, ω_i, ζ_i, k_{ij}, g_{ij}, and m_{ij}, a direct calculation of the two integrals appearing in eqn (2.164) is required. The integral of the denominator has already been encountered at the beginning of this chapter. Its value is

$$\int_{-\infty}^{\infty} m_i \omega^2 |H_i|^2 \, d\omega = \frac{\pi}{2\zeta_i m_i \omega_i} \tag{2.165}$$

For the integral of the numerator, it may be useful to transform it in a standard form. The integrand $\imath\omega \left[(k_{ij} + \omega^2 m_{ij})^2 + \omega^2 g_{ij}^2 \right] H_j |H_i|^2$ is a rational function. If we multiply the numerator and denominator by $m_j \omega_j^2 \left[1 - 2\imath\zeta_j \omega/\omega_j - \omega^2/\omega_j^2 \right]$; the integrand becomes

$$\imath\omega m_j \omega_j^2 \left[1 - 2\imath\zeta_j \omega/\omega_j - \omega^2/\omega_j^2 \right] \left[(k_{ij} + \omega^2 m_{ij})^2 + \omega^2 g_{ij}^2 \right] |H_j H_i|^2 \tag{2.166}$$

But since the function $\omega \mapsto \imath\omega \left[(k_{ij} + \omega^2 m_{ij})^2 + \omega^2 g_{ij}^2 \right] |H_j(\omega) H_i(\omega)|^2$ is odd, only the second monomial of the first bracket survives when performing the integration from $-\infty$ to ∞. This leads to

$$\beta_{ij} = \frac{4}{\pi} m_i m_j \zeta_i \zeta_j \omega_i \omega_j \int_{-\infty}^{\infty} \omega^2 \left[(k_{ij} + \omega^2 m_{ij})^2 + \omega^2 g_{ij}^2 \right] |H_j H_i|^2 \, d\omega \tag{2.167}$$

The advantage of this form of β_{ij} over (2.164) is that it reveals a symmetry under the permutation $i \leftrightarrow j$,

$$\beta_{ji} = \beta_{ij} \tag{2.168}$$

This is the so-called **reciprocity relationship**. Reciprocity will be generalized to multimodal subsystems in Chapter 3. The main consequence of reciprocity is that the exchanged power is found to be proportional to the difference of vibrational energies of adjacent subsystems,

$$\langle P_{ij} \rangle = \beta_{ij} \left(\langle E_i \rangle - \langle E_j \rangle \right) + o(\epsilon^2) \tag{2.169}$$

This equation is referred to as the **coupling power proportionality**. This result is valid up to order two in ϵ. The coupling power proportionality has been discussed by many authors and, as such, is certainly the most important result in statistical energy analysis. Let us remind ourselves of all the assumptions required to derive the coupling power proportionality. *In a set of resonators lightly coupled by conservative forces and excited* *by uncorrelated random white noise forces, the power exchanged in any pair of resonators is proportional to the difference in their energies.*

Let us finish this subsection by giving explicit relationships for the coefficient β_{ij}. By examining eqns (2.164) and (2.167), we see that the last step is to calculate the integrals $\int \iota\omega^{2n+1} H_j |H_i|^2 \, d\omega$ and $\int \omega^{2(n+1)} |H_j H_i|^2 \, d\omega$. This task is achieved in Appendix C by application of the residue theorem.

In the case of an elastic coupling alone, we have $g_{ij} = m_{ij} = 0$. The integral in the numerator of eqn (2.167) reduces to $k_{ij}^2 \int \iota\omega H_j |H_i|^2 \, d\omega$. This integral is given by eqn (C.56) when $\zeta_i, \zeta_j < 1$. It yields

$$\beta_{\text{stiff}, ij} = \frac{2k_{ij}^2 (\omega_i\zeta_i + \omega_j\zeta_j)}{m_i m_j \left[\left(\omega_i\sqrt{1-\zeta_i^2} + \omega_j\sqrt{1-\zeta_j^2}\right)^2 + (\omega_i\zeta_i + \omega_j\zeta_j)^2\right]\left[\left(\omega_i\sqrt{1-\zeta_i^2} - \omega_j\sqrt{1-\zeta_j^2}\right)^2 + (\omega_i\zeta_i + \omega_j\zeta_j)^2\right]} \tag{2.170}$$

This relationship was first derived by Newland (1966) in this exact form. However, an alternative form is possible. By using eqn (2.167) and the value of the integral $\int \omega^2 |H_j H_i|^2 \, d\omega$ given in eqn (C.85), we get

$$\beta_{\text{stiff}, ij} = \frac{2k_{ij}^2 (\zeta_i\omega_i + \zeta_j\omega_j)}{m_i m_j \left[(\omega_i^2 - \omega_j^2)^2 + 4\omega_i\omega_j(\zeta_i\omega_i + \zeta_j\omega_j)(\zeta_i\omega_j + \zeta_j\omega_i)\right]} \tag{2.171}$$

This is the formula derived by Lotz and Crandall (1973). The advantage of this form is that it is valid for any value of $\zeta_i \geq 0$, $\zeta_j \geq 0$ not both zero in opposition with eqn (2.170) which is limited to the range $0 < \zeta_i < 1$, $0 < \zeta_j < 1$. Of course, it may be checked after a tedious algebraic development that both numerators are equal.

In the case of a gyroscopic coupling, we write $k_{ij} = m_{ij} = 0$. Again, by eqn (2.167) and $\int \omega^4 |H_j H_i|^2 \, d\omega$ given in eqn (C.86),

$$\beta_{\text{gyr}, ij} = \frac{2g_{ij}^2 \omega_i\omega_j (\zeta_i\omega_j + \zeta_j\omega_i)}{m_i m_j \left[(\omega_i^2 - \omega_j^2)^2 + 4\omega_i\omega_j(\zeta_i\omega_i + \zeta_j\omega_j)(\zeta_i\omega_j + \zeta_j\omega_i)\right]} \tag{2.172}$$

This equation was derived by Fahy (1970).

The case of an inertial coupling is obtained by setting $k_{ij} = g_{ij} = 0$. Then we get, through eqns (2.167) and (C.87),

$$\beta_{\text{mass}, ij} = \frac{2m_{ij}^2 \omega_i\omega_j (\zeta_i\omega_j^3 + 4\zeta_i\zeta_j\omega_i\omega_j(\zeta_i\omega_j + \zeta_j\omega_i) + \zeta_j\omega_i^3)}{m_i m_j \left[(\omega_i^2 - \omega_j^2)^2 + 4\omega_i\omega_j(\zeta_i\omega_i + \zeta_j\omega_j)(\zeta_i\omega_j + \zeta_j\omega_i)\right]} \tag{2.173}$$

This formula also appears in Lotz and Crandall (1973).

Finally, the most general case is found by expanding the integral in eqn (2.167) as $\int [m_{ij}^2\omega^6 + (g_{ij}^2 + 2m_{ij}k_{ij})\omega^4 + k_{ij}^2\omega^2] \,|\, H_j H_i \,|^2 \, d\omega$ and combining eqns (2.171), (2.172), and (2.173),

$$\beta_{ij} = \frac{2m_{ij}^2\omega_i\omega_j(\zeta_i\omega_j^3 + 4\zeta_i\zeta_j\omega_i\omega_j(\zeta_i\omega_j + \zeta_j\omega_i) + \zeta_j\omega_i^3) + 2(g_{ij}^2 + 2k_{ij}m_{ij})\omega_i\omega_j\left(\omega_i\zeta_j + \omega_j\zeta_i\right) + 2k_{ij}^2(\zeta_i\omega_i + \zeta_j\omega_j)}{m_i m_j\left[(\omega_i^2 - \omega_j^2)^2 + 4\omega_i\omega_j(\zeta_i\omega_i + \zeta_j\omega_j)(\zeta_i\omega_j + \zeta_j\omega_i)\right]} \tag{2.174}$$

Mean dissipated power

Before closing this section on the energetics of weakly coupled resonators, we must examine the last term of the power balance (2.102): the mean dissipated power $P_{\text{diss},i}$. The instantaneous expression of dissipated power is given in eqn (2.94) by taking the mean value,

$$\langle P_{\text{diss},i}\rangle = c_i\langle \dot{x}_i^2\rangle \tag{2.175}$$

The hope is to find a relationship with the mean vibrational energy $\langle E_i\rangle$.

First, let us remark that the mean vibrational energy $\langle E_i\rangle$ may admit several expressions since we have proved that the mean Lagrangian $\langle L_i\rangle$ is zero. Therefore the mean energy remains unchanged by adding or subtracting the mean Lagrangian. Then, by adding eqns (2.98) and (2.99),

$$\langle E_i\rangle = m_i\langle \dot{x}_i^2\rangle + \sum_{j\neq i} m_{ij}\langle \dot{x}_i\dot{x}_j\rangle + \sum_{j\neq i} g_{ij}\langle \dot{x}_i x_j\rangle \tag{2.176}$$

where we have separated the internal kinetic energy $m_i\langle \dot{x}_i^2\rangle$. The above equality shows that if the coupling is realized by a spring, i.e. $m_{ij} = g_{ij} = 0$ when $i \neq j$, then the right-hand side reduces to $m_i\langle \dot{x}_i^2\rangle$ and consequently,

$$P_{\text{diss},\,i} = \frac{c_i}{m_i}\langle E_i\rangle \tag{2.177}$$

This equality shows that the mean power being dissipated is proportional to the mean vibrational energy. This is an exact equality but limited to the elastic coupling case.

In the presence of inertial or gyroscopic coupling, we must abandon the idea of finding an exact result but we may still employ the perturbation technique for a weaker result. By introducing the series (2.143) at order zero and $m_{ij} = \epsilon\alpha_{ij}''$, $g_{ij} = \epsilon\alpha_{ij}'$,

$$\langle E_i\rangle - m_i\langle \dot{x}_i^2\rangle = \epsilon\sum_{j\neq i} \alpha_{ij}''\langle \dot{x}_{i0}\dot{x}_{j0}\rangle + \alpha_{ij}'\langle \dot{x}_{i0}x_{j0}\rangle + o(\epsilon)$$

But by eqn (2.148) $\langle \dot{x}_{i0}\dot{x}_{j0}\rangle = \langle \dot{x}_{i0}x_{j0}\rangle = 0$ and the right-hand side is zero. We get

$$P_{\text{diss},\,i} = \frac{c_i}{m_i}\langle E_i\rangle + o(\epsilon) \tag{2.178}$$

Thus, the mean power dissipated is proportional to the mean vibrational energy at least to order one in ϵ in any case.

2.4 Two Coupled Resonators

Historically, the first system to be solved was not the case of N weakly coupled resonators but the special case of two resonators. In this section, we develop the theory of Scharton and Lyon (1968) in detail. Their results have not only an historical interest but are also important because a stronger result was obtained: the proportionality of exchanged power and difference of vibrational energies is valid for any coupling strength, weak or strong.

We have modified only a minor point compared with the original work: the definitions of subsystem energy and power flow are those of Section 2.2.

Governing equation

The governing equations of two coupled resonators are

$$\begin{cases} m_1\ddot{x}_1 + M(\ddot{x}_1 + \ddot{x}_2) + c_1\dot{x}_1 + G\dot{x}_2 + k_1x_1 + K(x_1 - x_2) = f_1 \\ m_2\ddot{x}_2 + M(\ddot{x}_1 + \ddot{x}_2) + c_2\dot{x}_2 - G\dot{x}_1 + k_2x_2 + K(x_2 - x_1) = f_2 \end{cases} \tag{2.179}$$

This is a special case of eqn (2.72) where we have noted M, G, K instead of m_{12}, g_{12}, k_{12} and the diagonal entries of the mass matrix is $m_{ii} = m_i + M$. From this set of equations, we can identify the mass, damping, gyroscopic, and stiffness matrices,

$$\mathbf{M} = \begin{pmatrix} m_1 + M & M \\ M & m_2 + M \end{pmatrix} \quad \mathbf{C} = \begin{pmatrix} c_1 & 0 \\ 0 & c_2 \end{pmatrix} \quad \mathbf{G} = \begin{pmatrix} 0 & G \\ -G & 0 \end{pmatrix} \quad \mathbf{K} = \begin{pmatrix} k_1 + K & -K \\ -K & k_2 + K \end{pmatrix}$$

The damping matrix is diagonal (no dashpot between resonators), the mass and stiffness matrices are symmetric and positive-definite, and the gyroscopic matrix is antisymmetric. Assumption 1 of conservative coupling is verified. Furthermore, to verify Assumption 2, we take f_i as uncorrelated white noises whose power spectral densities are noted S_i and cross-spectral density is zero.

Equation (2.179) is linear. We note $h_{ij}(\tau)$ the impulse response of resonator i with excitation on resonator j, and $H_{ij}(\omega)$ the related frequency response function obtained by Fourier transform. The system composed of two resonators is a linear time-invariant system with two inputs and two outputs.

By virtue of eqn (2.77), the dynamic stiffness matrix is

$$\mathbf{D}(\omega) = \begin{pmatrix} -(m_1 + M)\omega^2 + \imath c_1\omega + k_1 + K & -M\omega^2 + \imath\omega G - K \\ -M\omega^2 - \imath\omega G - K & -(m_2 + M)\omega^2 + \imath c_2\omega + k_2 + K \end{pmatrix} \tag{2.180}$$

Let us introduce the following parameters:

$$\begin{cases} \Omega_i &= \sqrt{(k_i + K)/(m_i + M)} \\ \Delta_i &= c_i/(m_i + M) \\ \kappa &= K/\sqrt{(m_1 + M)(m_2 + M)} \\ \gamma &= G/\sqrt{(m_1 + M)(m_2 + M)} \\ \mu &= M/\sqrt{(m_1 + M)(m_2 + M)} \end{cases} \tag{2.181}$$

The frequency response function matrix **H** is found by inverting **D**,

$$\mathbf{H}(\omega) = \frac{1}{Q(\omega)} \begin{pmatrix} \dfrac{-\omega^2 + \imath\omega\Delta_2 + \Omega_2^2}{m_1 + M} & \dfrac{\mu\omega^2 - \imath\omega\gamma + \kappa}{\sqrt{(m_1 + M)(m_2 + M)}} \\ \dfrac{\mu\omega^2 + \imath\omega\gamma + \kappa}{\sqrt{(m_1 + M)(m_2 + M)}} & \dfrac{-\omega^2 + \imath\omega\Delta_1 + \Omega_1^2}{m_2 + M} \end{pmatrix} \tag{2.182}$$

where

$$Q(\omega) = \left(-\omega^2 + \imath\omega\Delta_1 + \Omega_1^2\right)\left(-\omega^2 + \imath\omega\Delta_2 + \Omega_2^2\right) - \left|\mu\omega^2 + \imath\omega\gamma + \kappa\right|^2 \tag{2.183}$$

Let us remark that $Q(-\omega) = \overline{Q}(\omega)$ and therefore $\omega \mapsto |Q|^2(\omega)$ is an even function.

Energy and power

The definition of resonator energy was discussed in Section 2.2. We found that if the subsystems are defined in such a way that their vibrational energies are

$$E_i = \frac{1}{2}(m_i + M)\dot{x}_i^2 + \frac{1}{2}M\dot{x}_1\dot{x}_2 + \frac{1}{2}(k_i + K)x_i^2 - \frac{1}{2}Kx_1x_2 \tag{2.184}$$

then the power flow between the two resonators is

$$P_{12} = \frac{1}{2}M\left(\dot{x}_1\ddot{x}_2 - \ddot{x}_1\dot{x}_2\right) + G\dot{x}_1\dot{x}_2 + \frac{1}{2}K\left(x_1\dot{x}_2 - \dot{x}_1x_2\right) \tag{2.185}$$

and $P_{21} = -P_{12}$.

Mean vibrational energy

To calculate the mean vibrational energy $\langle E_i \rangle$, we must evaluate $\langle \dot{x}_i^2 \rangle$, $\langle \dot{x}_1\dot{x}_2 \rangle$, $\langle x_i^2 \rangle$, and $\langle x_1x_2 \rangle$ in terms of power spectral densities. Since eqn (2.179) is a MIMO system, application of eqn (1.83) and cancellation cross-product terms by virtue of Assumption 2 gives

$$\langle x_i x_j \rangle = \sum_{k=1}^{2} \frac{S_k}{2\pi} \int_{-\infty}^{\infty} \overline{H}_{ik}(\omega) H_{jk}(\omega)\, d\omega \tag{2.186}$$

and

$$\langle \dot{x}_i \dot{x}_j \rangle = \sum_{k=1}^{2} \frac{S_k}{2\pi} \int_{-\infty}^{\infty} \omega^2 \overline{H}_{ik}(\omega) H_{jk}(\omega)\, d\omega \tag{2.187}$$

Substitution into eqn (2.184) gives

$$\begin{pmatrix} \langle E_1 \rangle \\ \langle E_2 \rangle \end{pmatrix} = \begin{pmatrix} E_{11} & E_{12} \\ E_{21} & E_{22} \end{pmatrix} \begin{pmatrix} S_1/2\pi \\ S_2/2\pi \end{pmatrix} \tag{2.188}$$

where

$$E_{ik} = \frac{1}{2} \int_{-\infty}^{\infty} \left[(m_i + M)\omega^2 + (k_i + K) \right] |H_{ik}|^2(\omega) + \left(M\omega^2 - K \right) \overline{H}_{1k}(\omega) H_{2k}(\omega)\, d\omega \tag{2.189}$$

If the problem was to assess the vibrational energies from the knowledge of excitations, this set of equations would be the solution. But our purpose is rather to establish the coupling power proportionality (2.169) linking the exchanged power and vibrational energies but by a direct method which avoids the perturbation technique.

The set of equations (2.188) can be solved. The result is the power spectral densities S_i in terms of mean vibrational energies $\langle E_i \rangle$,

$$\begin{pmatrix} S_1/2\pi \\ S_2/2\pi \end{pmatrix} = \frac{1}{E_{11}E_{22} - E_{12}E_{21}} \begin{pmatrix} E_{22} & -E_{12} \\ -E_{21} & E_{11} \end{pmatrix} \begin{pmatrix} \langle E_1 \rangle \\ \langle E_2 \rangle \end{pmatrix} \tag{2.190}$$

In matrix notation,

$$E_{ik} = \frac{1}{2} \int_{-\infty}^{\infty} \overline{H}_{ik} \left[\omega^2 \mathbf{MH} + \mathbf{KH} \right]_{ik} d\omega \tag{2.191}$$

where $[.]_{ik}$ denotes the i, k-entry of the matrix. However, we have shown in eqn (2.123) that the equality $\langle L_i \rangle = 0$ is equivalent to

$$\frac{1}{2} \int_{-\infty}^{\infty} \overline{H}_{ik} \left[-\omega^2 \mathbf{MH} + \imath\omega \mathbf{GH} + \mathbf{KH} \right]_{ik} d\omega = 0 \tag{2.192}$$

So, by adding the two last equalities, we obtain

$$E_{ik} = \int_{-\infty}^{\infty} \overline{H}_{ik} \left[\frac{\imath\omega}{2} \mathbf{GH} + \mathbf{KH} \right]_{ik} d\omega \tag{2.193}$$

or alternatively by subtracting,

$$E_{ik} = \int_{-\infty}^{\infty} \overline{H}_{ik} \left[\omega^2 \mathbf{MH} - \frac{\imath\omega}{2}\mathbf{GH}\right]_{ik} \mathrm{d}\omega \tag{2.194}$$

Depending on the complexity of the mass and stiffness matrices, one of these equations for E_{ik} may be more suitable.

Mean exchanged power

To calculate the mean power flow $\langle P_{12}\rangle$ given in eqn (2.185), the mean products $\langle x_i\dot{x}_j\rangle$, $\langle \dot{x}_i\dot{x}_j\rangle$, and $\langle \dot{x}_i\ddot{x}_j\rangle$ are required. The second term has been calculated, while the two others are

$$\langle x_i\dot{x}_j\rangle = \sum_{k=1}^{2} \frac{S_k}{2\pi} \int_{-\infty}^{\infty} \imath\omega \overline{H}_{ik}(\omega) H_{jk}(\omega)\, \mathrm{d}\omega \tag{2.195}$$

and

$$\langle \dot{x}_i\ddot{x}_j\rangle = \sum_{k=1}^{2} \frac{S_k}{2\pi} \int_{-\infty}^{\infty} \imath\omega^3 \overline{H}_{ik}(\omega) H_{jk}(\omega)\, \mathrm{d}\omega \tag{2.196}$$

Bear in mind that when interchanging the dot position, the sign is modified. Hence, $\langle \dot{x}_i x_j\rangle = -\langle x_i\dot{x}_j\rangle$ and $\langle \ddot{x}_i\dot{x}_j\rangle = -\langle \dot{x}_i\ddot{x}_j\rangle$. Substitution of eqns (2.187), (2.195), and (2.196) into eqn (2.185) gives $\langle P_{12}\rangle$ as a linear combination whose matrix notation is

$$\langle P_{12}\rangle = \begin{pmatrix} A_1 & -A_2 \end{pmatrix} \begin{pmatrix} S_1/2\pi \\ S_2/2\pi \end{pmatrix} \tag{2.197}$$

with

$$\begin{cases} A_1 = \displaystyle\int_{-\infty}^{\infty} \imath\omega \left[\omega^2 M - \imath\omega G + K\right] \overline{H}_{11}(\omega) H_{21}(\omega)\, \mathrm{d}\omega \\ A_2 = -\displaystyle\int_{-\infty}^{\infty} \imath\omega \left[\omega^2 M - \imath\omega G + K\right] \overline{H}_{12}(\omega) H_{22}(\omega)\, \mathrm{d}\omega \end{cases} \tag{2.198}$$

Combining eqns (2.197) and (2.190) gives

$$\langle P_{12}\rangle = \frac{1}{E_{11}E_{22} - E_{12}E_{21}} \begin{pmatrix} A_1 & -A_2 \end{pmatrix} \begin{pmatrix} E_{22} & -E_{12} \\ -E_{21} & E_{11} \end{pmatrix} \begin{pmatrix} \langle E_1\rangle \\ \langle E_2\rangle \end{pmatrix} \tag{2.199}$$

After developing

$$\langle P_{12}\rangle = \beta_{12}\langle E_1\rangle - \beta_{21}\langle E_2\rangle \tag{2.200}$$

with

$$\begin{cases} \beta_{12} = (A_1 E_{22} + A_2 E_{21}) / (E_{11} E_{22} - E_{12} E_{21}) \\ \beta_{21} = (A_1 E_{12} + A_2 E_{11}) / (E_{11} E_{22} - E_{12} E_{21}) \end{cases} \tag{2.201}$$

In order to get an explicit relationship for β_{ij} in terms of the mechanical parameters, one must compute A_i and E_{ij}. But instead of doing it in the most general case, we solve separately the cases of elastic and inertial couplings.

Let us restrict the calculation to the particular case of an elastic coupling. We set $M = G = 0$. Firstly, we develop A_i in eqn (2.198) by using the expressions of H_{ij} given by eqn (2.182),

$$\begin{cases} A_1 = K \displaystyle\int_{-\infty}^{\infty} \imath\omega \overline{H}_{11}(\omega) H_{21}(\omega)\,\mathrm{d}\omega = \frac{\kappa^2}{m_1} \int_{-\infty}^{\infty} \imath\omega \frac{(-\omega^2 - \imath\omega\Delta_2 + \Omega_2^2)}{|Q(\omega)|^2}\,\mathrm{d}\omega \\ A_2 = -K \displaystyle\int_{-\infty}^{\infty} \imath\omega \overline{H}_{12}(\omega) H_{22}(\omega)\,\mathrm{d}\omega = -\frac{\kappa^2}{m_2} \int_{-\infty}^{\infty} \imath\omega \frac{(-\omega^2 + \imath\omega\Delta_1 + \Omega_1^2)}{|Q(\omega)|^2}\,\mathrm{d}\omega \end{cases} \tag{2.202}$$

Since $\omega \mapsto |Q|^2(\omega)$ is even, the odd terms of numerators do not contribute to the integrals and

$$\begin{cases} A_1 = \dfrac{\kappa^2 \Delta_2}{m_1} \displaystyle\int_{-\infty}^{\infty} \left| \frac{\imath\omega}{Q(\omega)} \right|^2 \mathrm{d}\omega \\ A_2 = \dfrac{\kappa^2 \Delta_1}{m_2} \displaystyle\int_{-\infty}^{\infty} \left| \frac{\imath\omega}{Q(\omega)} \right|^2 \mathrm{d}\omega \end{cases} \tag{2.203}$$

Secondly, to obtain explicit relationships for E_{ik} we have a choice between several expressions. Since **M** is diagonal and **G** is zero, the most appropriate equality is eqn (2.194). It reduces to

$$E_{ik} = \int_{-\infty}^{\infty} m_i \omega^2 |H_{ik}|^2(\omega)\,\mathrm{d}\omega \tag{2.204}$$

After developing with eqn (2.182),

$$\begin{cases} E_{11} = \dfrac{1}{m_1} \displaystyle\int_{-\infty}^{\infty} \omega^2 \left| \frac{-\omega^2 + \imath\omega\Delta_2 + \Omega_2^2}{Q(\omega)} \right|^2 \mathrm{d}\omega \\ E_{12} = \dfrac{\kappa^2}{m_2} \displaystyle\int_{-\infty}^{\infty} \left| \frac{\imath\omega}{Q(\omega)} \right|^2 \mathrm{d}\omega \\ E_{21} = \dfrac{\kappa^2}{m_1} \displaystyle\int_{-\infty}^{\infty} \left| \frac{\imath\omega}{Q(\omega)} \right|^2 \mathrm{d}\omega \\ E_{22} = \dfrac{1}{m_2} \displaystyle\int_{-\infty}^{\infty} \omega^2 \left| \frac{-\omega^2 + \imath\omega\Delta_1 + \Omega_1^2}{Q(\omega)} \right|^2 \mathrm{d}\omega \end{cases} \tag{2.205}$$

So, the problem reduces to the calculation of only two integrals: $\int |\imath\omega/Q|^2\,\mathrm{d}\omega$ and $\int |\imath\Omega_i^2\omega - \Delta_i\omega^2 - \imath\omega^3/Q|^2\,\mathrm{d}\omega$ where

$$Q(\omega) = \left(-\omega^2 + \imath\omega\Delta_1 + \Omega_1^2\right)\left(-\omega^2 + \imath\omega\Delta_2 + \Omega_2^2\right) - \kappa^2 \tag{2.206}$$

In both, we can recognize the standard form of the following quadrature formula,

$$\int_{-\infty}^{\infty} |f|^2\,\mathrm{d}\omega = \pi\frac{a_0b_3^2(a_1a_2 - a_0a_3) + a_0a_1a_4(b_2^2 - 2b_1b_3) + a_0a_3a_4(b_1^2 - 2b_0b_2) + a_4b_0^2(a_2a_3 - a_1a_4)}{a_0a_4(a_1a_2a_3 - a_0a_3^2 - a_1^2a_4)} \tag{2.207}$$

where

$$f(\omega) = \frac{b_0 + \imath\omega b_1 - \omega^2b_2 - \imath\omega^3b_3}{a_0 + \imath\omega a_1 - \omega^2a_2 - \imath\omega^3a_3 + \omega^4a_4} \tag{2.208}$$

Therefore, we consider in eqn (2.207) the set of values

$$\begin{aligned}
a_0 &= \Omega_1^2\Omega_2^2 - \kappa^2 & b_0 &= 0 & b_0' &= 0\\
a_1 &= \Delta_1\Omega_2^2 + \Delta_2\Omega_1^2 & b_1 &= 1 & b_1' &= \Omega_i^2\\
a_2 &= \Omega_1^2 + \Omega_2^2 + \Delta_1\Delta_2 & b_2 &= 0 & b_2' &= \Delta_i\\
a_3 &= \Delta_1 + \Delta_2 & b_3 &= 0 & b_3' &= 1\\
a_4 &= 1
\end{aligned} \tag{2.209}$$

The coefficients a_i are found by developing eqn (2.206) and ordering the monomials in increasing powers of ω. The coefficients b_i apply for the numerator $\imath\omega$ while b_i' apply for $\imath\Omega_i^2\omega - \Delta_i\omega^2 - \imath\omega^3$. We get successively

$$\int_{-\infty}^{\infty}\left|\frac{\imath\omega}{Q(\omega)}\right|^2\mathrm{d}\omega = \frac{n_1}{d} \tag{2.210}$$

and

$$\int_{-\infty}^{\infty}\left|\frac{\imath\Omega_i^2\omega - \Delta_i\omega^2 - \imath\omega^3}{Q(\omega)}\right|^2\mathrm{d}\omega = \frac{l_i}{d} \tag{2.211}$$

where

$$\begin{aligned}
d &= a_4(a_1a_2a_3 - a_0a_3^2 - a_1^2a_4)\\
n_1 &= a_3a_4 = \Delta_1 + \Delta_2\\
l_i &= a_1a_2 - a_0a_3a_1a_4(\Delta_i^2 - 2\Omega_i^2) + \Omega_i^4a_3a_4\\
&= (\Delta_1\Omega_2^2 + \Delta_2\Omega_1^2)(\Omega_1^2 + \Omega_2^2 + \Delta_1\Delta_2) - (\Omega_1^2\Omega_2^2 - \kappa^2)(\Delta_1 + \Delta_2) + (\Delta_1\Omega_2^2 + \Delta_2\Omega_1^2)(\Delta_i^2 - 2\Omega_i^2) + (\Delta_1 + \Delta_2)\Omega_i^4
\end{aligned}$$

Since $b_0 = b_0' = 0$, the coefficient a_0 simplifies in eqn (2.207)—this is why it does not appear in the above equations. It is not necessary to develop d since it simplifies in what follows. After some re-arrangements,

$$l_i = \Delta_i\left[(\Omega_1^2 - \Omega_2^2)^2 + (\Delta_1 + \Delta_2)(\Delta_1\Omega_2^2 + \Delta_2\Omega_1^2)\right] + \kappa^2(\Delta_1 + \Delta_2) \quad (2.212)$$

Now by combining eqns (2.201), (2.203), (2.205), (2.210), and (2.211), we get

$$\beta_{12} = \frac{\frac{\kappa^2\Delta_2 n_1}{m_1 d} \times \frac{l_1}{m_2 d} + \frac{\kappa^2\Delta_1 n_1}{m_2 d} \times \frac{\kappa^2 n_1}{m_1 d}}{\frac{l_2}{m_1 d} \times \frac{l_1}{m_2 d} - \frac{\kappa^2 n_1}{m_1 d} \times \frac{\kappa^2 n_1}{m_2 d}} \quad (2.213)$$

After simplifying the factor $m_1 m_2 d^2$,

$$\beta_{12} = \kappa^2 n_1 \times \frac{\Delta_2 l_1 + \kappa^2\Delta_1 n_1}{l_1 l_2 - \kappa^4 n_1^2} \quad (2.214)$$

Now, substituting the values of n_1, l_1, and l_2 gives

$$\beta_{12} = \kappa^2(\Delta_1 + \Delta_2) \times \frac{\Delta_2(\Delta_1[-] + \kappa^2(\Delta_1 + \Delta_2)) + \kappa^2\Delta_1(\Delta_1 + \Delta_2)}{(\Delta_1[-] + \kappa^2(\Delta_1 + \Delta_2))(\Delta_2[-] + \kappa^2(\Delta_1 + \Delta_2)) - \kappa^4(\Delta_1 + \Delta_2)^2} \quad (2.215)$$

where

$$[-] = (\Omega_1^2 - \Omega_2^2)^2 + (\Delta_1 + \Delta_2)(\Delta_1\Omega_2^2 + \Delta_2\Omega_1^2) \quad (2.216)$$

Factorization of $\Delta_1 + \Delta_2$ in the numerator and development of the denominator yields

$$\beta_{12} = \kappa^2(\Delta_1 + \Delta_2) \times \frac{\Delta_1\Delta_2[-] + \kappa^2(\Delta_1 + \Delta_2)^2}{\Delta_1\Delta_2[-]^2 + \kappa^2(\Delta_1 + \Delta_2)^2[-]} \quad (2.217)$$

After simplifying the term $\Delta_1\Delta_2[-] + \kappa^2(\Delta_1 + \Delta_2)^2$, the factor β_{12} finally reduces to

$$\beta_{\text{stiff},12} = \frac{K^2\,(\Delta_1 + \Delta_2)}{m_1 m_2\left[\left(\Omega_1^2 - \Omega_2^2\right)^2 + (\Delta_1 + \Delta_2)\left(\Delta_1\Omega_2^2 + \Delta_2\Omega_1^2\right)\right]} \quad (2.218)$$

where we have noted $\beta_{\text{stiff},12}$ instead of β_{12} to remind us that this value has been derived under the assumption of elastic coupling. This relationship was derived by Lyon and Maidanik (1962) for the first time. The symmetry

$$\beta_{21} = \beta_{12} \quad (2.219)$$

always holds, so that

$$\langle P_{12}\rangle = \beta_{12}\left(\langle E_1\rangle - \langle E_2\rangle\right) \quad (2.220)$$

This is the coupling power proportionality also derived by Lyon and Maidanik (1962). Let us remark that we have not assumed the coupling to be weak during the proof. The result is therefore valid for any strength of coupling but is limited to two interacting resonators as we shall see in Section 2.5.

We have obtained the coefficient β in two different ways in this section and in Section 2.3. Equation (2.171) has been derived under the weak coupling assumption by considering the single input–single output linear system H_i while eqn (2.218) has been derived in a more direct way by considering the multiple input–multiple output linear system H_{ij}. We may therefore wonder if both expressions agree. A simple examination of them shows that they match under the substitution

$$\begin{aligned} K &\leftrightarrow k_{ij} \\ \Omega_i &\leftrightarrow \omega_i \\ \Delta_i &\leftrightarrow 2\zeta_i\omega_i \end{aligned}$$

The first substitution is just a question of notation while the last two rest on the zero-order development $\Omega_i = \omega_i + o(1)$ in $\epsilon = k_{ij}/k_i$. Since the coefficient β is proportional to ϵ^2 by the presence of $k_{ij}^2 = K^2$ in the numerator, and the coupling power proportionality is valid at order two in ϵ, any change in β by a third-order term does not modify the coupling power proportionality. By this argument, eqns (2.171) and (2.218) are equivalent.

We now turn to the last case of coupling. We set $G = K = 0$ and $M \neq 0$ or equivalently $\gamma = \kappa = 0$ and $\mu \neq 0$. By eqns (2.182) and (2.198),

$$\begin{cases} A_1 = \dfrac{\mu^2\Delta_2}{(m_1+M)} \displaystyle\int_{-\infty}^{\infty} \left| \dfrac{-\imath\omega^3}{Q(\omega)} \right|^2 \mathrm{d}\omega \\ A_2 = \dfrac{\mu^2\Delta_1}{(m_2+M)} \displaystyle\int_{-\infty}^{\infty} \left| \dfrac{-\imath\omega^3}{Q(\omega)} \right|^2 \mathrm{d}\omega \end{cases} \tag{2.221}$$

Furthermore, we have seen that by using the property of null Lagrangian, the mean vibrational energy may take several forms. The more appropriate expression of the matrix E_{ik} is that for which the matrices are diagonal or null. We choose eqn (2.193),

$$E_{ik} = \int_{-\infty}^{\infty} k_i |H_{ik}|^2(\omega)\, \mathrm{d}\omega \tag{2.222}$$

After developing the entries of the matrix $\mathbf{H}$ given in eqn (2.182),

$$\begin{cases} E_{11} = \dfrac{\Omega_1^2}{m_1+M} \displaystyle\int_{-\infty}^{\infty} \left| \dfrac{-\omega^2 + \imath\omega\Delta_2 + \Omega_2^2}{Q(\omega)} \right|^2 \mathrm{d}\omega \\ E_{12} = \dfrac{\mu^2\Omega_1^2}{(m_2+M)} \displaystyle\int_{-\infty}^{\infty} \left| \dfrac{-\omega^2}{Q(\omega)} \right|^2 \mathrm{d}\omega \\ E_{21} = \dfrac{\mu^2\Omega_2^2}{(m_1+M)} \displaystyle\int_{-\infty}^{\infty} \left| \dfrac{-\omega^2}{Q(\omega)} \right|^2 \mathrm{d}\omega \\ E_{22} = \dfrac{\Omega_2^2}{m_2+M} \displaystyle\int_{-\infty}^{\infty} \left| \dfrac{-\omega^2 + \imath\omega\Delta_1 + \Omega_1^2}{Q(\omega)} \right|^2 \mathrm{d}\omega \end{cases} \tag{2.223}$$

where we have substituted $\Omega_i^2 = k_i/(m_i + M)$. Each of the four integrals appearing in eqns (2.221) and (2.223) may be calculated by the quadrature formula (2.207). The common denominator is the polynomial

$$Q(\omega) = \left(-\omega^2 + \imath\omega\Delta_1 + \Omega_1^2\right)\left(-\omega^2 + \imath\omega\Delta_2 + \Omega_2^2\right) - \mu^2\omega^4 \tag{2.224}$$

One takes the following set of parameters for eqn (2.207),

$$\begin{array}{llll} a_0 = \Omega_1^2\Omega_2^2 & b_0 = 0 & b'_0 = 0 & b''_0 = \Omega_i^2 \\ a_1 = \Delta_1\Omega_2^2 + \Delta_2\Omega_1^2 & b_1 = 0 & b'_1 = 0 & b''_1 = \Delta_i \\ a_2 = \Omega_1^2 + \Omega_2^2 + \Delta_1\Delta_2 & b_2 = 1 & b'_2 = 0 & b''_2 = 1 \\ a_3 = \Delta_1 + \Delta_2 & b_3 = 0 & b'_3 = 1 & b''_3 = 0 \\ a_4 = 1 - \mu^2 & & & \end{array} \tag{2.225}$$

The parameters a_i are obtained by developing Q and identifying the coefficient in powers of $\imath\omega$ whereas the parameters b_k, b'_k, and b''_k respectively apply to $\int |-\omega^2/Q|^2\,\mathrm{d}\omega$, $\int |-\imath\omega^3/Q|^2\,\mathrm{d}\omega$, and $\int |(-\omega^2 + \imath\omega\Delta_i + \Omega_i^2)/Q|^2\,\mathrm{d}\omega$. One obtains

$$\int_{-\infty}^{\infty} \left|\frac{(\imath\omega)^k}{Q(\omega)}\right|^2 \mathrm{d}\omega = \frac{n_k}{d}, \quad k = 2, 3 \tag{2.226}$$

and

$$\int_{-\infty}^{\infty} \left|\frac{\Omega_i^2 + \imath\omega\Delta_i - \omega^2}{Q(\omega)}\right|^2 \mathrm{d}\omega = \frac{l_i}{d}, \quad i = 1, 2 \tag{2.227}$$

where

$$\begin{aligned} d &= a_0a_4(a_1a_2a_3 - a_0a_3^2 - a_1^2a_4) \\ n_2 &= a_0a_1a_4 \\ n_3 &= a_0(a_1a_2 - a_0a_3) \\ l_i &= a_0a_1a_4 + a_0a_3a_4(\Delta_i^2 - 2\Omega_i^2) + a_4\Omega_i^4(a_2a_3 - a_1a_4) \end{aligned}$$

After developing and simplifying,

$$\begin{aligned} n_2 &= (1-\mu^2)\Omega_1^2\Omega_2^2(\Delta_1\Omega_2^2+\Delta_2\Omega_1^2) \\ n_3 &= \Omega_1^2\Omega_2^2\left[\Delta_1\Omega_2^4+\Delta_2\Omega_1^4+\Delta_1\Delta_2(\Delta_1\Omega_2^2+\Delta_2\Omega_1^2)\right] \\ l_i &= (1-\mu^2)\Omega_i^2\left\{\Delta_i\left[(\Omega_1^2-\Omega_2^2)^2+(\Delta_1+\Delta_2)(\Delta_1\Omega_2^2+\Delta_2\Omega_1^2)\right]+\mu^2\Omega_i^2(\Delta_1\Omega_2^2+\Delta_2\Omega_1^2)\right\} \end{aligned} \tag{2.228}$$

By eqn (2.201),

$$\beta_{12} = \frac{\dfrac{\mu^2\Delta_2 n_3}{(m_1+M)d} \times \dfrac{\Omega_2^2 l_1}{(m_2+M)d} + \dfrac{\mu^2\Delta_1 n_3}{(m_2+M)d} \times \dfrac{\mu^2\Omega_2^2 n_2}{(m_1+M)d}}{\dfrac{\Omega_1^2 l_1}{(m_1+M)d} \times \dfrac{\Omega_2^2 l_2}{(m_2+M)d} - \dfrac{\mu^2\Omega_1^2 n_2}{(m_2+M)d} \times \dfrac{\mu^2\Omega_2^2 n_2}{(m_1+M)d}} \tag{2.229}$$

After simplifying the factor $\Omega_2^2/m_1 m_2 d^2$,

$$\beta_{12} = \frac{\mu^2 n_3}{\Omega_1^2} \times \frac{\Delta_2 l_1 + \mu^2 \Delta_1 n_2}{l_1 l_2 - \mu^4 n_2^2} \tag{2.230}$$

By a direct substitution,

$$\begin{aligned} \Delta_2 l_1 + \mu^2 \Delta_1 n_2 &= (1-\mu^2)\Omega_1^2 \left\{ \Delta_1 \Delta_2 [-] + \mu^2 (\Delta_1 \Omega_2^2 + \Delta_2 \Omega_1^2)^2 \right\} \\ l_1 l_2 - \mu^4 n_2^2 &= (1-\mu^2)^2 \Omega_1^2 \Omega_2^2 [-] \left\{ \Delta_1 \Delta_2 [-] + \mu^2 (\Delta_1 \Omega_2^2 + \Delta_2 \Omega_1^2)^2 \right\} \end{aligned} \tag{2.231}$$

where

$$[-] = (\Omega_1^2 - \Omega_2^2)^2 + (\Delta_1 + \Delta_2)(\Delta_1 \Omega_2^2 + \Delta_2 \Omega_1^2) \tag{2.232}$$

Therefore after substitution and simplification,

$$\beta_{\text{mass},12} = \frac{\mu^2 \left[\Delta_1 \Omega_2^4 + \Delta_2 \Omega_1^4 + \Delta_1 \Delta_2 (\Delta_1 \Omega_2^2 + \Delta_2 \Omega_1^2) \right]}{(1-\mu^2) \left[\left(\Omega_1^2 - \Omega_2^2 \right)^2 + (\Delta_1 + \Delta_2) \left(\Delta_1 \Omega_2^2 + \Delta_2 \Omega_1^2 \right) \right]} \tag{2.233}$$

This equation has been derived by Lyon and Maidanik (1962) under the weak coupling assumption ($\mu \ll 1$) and by Lyon and Scharton (1968) in this exact form. This expression verifies the symmetry $\beta_{12} = \beta_{21}$. Furthermore, it may be compared with eqn (2.173) obtained under the weak coupling assumption for two or more oscillators.

2.5 Strongly Coupled Resonators

The method employed in Section 2.4 can be straightforwardly generalized to N resonators. We again do *not* assume the coupling to be weak (in opposition with what was done in Section 2.3). But we still assume that excitations are uncorrelated white noises.

We have seen in eqn (2.99) the general expression for the mean vibrational energy,

$$\langle E_i \rangle = \frac{1}{2} \left[\sum_j m_{ij} \langle \dot{x}_i \dot{x}_j \rangle + k_{ii} \langle x_i^2 \rangle - \sum_{j \neq i} k_{ij} \langle x_i x_j \rangle \right]$$

The evaluation of $\langle E_i \rangle$ requires the calculation of the products $\langle x_i x_j \rangle$, $\langle \dot{x}_i \dot{x}_j \rangle$. We have calculated them in eqns (2.186), (2.187) when considering the special case of two resonators. But these expressions remain valid for any N. So, if H_{ij} denotes the frequency response function of resonator i with excitation on j, then the mean vibrational energy becomes

$$\langle E_i \rangle = \sum_k E_{ik} \frac{S_k}{2\pi} \tag{2.234}$$

where

$$E_{ik} = \int_{-\infty}^{\infty} \sum_j m_{ij}\omega^2 \overline{H}_{ik}(\omega) H_{jk}(\omega) + k_{ii} \, |H_{ik}|^2(\omega) - \sum_j k_{ij} \overline{H}_{ik}(\omega) H_{jk}(\omega) \, \mathrm{d}\omega \tag{2.235}$$

This is simply the mean square response of a linear system to multiple inputs mutually uncorrelated that we have encountered several times in this chapter. In matrix notation, this linear relationship reads

$$\begin{pmatrix} \langle E_1 \rangle \\ \vdots \\ \langle E_n \rangle \end{pmatrix} = \begin{pmatrix} E_{11} & \dots & E_{1n} \\ \vdots & & \vdots \\ E_{n1} & \dots & E_{nn} \end{pmatrix} \begin{pmatrix} S_1/2\pi \\ \vdots \\ S_n/2\pi \end{pmatrix} \tag{2.236}$$

The coefficients E_{ij} are referred to as **energy influence coefficients**. Their properties are investigated by Guyader et al. (1986), Lesueur (1988), and Mace (2003, 2005).

On the other hand, the mean power exchanged between the resonators i and j is, according to eqn (2.100),

$$\langle P_{ij} \rangle = m_{ij} \langle \dot{x}_i \ddot{x}_j \rangle + g_{ij} \langle \dot{x}_i \dot{x}_j \rangle + k_{ij} \langle x_i \dot{x}_j \rangle$$

Again, the products $\langle \dot{x}_i \ddot{x}_j \rangle$ and $\langle x_i \dot{x}_j \rangle$ have already been calculated in eqns (2.195) and (2.196) by considering the mean response of a linear system to multiple inputs mutually uncorrelated. We obtain

$$\langle P_{ij} \rangle = \sum_k A_{ijk} \frac{S_k}{2\pi} \tag{2.237}$$

where

$$A_{ijk} = \int_{-\infty}^{\infty} \iota\omega \left[m_{ij}\omega^2 - \iota\omega g_{ij} + k_{ij} \right] \overline{H}_{ik}(\omega) H_{jk}(\omega) \, \mathrm{d}\omega \tag{2.238}$$

In matrix notation,

$$\langle P_{ij} \rangle = \begin{pmatrix} A_{ij1} & \dots & A_{ijn} \end{pmatrix} \begin{pmatrix} S_1/2\pi \\ \vdots \\ S_n/2\pi \end{pmatrix} \tag{2.239}$$

Hence, by combining eqns (2.236) and (2.239), we obtain $\langle P_{ij} \rangle$ as a linear combination of the mean vibrational energies $\langle E_i \rangle$,

$$\langle P_{ij} \rangle = \begin{pmatrix} A_{ij1} \dots A_{ijn} \end{pmatrix} \begin{pmatrix} E_{11} \dots E_{1n} \\ \vdots \quad\quad \vdots \\ E_{n1} \dots E_{nn} \end{pmatrix}^{-1} \begin{pmatrix} \langle E_1 \rangle \\ \vdots \\ \langle E_2 \rangle \end{pmatrix} \tag{2.240}$$

that may also be written as

$$\langle P_{ij} \rangle = \sum_k \beta_{ijk} \langle E_k \rangle \tag{2.241}$$

where β_{ijk} is a coefficient which depends on all A_{ijk} and E_{kl}. This equation highlights that in general, when resonators are strongly coupled, the mean exchanged power is a linear combination of *all* vibrational energies and not only the *two* vibrational energies of adjacent subsystems as for lightly coupled resonators.

In the case of only two resonators, this linear combination reduces to a linear combination with the only two energies $\langle E_1 \rangle$ and $\langle E_2 \rangle$,

$$\langle P_{12} \rangle = \beta_{121} \langle E_1 \rangle + \beta_{122} \langle E_2 \rangle \tag{2.242}$$

And the effective calculation of the coefficients β_{12i} done in the previous subsection has revealed the antisymmetry

$$\beta_{121} = -\beta_{122} \tag{2.243}$$

Therefore, the exchanged power has been reduced to a proportionality with the *difference* of adjacent energies

$$\langle P_{12} \rangle = \beta \left(\langle E_1 \rangle - \langle E_2 \rangle \right) \tag{2.244}$$

This is what we have called the coupling power proportionality.

Another situation where eqn (2.241) simplifies is the case of weak coupling. We have extensively discussed this situation in Section 2.3 and the main conclusion was that once again, eqn (2.241) reduces to the coupling power proportionality (2.244).

It becomes clear now why the assumption of light coupling was not a requirement for two resonators but remains unavoidable in the general case of more than two resonators. Both situations—two resonators or more resonators but weakly coupled—lead to the same reduction of eqn (2.241) and to the coupling power proportionality eqn (2.244), which does not hold in the general case or $N > 2$ strongly coupled resonators.

2.6 Statistical Energy Analysis of Coupled Oscillators

The energetic analysis introduced in the simple case of a single oscillator can now be developed for the less trivial system of N coupled oscillators. The global system is the set of all oscillators. This system is subjected to random external forces supplying energy to the system. Besides, the system may dissipate energy in dashpots. A thermodynamical reasoning based on a balance of the vibrational energy in steady-state condition is therefore applicable.

But to analyse the path of energy inside the system, we may divide it into subsystems. Each subsystem consists of a single oscillator as shown in Fig. 2.7. In accordance with Section 2.2, subsystems are defined in such a way that their Lagrangian and energy are given by eqns (2.89) and (2.90).

The list of assumptions is the following,

- conservative coupling,
- random, stationary white noise and uncorrelated forces,
- weak coupling (only when $N > 2$).

The coupling power proportionality has been derived in two different situations. In Section 2.2 the weak coupling assumption was necessary. But in Section 2.4 the same result has been proved for any strength of coupling in the special case $N = 2$. So, in the above list of assumptions, the weak coupling assumption is required only when $N > 2$.

Forces are stationary random processes. Therefore, in steady-state conditions the power balance for subsystem i is

$$\langle P_{\text{diss},i} \rangle + \sum_{j \neq i} \langle P_{ij} \rangle = \langle P_i \rangle \tag{2.245}$$

The injected power is either assumed to be known or assessed with the power spectral density of excitation force,

$$\langle P_i \rangle = \frac{S_i}{2m_i} \tag{2.246}$$

If the mass matrix is not diagonal, eqn (2.112) provides a better estimation.

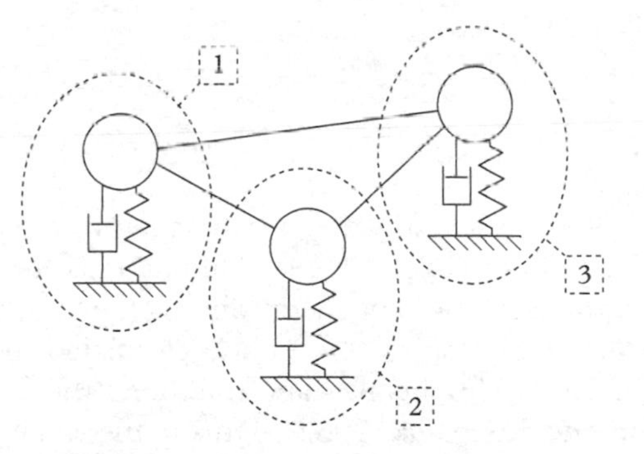

Figure 2.7 *Statistical energy analysis of coupled oscillators. A subsystem reduces to a single oscillator, the coupling energy being shared between adjacent subsystems.*

The power being dissipated in a dashpot is proportional to the kinetic energy of the isolated oscillator. But we have furthermore established that this internal kinetic energy is equal to the vibrational energy $\langle E_i \rangle$. We found

$$\langle P_{\text{diss},i} \rangle = \eta_i \omega_i \langle E_i \rangle \tag{2.247}$$

where $\eta_i \omega_i = c_i/m_i$. This is the dissipation law usually adopted in statistical energy analysis.

The coupling power proportionality states that the exchanged power $\langle P_{ij} \rangle$ is proportional to the difference of vibrational energies $\langle E_i \rangle$ and $\langle E_j \rangle$,

$$\langle P_{ij} \rangle = \beta_{ij} \left(\langle E_i \rangle - \langle E_j \rangle \right) \tag{2.248}$$

The coefficients β_{ij} are symmetric.

Introducing these two relationships into the power balance leads to

$$\eta_i \omega_i \langle E_i \rangle + \sum_{j \neq i} \beta_{ij} \left(\langle E_i \rangle - \langle E_j \rangle \right) = \langle P_i \rangle \tag{2.249}$$

In a matrix form, this equality reads

$$\begin{pmatrix} \sum_{k \neq 1} \beta_{1k} + \eta_1 \omega_1 & & -\beta_{ji} \\ & \ddots & \\ -\beta_{ji} & & \sum_{k \neq N} \beta_{Nk} + \eta_N \omega_N \end{pmatrix} \begin{pmatrix} \langle E_1 \rangle \\ \vdots \\ \langle E_N \rangle \end{pmatrix} = \begin{pmatrix} \langle P_1 \rangle \\ \vdots \\ \langle P_N \rangle \end{pmatrix} \tag{2.250}$$

This is the statistical energy analysis equation for a system of N coupled oscillators, each subsystem being a single oscillator. All coefficients β_{ij} are given in terms of mechanical parameters m_i, c_i, k_i, and k_{ij}.

2.7 Comments

R. H. Lyon began, in 1960, to investigate the interaction of a mechanical resonator in equilibrium with a thermal bath (Lyon, 2003). Although his former motivation was to determine the equilibrium distribution of energy among the 'modes' in a turbulent fluid, he turned to a problem of vibrating structures and calculated the power flow between two modes randomly excited. He obtained and published the coupling power proportionality (2.200) (Lyon and Maidanik, 1962). During the same period, Smith carried out a calculation of the interaction of a structural mode excited by a diffuse sound field. As a result, he showed that the response reaches a limit when the coupling loss factor due to acoustic radiation dominates the internal damping loss factor of the mode (Smith, 1962). This result is today clearly related to statistical energy analysis.

On the factor β

In Lyon and Maidanik (1962), the system considered was two resonators of unit total mass $m_i + M = 1$ with coupling stiffness K, coupling inertia M, and gyroscopic coupling G. The factor β was

$$\beta = \frac{M^2(\Delta_1\Omega_2^4 + \Delta_2\Omega_1^4 + \Delta_1\Delta_2(\Delta_1\Omega_2^2 + \Delta_2\Omega_1^2)) + (G^2 + 2MK)(\Delta_1\Omega_2^2 + \Delta_2\Omega_1^2) + K^2(\Delta_1 + \Delta_2)}{\left(\Omega_1^2 - \Omega_2^2\right)^2 + (\Delta_1 + \Delta_2)\left(\Delta_1\Omega_2^2 + \Delta_2\Omega_1^2\right)} \tag{2.251}$$

where Ω_i and Δ_i are defined by

$$\Omega_i^2 = (k_i + K)/(m_i + M)$$
$$\Delta_i = c_i/(m_i + M)$$

However, in Lyon and Maidanik (1962) the coupling power proportionality was derived under the assumption of light coupling, i.e. $M \ll m_i$, $G \ll \Delta_i$, and $K \ll m_i\Omega_i^2$. But in Scharton and Lyon (1968) the light coupling assumption was relaxed without any trouble for the coupling power proportionality. The final expression of β is

$$\beta = \frac{\mu^2(\Delta_1\Omega_2^4 + \Delta_2\Omega_1^4 + \Delta_1\Delta_2(\Delta_1\Omega_2^2 + \Delta_2\Omega_1^2)) + (\gamma^2 + 2\mu\kappa)(\Delta_1\Omega_2^2 + \Delta_2\Omega_1^2) + \kappa^2(\Delta_1 + \Delta_2)}{(1-\mu^2)\left[\left(\Omega_1^2 - \Omega_2^2\right)^2 + (\Delta_1 + \Delta_2)\left(\Delta_1\Omega_2^2 + \Delta_2\Omega_1^2\right)\right]} \tag{2.252}$$

where

$$\mu = M/\sqrt{(m_1 + M)(m_2 + M)}$$
$$\gamma = G/\sqrt{(m_1 + M)(m_2 + M)}$$
$$\kappa = K/\sqrt{(m_1 + M)(m_2 + M)}$$

This is also the expression which appears in Lyon (1975), Lyon and DeJong (1995), and Ungar (1966) in the special case $\mu = 0$. This equation generalizes eqns (2.218) and (2.233).

Newland (1966) generalized the calculation of power flow between two resonators to a set of arbitrary number of resonators. To that purpose he introduced the assumption of weak coupling and used the perturbation technique. He obtained that the coupling power proportionality always holds but only up to order two in ϵ (see eqn (2.169)). The factor β derived in Newland (1966) is exactly the one given by eqn (2.170). In Newland (1968), the result is extended to the case of both elastic and inertial coupling with the expression

$$\beta = \frac{2(K - M\omega_i^2)(K - M\omega_j^2)\left(\omega_i\zeta_i + \omega_j\zeta_j\right)}{m_i m_j\left[\left(\omega_i\sqrt{1-\zeta_i^2} + \omega_j\sqrt{1-\zeta_j^2}\right)^2 + (\omega_i\zeta_i + \omega_j\zeta_j)^2\right]\left[\left(\omega_i\sqrt{1-\zeta_i^2} - \omega_j\sqrt{1-\zeta_j^2}\right)^2 + (\omega_i\zeta_i + \omega_j\zeta_j)^2\right]} \tag{2.253}$$

However, due to the particular method used for the calculation of the involved integral, these expressions of the factor β are limited to ζ_i, $\zeta_j < 1$. Fahy (1970) solved the case of a gyroscopic coupling for any value of ζ_i, ζ_j. Finally, Lotz and Crandall (1973) obtained the factor β by Newland's method in the special cases of elastic and inertial couplings,

$$\beta = \frac{2}{m_i m_j} \times \frac{\zeta_i\omega_i(K - M\omega_j^2)^2 + \zeta_j\omega_j(K - M\omega_i^2)^2 + 4\zeta_i\omega_i\zeta_j\omega_j M^2(\zeta_i\omega_i\omega_j^2 + \zeta_j\omega_j\omega_i^2)}{(\omega_i^2 - \omega_j^2)^2 + 4\omega_i\omega_j(\zeta_i\omega_i + \zeta_j\omega_j)(\zeta_i\omega_j + \zeta_j\omega_i)} \tag{2.254}$$

This equation embodies eqns (2.171) and (2.173) as special cases.

Let us remark that the natural frequencies ω_i involved in the above equations and eqns (2.170)–(2.173) are defined as the 'uncoupled' natural frequencies of resonators $\omega_i^2 = k_i/m_i$ in opposition with the 'blocked' natural frequencies $\Omega_i^2 = (k_i + K)/(m_i + M)$ used by Lyon. It may be checked, however, that the substitution

$$\begin{aligned}
\Omega_i &\leftrightarrow \omega_i \\
\Delta_i &\leftrightarrow 2\zeta_i\omega_i \\
\mu &\leftrightarrow m_{ij}/\sqrt{(m_1 + M)(m_2 + M)} \\
\gamma &\leftrightarrow g_{ij}/\sqrt{(m_1 + M)(m_2 + M)} \\
\kappa &\leftrightarrow k_{ij}/\sqrt{(m_1 + M)(m_2 + M)}
\end{aligned}$$

in eqn (2.252) which is valid for light couplings gives successively the three coefficients (2.171), (2.172), and (2.173) as well as eqn (2.254) up to order two in ϵ.

The complete calculations of these coefficients β are only sometimes given in the original papers by mentioning the main steps of the proof. Most often the subsequent references require one to content oneself with the reporting of the result without the proof. But even in the original papers, the mathematical developments are sometimes difficult to follow due to a lack of detail. Thus, the palm of laconism falls to Fahy (1970), who simply wrote 'The result of a large algebraic book-keeping operation is' and gave the right coefficient β (2.172) and much more besides! We have tried to restore the details in this chapter and Appendix C.

On uncoupled energies

Originally, the coupling power proportionality stated the proportionality between the exchanged power and the difference of **uncoupled energies**. Uncoupled energies are not the actual energies of resonators but the energies that resonators would have in the absence of coupling, i.e. when $M = G = K = 0$ with the same levels of excitation S_i. With the notations of Section 2.3, uncoupled energies are the zero-order development of actual energies. According to eqn (2.161), the uncoupled energies are

$$\theta_i = \frac{S_i}{2\pi}\int_{-\infty}^{\infty} k_i |H_i|^2(\omega)\,\mathrm{d}\omega = \frac{S_i}{2\pi} \times \frac{\pi k_i}{2\zeta_i m_i^2 \omega_i^3} = \frac{S_i}{2c_i} \tag{2.255}$$

where $c_i = 2m_i\zeta_i\omega_i$. Thus, in Lyon and Maidanik (1962) the coupling power proportionality is

$$\langle P_{12}\rangle = \beta\,(\theta_1 - \theta_2) \tag{2.256}$$

But in Scharton and Lyon (1968) it was proved that the coupling power is also proportional to the difference of **actual energies**. That is the latter result that has been reported in Lyon (1975), Lyon and DeJong (1995), and in Section 2.3. Since the concept of uncoupled energies seems no longer useful, we have avoided introducing it in this text.

On dissipative couplings

The coupling power proportionality fails in some situations. For instance when the coupling is dissipative, Fahy (1987) found that the exchanged power between two resonators is a linear combination of adjacent energies which does not reduce to a proportionality with the difference of energies. The general relationship is

$$\langle P_{12}\rangle = \beta\,(\langle E_1\rangle - \langle E_2\rangle) + \alpha\langle E_1\rangle + \gamma\langle E_2\rangle \tag{2.257}$$

where β of the above slightly differs from those of eqn (2.218). This result provides a justification of the necessity of Assumption 1.

On energy sharing

The equality of kinetic and elastic energies is a well-known result for a single resonator excited by a white noise random force. This result is demonstrated in many elementary courses (Soize, 2001) in random vibration. A similar result is introduced in Scharton and Lyon (1968), Lyon (1975), and Lyon and DeJong (1995) in the case of two resonators. However, this is not the equality of *individual* kinetic and elastic energies $\langle K_i\rangle$ and $\langle V_i\rangle$ but only the equality of their difference,

$$\langle K_1\rangle - \langle K_2\rangle = \langle V_1\rangle - \langle V_2\rangle \tag{2.258}$$

See eqn (15) in Scharton and Lyon (1968) and eqn (3.1.124) in Lyon (1975). The fact that $\langle K_i\rangle \neq \langle V_i\rangle$ even in the absence of gyroscopic coupling comes from the particular expressions adopted for energies. In Scharton and Lyon (1968), these are

$$\langle K_i\rangle = \frac{1}{2}(m_i + M)\langle \dot{x}_i^2\rangle \tag{2.259}$$

$$\langle V_i\rangle = \frac{1}{2}(k_i + K)\langle x_i^2\rangle \tag{2.260}$$

But these expressions violate the constraint that their sum must give the total vibrational energy (the cross-product terms of the coupling energy are missing, see eqn (2.184)).

By imposing that the whole is the sum of the parts and sharing the coupling energy and Lagrangian as in eqns (2.89) and (2.90), we have restored the equality of individual kinetic and elastic energies $\langle K_i \rangle = \langle V_i \rangle$ (in the absence of gyroscopic coupling) and $\langle L_i \rangle = 0$ in the most general case. It is remarkable that no modification is induced on the final expression of the factor β.

On exchanged power

In most of references, the expression of exchanged power between oscillators is derived from considerations of force and velocity. All authors agree with the following relationship:

$$\langle P_{ij} \rangle = M\langle \dot{x}_i \ddot{x}_j \rangle + G\langle \dot{x}_i \dot{x}_j \rangle + K\langle x_i \dot{x}_j \rangle \tag{2.261}$$

See for instance eqn (6) in Scharton and Lyon (1968) and eqn (31) in Gersh (1969). In steady-state conditions, $\langle \ddot{x}_i \dot{x}_j \rangle = -\langle \dot{x}_i \ddot{x}_j \rangle$, $\langle \dot{x}_i x_j \rangle = -\langle x_i \dot{x}_j \rangle$, and substituting $i \leftrightarrow j$ and $G \leftrightarrow -G$ leads to $\langle P_{ij} \rangle = -\langle P_{ji} \rangle$. But the corresponding expression of the *instantaneous* exchanged power P_{ij} is generally not given. A simple transposition by removing the brackets would lead to a violation of the condition $P_{ij} = -P_{ji}$. In this chapter, we have derived a more symmetric expression for P_{ij} in eqn (2.96) under the constraint that the power balance is satisfied. The symmetry $P_{ij} = -P_{ji}$ then simply means that the energy entering subsystem j equals that leaving subsystem i, that is no energy is stored in the coupling. In other words, the sum of subsystem energies is also the energy of the whole system.

3

Ensemble of Resonators

So far, we have considered the response of *deterministic* resonators to random forces. In this case resonators behave like linear filters, which allows us to apply the theory of random vibration exposed in Chapter 1. We now introduce a second cause of randomness: resonators themselves are random. They are chosen at random from a set of resonators with different characteristics m_i, c_i, k_i, m_{ij}, g_{ij}, and k_{ij}. This chapter focuses on the statistical description of the behaviour of such a population of random resonators. When dealing with an actual system which is unique and therefore well determined, the hope is that its behaviour will not deviate significantly from the mean behaviour of the population. The confidence in this outline is naturally better when the system is disordered and has a great complexity.

3.1 Multi-Degree of Freedom Subsystems

The first step is to find the way from a deterministic description of a complex system to a random description. In practice, a subsystem is a mechanical structure which, in general, contains a large number of modes. We must therefore address the problem of the equivalence between a large set of actual resonators and a random ensemble of resonators.

Definition of subsystem

As in Chapter 2, we consider a set of coupled resonators. These resonators verify Assumptions 1–3, that is conservative and weak couplings, and wide-band uncorrelated random forces. But in addition, we constitute groups of resonators that will be called **subsystems**. They are referred to by a subscript i. Each subsystem contains N_i resonators, also called modes. A mode of a subsystem is then referred to with two subscripts i for its subsystem and α for its number. For instance, the stiffness of mode α in subsystem i is noted $k_{i\alpha}$, its natural frequency $\omega_{i\alpha}$, and so on. The inter-modal coupling terms between mode α of subsystem i and mode β of subsystem j are noted $m_{i\alpha,j\beta}$,

Foundation of Statistical Energy Analysis in Vibroacoustics. First Edition. A. Le Bot.

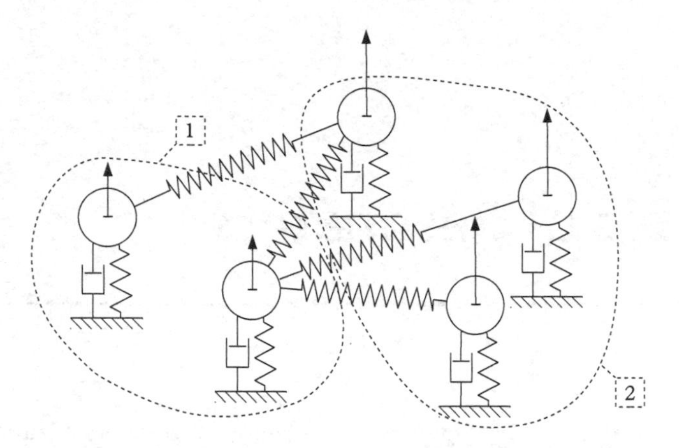

Figure 3.1 *Definition of subsystems as groups of uncoupled resonators excited with same power spectral density.*

$g_{i\alpha,j\beta}$, and $k_{i\alpha,j\beta}$. The internal variable of modes is noted $X_{i\alpha}$ and is usually interpreted as the mass position although it can have different interpretation depending on the nature of the system.

Subsystems are defined as groups of resonators which verify the following conditions (see Fig. 3.1),

- Modes of subsystem i have the same modal mass m_i and damping coefficient c_i.
- Modes are uncoupled in subsystem i, $m_{i\alpha,i\beta} = g_{i\alpha,i\beta} = k_{i\alpha,i\beta} = 0$ for any α, β.
- The modal forces $F_{i\alpha}$ have same power spectral density S_i in subsystem i.

Let us make some remarks on these conditions. The first point is not really a restricting condition since it is always possible to define the mode magnitude $X_{i\alpha}$ in such a way that the equality of modal mass is verified. The constancy of the damping coefficient is well verified by viscous forces. This particular choice of dissipation force will result in drastic simplifications in forthcoming calculations. The second point will be naturally interpreted in Chapter 4 as being the orthogonality of modes, a property always verified. The last point goes further than Assumption 2 which states that forces are white noises mutually uncorrelated. It is assumed that all power spectral densities are identical within a subsystem. This constitutes a true assumption of statistical energy analysis and we may now re-formulate Assumption 2.

Assumption 2b *Modal forces are random, uncorrelated, white noises, and have the same power spectral density in each subsystem.*

This kind of excitation is often named a 'rain-on-the-roof' excitation.

Canonical problem

Modes are henceforth grouped into subsystems. To perform an energy analysis of these subsystems, we must first introduce the energy and power of subsystems. For this, we adopt the axiomatic method. We introduce a set of equations a priori that we shall call a canonical problem. It will be solved by the method of Chapter 2. This canonical problem will turn out to be useful in Chapter 4 where continuous or discrete structures will be reduced to the canonical problem.

Let us now introduce the equations of the canonical problem. The governing equation of mode i, α is

$$m_i \ddot{X}_{i\alpha} + c_i \dot{X}_{i\alpha} + k_{i\alpha} X_\alpha = F_{i\alpha} + \sum_{j \neq i} \sum_{\beta=1}^{N_j} k_{i\alpha,j\beta} X_{j\beta} - g_{i\alpha,j\beta} \dot{X}_{j\beta} - m_{i\alpha,j\beta} \ddot{X}_{j\beta} \tag{3.1}$$

where mode i, α is coupled with all modes j, β with $j \neq i$ but not with modes of subsystem i.

The vibrational energy E_i of a subsystem is the sum,

$$E_i = \sum_{\alpha=1}^{N_i} E_{i\alpha} \tag{3.2}$$

where the modal energies are

$$E_{i\alpha} = \frac{1}{2} \left[m_i \dot{X}_{i\alpha}^2 + k_{i\alpha} X_{i\alpha}^2 + \sum_{j \neq i} \sum_{\beta=1}^{N_j} m_{i\alpha,j\beta} \dot{X}_{i\alpha} \dot{X}_{j\beta} - k_{i\alpha,j\beta} X_{i\alpha} X_{j\beta} \right] \tag{3.3}$$

The power being exchanged between groups i and j is the sum of powers exchanged between individual modes,

$$P_{ij} = \sum_{\alpha,\beta} P_{i\alpha,j\beta} \tag{3.4}$$

where the inter-modal exchanged power is

$$P_{i\alpha,j\beta} = \frac{1}{2} m_{i\alpha,j\beta} \left[\dot{X}_{i\alpha} \ddot{X}_{j\beta} - \ddot{X}_{i\alpha} \dot{X}_{j\beta} \right] + g_{i\alpha,j\beta} \dot{X}_{i\alpha} \dot{X}_{j\beta} + \frac{1}{2} k_{i\alpha,j\beta} \left[X_{i\alpha} \dot{X}_{j\beta} - \dot{X}_{i\alpha} X_{j\beta} \right] \tag{3.5}$$

Let us remark that the above expressions are consistent with those derived in Chapter 2.

Equations (3.1–3.5) constitute the canonical problem of the modal approach of statistical energy analysis. It is representative of a wide class of problems. The situation is sketched in Fig. 3.2. Modes are grouped in subsystems $1, \ldots, n$ and are coupled with modes of other subsystems. Energy of subsystems is the sum of modal energies and the net exchanged power the sum of inter-modal powers.

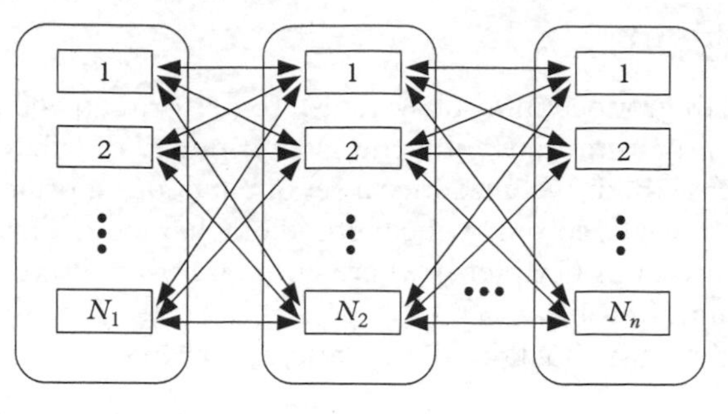

Figure 3.2 *Power exchanged between n multi-degree-of-freedom subsystems.*

Deterministic exchanged power

Let us provisionally forget that modes belong to groups and consider that we have a single system composed of a large number $\sum_i N_i$ of resonators. The fact that some of them are uncoupled is not important, but all existing couplings are light and conservative. Furthermore, the forces $F_{i\alpha}$ are white noises mutually uncorrelated. We are therefore in the situation described in Chapter 2 and we can apply its results. In particular, the power exchanged between two modes $i\alpha, j\beta$ is given by eqn (2.157),

$$\begin{aligned}\langle P_{i\alpha,j\beta}\rangle = \Bigg[& \frac{S_i}{2\pi}\int_{-\infty}^{\infty} \imath\omega\left[\left(k_{i\alpha,j\beta}+\omega^2 m_{i\alpha,j\beta}\right)^2+\omega^2 g_{i\alpha,j\beta}^2\right] H_{j\beta}\,|H_{i\alpha}|^2\,\mathrm{d}\omega \\ & -\frac{S_j}{2\pi}\int_{-\infty}^{\infty} \imath\omega\left[\left(k_{i\alpha,j\beta}+\omega^2 m_{i\alpha,j\beta}\right)^2+\omega^2 g_{i\alpha,j\beta}^2\right] H_{i\alpha}\,|H_{j\beta}|^2\,\mathrm{d}\omega\Bigg]\end{aligned} \tag{3.6}$$

In the above equation, we have introduced the power spectral density S_i of the random force applied to resonator $i\alpha$. The uncoupled frequency response function of resonator $i\alpha$ has been noted $H_{i\alpha}(\omega)$. In accordance with eqn (3.4), the net power exchanged between subsystems i and j is

$$\begin{aligned}\langle P_{ij}\rangle = \sum_{\alpha,\beta}\Bigg[& \frac{S_i}{2\pi}\int_{-\infty}^{\infty} \imath\omega\left[\left(k_{i\alpha,j\beta}+\omega^2 m_{i\alpha,j\beta}\right)^2+\omega^2 g_{i\alpha,j\beta}^2\right] H_{j\beta}\,|H_{i\alpha}|^2\,\mathrm{d}\omega \\ & -\frac{S_j}{2\pi}\int_{-\infty}^{\infty} \imath\omega\left[\left(k_{i\alpha,j\beta}+\omega^2 m_{i\alpha,j\beta}\right)^2+\omega^2 g_{i\alpha,j\beta}^2\right] H_{i\alpha}\,|H_{j\beta}|^2\,\mathrm{d}\omega\Bigg]\end{aligned} \tag{3.7}$$

On the other hand, eqn (2.161) gives the vibrational energy of resonator $i\alpha$ in terms of the power spectral density S_i.

$$\langle E_{i\alpha}\rangle = \frac{S_i}{2\pi}\int_{-\infty}^{\infty} m_i\omega^2\,|H_{i\alpha}|^2(\omega)\,\mathrm{d}\omega \tag{3.8}$$

By applying (3.2), the total energy of subsystem i is

$$\langle E_i \rangle = \sum_{\alpha=1}^{N_i} \frac{S_i}{2\pi} \int_{-\infty}^{\infty} m_i \omega^2 \,|\, H_{i\alpha} \,|^2 (\omega) \, \mathrm{d}\omega \tag{3.9}$$

By remarking that S_i and S_j do not depend α and β by virtue of Assumption 2b, and combining eqns (3.7) and (3.9), we obtain the coupling power proportionality

$$\langle P_{ij} \rangle = \beta_{ij} \langle E_i \rangle - \beta_{ji} \langle E_j \rangle \tag{3.10}$$

with the following factor,

$$\beta_{ij} = \frac{\sum_{\alpha,\beta} \int_{-\infty}^{\infty} \iota\omega \left[\left(k_{i\alpha,j\beta} + \omega^2 m_{i\alpha,j\beta} \right)^2 + \omega^2 g_{i\alpha,j\beta}^2 \right] H_{j\beta} \,|\, H_{i\alpha} \,|^2 \, \mathrm{d}\omega}{\sum_{\alpha} \int_{-\infty}^{\infty} m_i \omega^2 \,|\, H_{i\alpha} \,|^2 (\omega) \, \mathrm{d}\omega} \tag{3.11}$$

Equation (3.11) is established under the assumption of conservative and weak coupling. Although we have not explicitly written the series in terms of powers of ϵ, we know from Section 2.3 that the result is valid up to order 2 in ϵ. Formally, eqn (3.11) is similar to eqn (2.164) except for the presence of sums in both numerator and denominator. But in practice there is a strong difference. Applying equation (2.164) requires knowledge of mechanical parameters of only two resonators m_i, k_i, c_i, m_j, k_j, c_j while eqn (3.11) requires a large number of mechanical parameters m_i, c_i, $k_{i\alpha}$, $\alpha = 1, \ldots, N_i$, and m_j, c_j, $k_{j\beta}$, $\beta = 1, \ldots, N_j$ to be known. In practice, the number of modes N_i, N_j can be very large and it is often illusive to try to compute all of them. This is exactly what we want to avoid in statistical energy analysis.

Before interpreting eqn (3.11) in terms of probability, we need a technical result.

Monte Carlo method

The Monte Carlo method is a numerical method to assess an integral. In the theory of probability, an integral can be interpreted as the expectation of a random variable. Let us consider a random variable x with a uniform probability density function within the interval $[a, b]$. Then, for any deterministic function $f(x)$,

$$\langle f \rangle = \frac{1}{b-a} \int_a^b f(x) \, \mathrm{d}x \simeq \frac{1}{N} \sum_{\alpha=1}^{N} f(x_\alpha) \tag{3.12}$$

provided that the collection x_α is taken randomly and uniformly over the interval $[a, b]$. Of course, the equality is not exact but only gives an approximation of $\langle f \rangle$. The larger the number of samples x_α the better the approximation.

In the case of a non-uniform probability density function $p(x)$, the result always holds provided that the collection x_α is taken randomly with the probability density $p(x)$.

$$\langle f \rangle = \int_{-\infty}^{\infty} f(x)p(x)\,\mathrm{d}x \simeq \frac{1}{N}\sum_{\alpha=1}^{N} f(x_\alpha) \tag{3.13}$$

This result admits a direct generalization for multi-dimensional random variables. If $f(x, y)$ is a deterministic function of two variables and if X and Y are two random variables with mutual probability density $p(x, y)$, then

$$\langle f \rangle = \int_{-\infty}^{\infty}\int_{-\infty}^{\infty} f(x, y)p(x, y)\,\mathrm{d}x\mathrm{d}y \simeq \frac{1}{N}\sum_{\alpha=1}^{N} f(x_\alpha, y_\alpha) \tag{3.14}$$

provided that the collection x_α, y_α is chosen with respect to the joint probability density $p(x, y)$.

The special case where x and y are statistically independent is of particular interest. The joint probability function is of the form $p(x)p'(y)$. The double integral then splits into two successive integrals that can be approximated by two applications of eqn (3.13). It yields

$$\langle f \rangle = \int_{-\infty}^{\infty}\int_{-\infty}^{\infty} f(x, y)p(x)p'(y)\,\mathrm{d}x\mathrm{d}y \simeq \frac{1}{NN'}\sum_{\alpha=1}^{N}\sum_{\beta=1}^{N'} f(x_\alpha, y_\beta) \tag{3.15}$$

Conversely, any discrete sum with two dummy variables as above may be approximated by a probabilistic integral with two independent random variables.

Reduction to a probability problem

We are now in a position to introduce a probabilistic interpretation of the factor β_{ij} of eqn (3.11). We shall replace the discrete sums by integrals following the Monte Carlo method. Of course, this numerical approximation works better when the number of samples is large. In the present context, samples are modes, so we must introduce the next assumption of statistical energy analysis.

Assumption 4 *The number of modes is large in all subsystems;* $N_i \gg 1$ *for all* i.

Let us examine the sums in eqn (3.11). They run over the subscripts α, β on which depend the values of $\omega_{i\alpha}, \omega_{j\beta}, k_{i\alpha,j\beta}, g_{i\alpha,j\beta}, m_{i\alpha,j\beta}$. The modal mass m_i and the half-power bandwidth $\Delta_i = c_i/m_i$ do not depend on the mode index as specified in the canonical problem. Instead of summing over α and β, we may forget the sequence of exact values $\omega_{i\alpha}, \omega_{j\beta}, k_{i\alpha,j\beta}, g_{i\alpha,j\beta}, m_{i\alpha,j\beta}$ where $\alpha = 1, \ldots, N_i$, $\beta = 1, \ldots, N_j$ and replace them by a well-suited probability density function $p(\omega_i, \omega_j, k_{ij}, g_{ij}, m_{ij})$. Thus, the sums in the numerator and denominator of eqn (3.11) may be interpreted as resulting from the application of the Monte Carlo method to the random variables ω_i, ω_j, m_{ij}, g_{ij}, and k_{ij}. The joint probability density function $p(\omega_i, \omega_j, k_{ij}, g_{ij}, m_{ij})$ must be inferred from the actual population of numbers $\omega_{i\alpha}$, $\omega_{j\beta}$, $k_{i\alpha,j\beta}$, $g_{i\alpha,j\beta}$, and $m_{i\alpha,j\beta}$.

The sum of the numerator in eqn (3.11) may be written as a probability integral,

$$\sum_{\alpha,\beta} \int_{-\infty}^{\infty} \imath\omega \left[\left(k_{i\alpha,j\beta} + \omega^2 m_{i\alpha,j\beta}\right)^2 + \omega^2 g_{i\alpha,j\beta}^2 \right] H_{j\beta} \, | H_{i\alpha} |^2 \, \mathrm{d}\omega = N_i N_j$$
$$\times \int \int_{-\infty}^{\infty} \imath\omega \left[\left(k_{ij} + \omega^2 m_{ij}\right)^2 + \omega^2 g_{ij}^2 \right] H_{j\beta} \, | H_{i\alpha} |^2 \, \mathrm{d}\omega \times p \, \mathrm{d}\omega_i \mathrm{d}\omega_j \mathrm{d}m_{ij} \mathrm{d}g_{ij} \mathrm{d}k_{ij} \tag{3.16}$$

where $p = p(\omega_i, \omega_j, k_{ij}, g_{ij}, m_{ij})$ is the joint probability density function. For the sum of the denominator of eqn (3.11), this gives

$$\sum_{\alpha} \int_{-\infty}^{\infty} m_i \omega^2 \, | H_{i\alpha} |^2 \, \mathrm{d}\omega = N_i \int \int_{-\infty}^{\infty} m_i \omega^2 \, | H_i |^2 \, \mathrm{d}\omega \times p_{\omega_i}(\omega_i) \, \mathrm{d}\omega_i \tag{3.17}$$

where $p_{\omega_i}(\omega_i)$ is the marginal probability density function of the random variables ω_i.

We may therefore introduce the bracket $\langle . \rangle$ to denote statistical average over the probability density function p,

$$\langle . \rangle = \int . \, p(\omega_i, \omega_j, k_{ij}, g_{ij}, m_{ij}) \, \mathrm{d}\omega_i \mathrm{d}\omega_j \mathrm{d}m_{ij} \mathrm{d}g_{ij} \mathrm{d}k_{ij} \tag{3.18}$$

We also introduce the double brackets $\langle\langle . \rangle\rangle$ used for the quantities E_i, P_{ij} to denote the statistical average over both the random excitations and the probability density function p.

We may now give the probabilistic form of eqns (3.10) and (3.11). The mean exchanged power between subsystems i and j is noted $\langle\langle P_{ij} \rangle\rangle$ where the double bracket $\langle\langle . \rangle\rangle$ reminds us that this is a probability-based estimation of the mean exchanged power between two groups of oscillators containing a large number of modes randomly excited. Similarly $\langle\langle E_i \rangle\rangle$ is the probability-based estimation of the mean vibrational energy of subsystem i. Both are related by the coupling power proportionality,

$$\langle\langle P_{ij} \rangle\rangle = \omega_0 \eta_{ij} \langle\langle E_i \rangle\rangle - \omega_0 \eta_{ji} \langle\langle E_j \rangle\rangle \tag{3.19}$$

where

$$\omega_0 \eta_{ij} = N_j \frac{\left\langle \int_{-\infty}^{\infty} \imath\omega \left[\left(k_{ij} + \omega^2 m_{ij}\right)^2 + \omega^2 g_{ij}^2 \right] H_j \, | H_i |^2 \, \mathrm{d}\omega \right\rangle}{\left\langle \int_{-\infty}^{\infty} m_i \omega^2 \, | H_i |^2 (\omega) \, \mathrm{d}\omega \right\rangle} \tag{3.20}$$

The coefficients η_{ij} are called **coupling loss factors**. They are introduced in place of the previous factors β_{ij} traditionally used for the deterministic form of the coupling power proportionality. In eqn (3.20) also appears the **centre frequency** ω_0 whose significance will be clarified in Section 3.3.

The calculation of the deterministic coupling loss factor by eqn (3.11) requires the knowledge of modal parameters of all resonators. In practice this is a very large number of parameters since the number of modes may be of the order of millions. We have reduced the question of the exchanged power between two sets containing a large number of deterministic resonators to a problem of theory of probability. The probabilistic coupling loss factor of eqn (3.20) only requires knowing the statistical properties of a random resonator which are specified by a unique probability density function; this means that the process is simplified.

3.2 Random Resonator

In this section we clarify the notion of the random resonator, and, in particular, we specify how to determine the probability density function p.

Random natural frequency

The question reduces to the determination of the marginal probability density function p_{ω_i}. The sequence of actual values of the natural frequency,

$$\omega_{i1},\ \omega_{i2},\ \ldots,\omega_{i\alpha},\ldots,\omega_{iN_i}$$

may be interpreted as N_i samples of the same random variables ω_i. We may introduce the 'empirical' cumulative distribution function,

$$F_{\omega_i}(\omega) = \frac{1}{N_i}\text{Card}\{\alpha/\omega_{i\alpha} \leq \omega\} \tag{3.21}$$

defined as the ratio of natural frequencies below the value ω. This is a step function and as such is highly irregular and therefore not derivable. The 'empirical' cumulative distribution function may be replaced by an 'idealized' cumulative distribution function that may be conveniently chosen as derivable. The marginal probability density function p_{ω_i} is then

$$p_{\omega_i} = \frac{\mathrm{d}F_{\omega_i}}{\mathrm{d}\omega} \tag{3.22}$$

Naturally, all these considerations make sense only when the number of modes is large enough.

The probability density function p_{ω_i} is closely related to a very common concept in high-frequency vibrations: the modal density. The modal density is defined as the mean number of modes per unit frequency bandwidth and is usually noted $n(\omega)$. The link between the probability density function p_{ω_i} and n is a simple proportionality,

$$p_{\omega_i}(\omega) = \frac{n(\omega)}{N_i} \tag{3.23}$$

where the proportionality constant is imposed by the normality condition. The question of asymptotic values of modal densities for various subsystems will be discussed in Chapter 6.

For the joint probability density function $p_{\omega_i,\omega_j}(\omega,\omega')$ one must examine the two sequences,

$$\begin{aligned}&\omega_{i1}, \omega_{i2}, \ldots, \omega_{i\alpha}, \ldots, \omega_{iN_i}\\&\omega_{j1}, \omega_{j2}, \ldots, \omega_{j\alpha}, \ldots, \omega_{jN_j}\end{aligned}$$

and introduce the joint cumulative distribution function,

$$F_{\omega_i,\omega_j}(\omega,\omega') = \frac{1}{N_iN_j}\text{Card}\{(\alpha,\beta)/\omega_{i\alpha} \leq \omega \,\&\, \omega_{j\beta} \leq \omega'\} \tag{3.24}$$

It is clear that the above cardinal may be split into a product of two cardinals,

$$F_{\omega_i,\omega_j}(\omega,\omega') = \frac{1}{N_i}\text{Card}\{\alpha/\omega_{i\alpha} \leq \omega\} \times \frac{1}{N_j}\text{Card}\{\beta/\omega_{j\beta} \leq \omega'\} \tag{3.25}$$

Thus, the joint probability density function obtained by deriving the cumulative distribution function becomes

$$p_{\omega_i,\omega_j}(\omega,\omega') = \frac{\partial F_{\omega_i,\omega_j}}{\partial\omega\partial\omega'} = p_{\omega_i}(\omega) \times p_{\omega_j}(\omega') \tag{3.26}$$

The joint probability density function is found to be the product of the marginal probability density functions. The random variables ω_i, ω_j are therefore statistically independent. This is not a physical assumption on the mechanical system but only a consequence of the approximation formula (3.15).

Random mode

Let us examine the random variables m_{ij}, g_{ij}, and k_{ij}. We need to introduce the notion of the random mode. A mode is a spatial function $\psi_i(x)$ which has certain properties that will be detailed in Chapter 4. But for the moment, we simply admit that a mode is a spatial random process noted $\psi_i(x)$.

The way by which inter-modal coupling parameters depend on the mode shape can be complex. But the simplest situation is when two subsystems are coupled by a mechanical spring. If the spring of stiffness K is attached at points x_i and x_j, the inter-modal stiffness between mode α of subsystem i and mode β of subsystem j is $k_{i\alpha,j\beta} = K\psi_{i\alpha}(x_i)\psi_{j\beta}(x_j)$. Other situations may arise. For instance, if the subsystems are coupled by a torsional spring of stiffness K, the mode shapes are involved in the inter-modal stiffness through their spatial derivatives $k_{i\alpha,j\beta} = K\psi'_{i\alpha}(x_i)\psi'_{j\beta}(x_j)$. In the case of fluid structure interaction, subsystems are coupled by a gyroscopic term which depends on mode shapes through the integral $g_{i\alpha,j\beta} = \rho_0 \int \psi_{i\alpha}\psi_{j\beta}\, \mathrm{d}S$ where ρ_0 is the air density. In all cases, the inter-modal coupling parameter depends on mode shapes through a functional dependence,

$$k_{i\alpha,j\beta},\ g_{i\alpha,j\beta},\ m_{i\alpha,j\beta} = f(\psi_{i\alpha}, \psi_{j\beta}) \tag{3.27}$$

where f may be any operator.

To infer the joint probability density function $p(\omega_i, \omega_j, k_{ij}, g_{ij}, m_{ij})$, the double subscript sequences $\omega_{i\alpha}$, $\omega_{j\beta}$, $k_{i\alpha,j\beta}$, $g_{i\alpha,j\beta}$, $m_{i\alpha,j\beta}$ must be examined. We then define the 'empirical' cumulative distribution function,

$$F(\omega, \omega', k, g, m) = \text{Card}\left\{(\alpha, \beta)/\omega_{i\alpha} \leq \omega \ \&\ \omega_{j\beta} \leq \omega' \ \&\ m_{i\alpha,j\beta} \leq m \ \&\ g_{i\alpha,j\beta} \leq g \ \&\ k_{i\alpha,j\beta} \leq k\right\}$$

from which an 'idealized' cumulative distribution function may be proposed by any interpolation process. The associated probability density function is

$$p(\omega, \omega', k, g, m) = \frac{1}{N_i N_j} \times \frac{\partial^5 F}{\partial\omega\partial\omega'\partial m\partial g\partial k} \tag{3.28}$$

Most often, natural frequencies and mode shapes will be found to be independent. This property must be checked on a case-by-case basis. But when this is achieved, the joint probability density function becomes

$$p(\omega, \omega', k, g, m) = p_{\omega_i}(\omega)p_{\omega_j}(\omega')p_{\psi}(k, g, m) \tag{3.29}$$

and evaluating expectation on natural frequencies and modes may be done in two separated steps.

Averaging over ensemble

We can now interpret the results of Section 3.1. One considers a statistical ensemble of similar systems but with varying natural frequencies and modes. The variability of modal properties is usually associated with unavoidable uncertainty that results from manufacturing processes. For instance, this ensemble may be realized by a collection of cars produced on an assembly line. All members are nominally identical but differ in structural detail and material properties due to manufacturing tolerances. The double bracket $\langle\langle . \rangle\rangle$ then gives a prediction of the average behaviour of the population but is in any case a prediction of a member. However, if the structural complexity of the system is high enough (see Shorter (2011) for a discussion on complexity in vibroacoustics), namely regarding the number of modes, it may be expected that the behaviour of any selected member does not significantly differ from the mean behaviour. In this case the double bracket $\langle\langle . \rangle\rangle$ prediction may be applied to a unique member.

3.3 Uniform Probability

In order to perform explicit calculations on expectation, it is necessary to give the probability density function $p_{\omega_i}(\omega)$. Several models are possible, depending on the system being considered. Among several possibilities, the most common is the uniform probability density function $p_{\omega_i}(\omega) = \text{cste}$. This is the model adopted in Newland (1968), Lotz and Crandall (1973), and Mace and Ji (2007). For other distributions such as Poisson or Gaussian orthogonal ensemble, see Lyon (1969), and Langley and Brown (2004a, 2004b). See also Wright and Weaver (2010) for a review on random matrix theory.

Probability density function

In the uniform model, the natural frequencies ω_i and ω_j are statistically independent and have a uniform probability density function (see Fig. 3.3). The natural frequencies lie between extreme values $\omega_{\min}$ and $\omega_{\max}$,

$$p(\omega_i) = \begin{cases} 1/\Delta\omega & \text{if } \omega_i \in [\omega_{\min}, \omega_{\max}] \\ 0 & \text{otherwise} \end{cases} \tag{3.30}$$

where $\Delta\omega = \omega_{\max} - \omega_{\min}$ is the frequency bandwidth. We shall also introduce the centre frequency $\omega_0 = (\omega_{\max}\omega_{\min})^{1/2}$. We have

$$\omega_{\min}, \omega_{\max} = \omega_0 \left[\sqrt{1 + \left(\frac{\Delta\omega}{2\omega_0}\right)^2} \mp \frac{\Delta\omega}{2\omega_0} \right] \tag{3.31}$$

In particular, it may be observed that $\omega_{\min}/\omega_0$ and $\omega_{\max}/\omega_0$ depend on $\Delta\omega/\omega_0$ solely.

Calculation of $\langle \int m_i\omega^2 \,|\, H_i \,|^2 \, d\omega \rangle$

The integral appearing in the denominator of eqn (3.20) has been calculated already in Chapter 2,

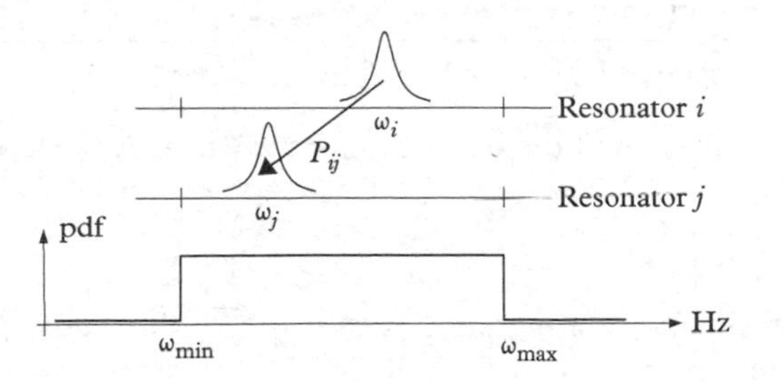

Figure 3.3 *Power exchanged between two random resonators.*

$$\int_{-\infty}^{\infty} m_i\omega^2 |H_i|^2 \, d\omega = \frac{\pi}{2\zeta_i m_i \omega_i} = \frac{\pi}{m_i \Delta_i} \tag{3.32}$$

where we have substituted the half-power bandwidth $\Delta_i = 2\zeta_i\omega_i$. Since both modal mass and half-power bandwidth have been assumed to be deterministic, the bracket $\langle . \rangle$ has no effect on the integral. In consequence, the brackets $\langle . \rangle$ of the denominator of eqn (3.20) vanish and those of the numerator may be conveniently applied to the entire ratio. So, in the particular case of a constant half-power bandwidth, i.e. viscous-type dissipation, the coupling loss factor $\omega_0\eta_{ij}$ is the mean value of the deterministic factors β_{ij} times N_j,

$$\omega_0\eta_{ij} = N_j \left\langle \frac{\int_{-\infty}^{\infty} \imath\omega \left[(k_{ij} + \omega^2 m_{ij})^2 + \omega^2 g_{ij} \right] H_j |H_i|^2 \, d\omega}{\int_{-\infty}^{\infty} m_i \omega^2 |H_i|^2 \, d\omega} \right\rangle = N_j \langle \beta_{ij} \rangle \tag{3.33}$$

Light damping

Several expressions of the factor β_{ij} were obtained in Chapter 2. The most appropriate for the present purpose is

$$\beta_{ij} = \frac{2m_{ij}^2\omega_i\omega_j\left(\zeta_i\omega_j^3+4\zeta_i\zeta_j\omega_i\omega_j(\zeta_i\omega_j+\zeta_j\omega_i)+\zeta_j\omega_i^3\right)+2\left(g_{ij}^2+2k_{ij}m_{ij}\right)\omega_i\omega_j(\omega_i\zeta_j+\omega_j\zeta_i)+2k_{ij}^2(\zeta_i\omega_i+\zeta_j\omega_j)}{m_i m_j\left[\left(\omega_i\sqrt{1-\zeta_i^2}+\omega_j\sqrt{1-\zeta_j^2}\right)^2+(\omega_i\zeta_i+\omega_j\zeta_j)^2\right]\left[\left(\omega_i\sqrt{1-\zeta_i^2}-\omega_j\sqrt{1-\zeta_j^2}\right)^2+(\omega_i\zeta_i+\omega_j\zeta_j)^2\right]} \tag{3.34}$$

Now, we introduce a further assumption. We shall assume that the damping is light. More exactly we shall assume that the quality factor of the peak is sufficiently high.

Assumption 5 *Damping loss factors of subsystems are low;* $\zeta_j \ll \Delta\omega/2\omega_0$.

But $\Delta\omega/2\omega_0$ is always one order of unity or less. For instance, $\Delta\omega/2\omega_0 = \sqrt{2}/4$ for an octave band. Therefore the above assumption implies that $\zeta_j \ll 1$. In particular we can neglect all terms of second order in damping loss factors, $\sqrt{1-\zeta_i^2} \sim 1$. Equation (3.34) becomes

$$\beta_{ij} = \frac{m_{ij}^2\left(\Delta_i\omega_j^4+\Delta_i\Delta_j\left(\Delta_i\omega_j^2+\Delta_j\omega_i^2\right)+\Delta_j\omega_i^4\right)+\left(g_{ij}^2+2k_{ij}m_{ij}\right)\left(\Delta_j\omega_i^2+\Delta_i\omega_j^2\right)+k_{ij}^2(\Delta_i+\Delta_j)}{m_i m_j\left[(\omega_i+\omega_j)^2+(\Delta_i+\Delta_j)^2/4\right]\left[(\omega_i-\omega_j)^2+(\Delta_i+\Delta_j)^2/4\right]} \tag{3.35}$$

where we have again substituted the half-power bandwidth $\Delta_i = 2\zeta_i\omega_i$.

Calculation of $\langle \beta_{ij} \rangle$

The brackets $\langle . \rangle$ correspond to integration with respect to ω_i, ω_j, m_{ij}, g_{ij}, k_{ij} and may be separated in three steps,

$$\langle \beta_{ij} \rangle = \langle \langle \langle \beta_{ij} \rangle_{\omega_i} \rangle_{\omega_j} \rangle_{\psi} = \frac{1}{\Delta\omega^2} \left\langle \int_{\omega_{\min}}^{\omega_{\max}} \int_{\omega_{\min}}^{\omega_{\max}} \beta_{ij} \, d\omega_i d\omega_j \right\rangle_{\psi} \tag{3.36}$$

where $\langle . \rangle_{\omega_i}$ denotes integration with respect to ω_i and $\langle . \rangle_\psi$ with respect to m_{ij}, g_{ij}, k_{ij}. We are now going to calculate them separately.

Let us remark that β_{ij} may be split into the sum of four terms,

$$\beta_{ij} = \frac{m_{ij}^2 Q_2 + m_{ij}^2 \Delta_i \Delta_j Q_1 + (g_{ij}^2 + 2k_{ij}m_{ij})Q_1 + k_{ij}^2 Q_0}{m_i m_j} \tag{3.37}$$

where the rational function Q_n is

$$Q_n(\omega_i, \omega_j) = \frac{\Delta_i \omega_j^{2n} + \Delta_j \omega_i^{2n}}{\left[(\omega_i + \omega_j)^2 + (\Delta_i + \Delta_j)^2/4\right]\left[(\omega_i - \omega_j)^2 + (\Delta_i + \Delta_j)^2/4\right]} \tag{3.38}$$

The rational function $Q_n = GF$ is in turn the product of two terms; a smooth term,

$$G(\omega_i, \omega_j) = \frac{\Delta_i \omega_j^{2n} + \Delta_j \omega_i^{2n}}{\left[(\omega_i + \omega_j)^2 + (\Delta_i + \Delta_j)^2/4\right]} \tag{3.39}$$

and a fluctuating part F,

$$F(\omega_i, \omega_j) = \frac{1}{\left[(\omega_i - \omega_j)^2 + (\Delta_i + \Delta_j)^2/4\right]} \tag{3.40}$$

Figure 3.4 shows a typical evolution of the two terms G and F versus ω_i. It can be seen that G is slowly varying while F has a strong peak about $\omega_i \sim \omega_j$. The function F looks like a δ-function and therefore it can be expected that it behaves like $\delta(\omega_i - \omega_j)$ during integration. It yields

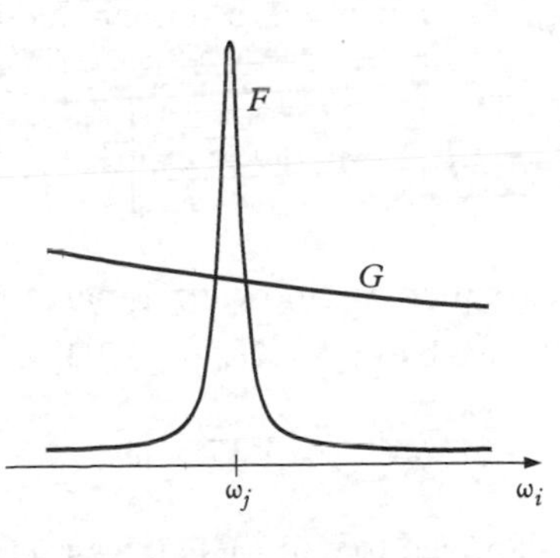

Figure 3.4 *Typical evolution of the smooth term $\omega_i \mapsto G(\omega_i, \omega_j)$ and the fluctuating term $\omega_i \mapsto F(\omega_i, \omega_j)$ over an octave band.*

$$\langle Q_n \rangle_{\omega_i} = \frac{1}{\Delta\omega} \int_{\omega_{\min}}^{\omega_{\max}} GF \, \mathrm{d}\omega_i = \frac{G(\omega_j, \omega_j)}{\Delta\omega} \int_{\omega_{\min}}^{\omega_{\max}} F \, \mathrm{d}\omega_i = \frac{G(\omega_j, \omega_j)}{\Delta\omega} \int_{-\infty}^{\infty} F \, \mathrm{d}\omega_i \tag{3.41}$$

Since the quality factor is high by virtue of Assumption 5, the integration limits can be extended to infinity without significantly modifying the value of the integral. Furthermore,

$$\int_{-\infty}^{\infty} \frac{\mathrm{d}\omega_i}{\left[(\omega_i - \omega_j)^2 + (\Delta_i + \Delta_j)^2/4\right]} = \frac{2}{\Delta_i + \Delta_j} \int_{-\infty}^{\infty} \frac{\mathrm{d}u}{1 + u^2} = \frac{2\pi}{\Delta_i + \Delta_j} \tag{3.42}$$

with the change of variable $u = 2(\omega_i - \omega_j)/(\Delta_i + \Delta_j)$. Therefore, in the limit of light damping $\Delta_i + \Delta_j \ll 2\omega_j$, combining eqns (3.39), (3.41), and (3.42) yields

$$\langle Q_n \rangle_{\omega_i} = \frac{\pi \omega_j^{2n-2}}{2\Delta\omega} \tag{3.43}$$

Finally, substituting eqn (3.43) into eqn (3.37) and neglecting the second-order term $m_{ij}\Delta_i\Delta_j Q_1$ gives

$$\langle \beta_{ij} \rangle_{\omega_i} = \pi \frac{m_{ij}^2 \omega_j^4 + (g_{ij}^2 + 2k_{ij}m_{ij})\omega_j^2 + k_{ij}^2}{2m_i m_j \omega_j^2 \Delta\omega} \tag{3.44}$$

To perform the second integration, we can apply the brackets $\langle . \rangle_{\omega_j}$ separately to the three terms of eqn (3.44). The problem therefore reduces to the calculation of $\langle \omega^{2n-2} \rangle_{\omega_j}$ for $n = 0, 1, 2$. First, for $n = 0$,

$$\left\langle \frac{1}{\omega_j^2} \right\rangle_{\omega_j} = \frac{1}{\Delta\omega} \int_{\omega_{\min}}^{\omega_{\max}} \frac{\mathrm{d}\omega_j}{\omega_j^2} = \frac{1}{\omega_{\max}\omega_{\min}} = \frac{1}{\omega_0^2} \tag{3.45}$$

The case $n = 1$ is trivial $\langle 1 \rangle_{\omega_j} = 1$. For $n = 2$, a similar calculation gives

$$\langle \omega_j^2 \rangle_{\omega_j} = \frac{1}{\Delta\omega} \int_{\omega_{\min}}^{\omega_{\max}} \omega_j^2 \, \mathrm{d}\omega_j = \omega_0^2 \left[1 + \frac{4}{3} \left(\frac{\Delta\omega}{2\omega_0} \right)^2 \right] \tag{3.46}$$

If we further assume that the frequency bandwidth is narrow, $\langle \omega_j^2 \rangle_{\omega_j}$ is well approximated by ω_0^2. For instance, for an octave band $\Delta\omega/2\omega_0 = \sqrt{2}/4$,

$$\langle \omega_j^2 \rangle_{\omega_j} = 1.17 \times \omega_0^2 \tag{3.47}$$

With the above relationships applied to eqn (3.44), we readily find

$$\langle \langle \beta_{ij} \rangle_{\omega_i} \rangle_{\omega_j} = \pi \frac{m_{ij}^2 \omega_0^4 + (g_{ij}^2 + 2k_{ij}m_{ij})\omega_0^2 + k_{ij}^2}{2m_i m_j \omega_0^2 \Delta\omega} \tag{3.48}$$

Since the only remaining random variables are m_{ij}, g_{ij}, and k_{ij}, the last bracket $\langle . \rangle_\psi$ applies trivially to the numerator only,

$$\langle\langle\langle \beta_{ij}\rangle_{\omega_i}\rangle_{\omega_j}\rangle_\psi = \pi \frac{\left\langle m_{ij}^2\omega_0^4 + (g_{ij}^2 + 2k_{ij}m_{ij})\omega_0^2 + k_{ij}^2 \right\rangle_\psi}{2m_i m_j \omega_0^2 \Delta\omega} \tag{3.49}$$

Coupling loss factor

Two special cases are of interest; two sets of resonators in weak interaction and a set of resonators coupled to a single resonator.

When two sets of resonators are in interaction, both ω_i and ω_j are random. If their probability density functions are also uniform, the relevant mean value of $\langle\langle\langle \beta_{ij}\rangle_{\omega_i}\rangle_{\omega_j}\rangle_\psi$ is given in eqn (3.49). In accordance with eqns (3.33) and (3.49),

$$\omega_0 \eta_{ij} = \pi N_j \frac{\left\langle \left(k_{ij} + \omega_0^2 m_{ij}\right)^2 + \omega_0^2 g_{ij}^2 \right\rangle_\psi}{2m_i m_j \omega_0^2 \Delta\omega} \tag{3.50}$$

where we have factorized the numerator. By permuting the subscripts i and j and remarking that $m_{ij} = m_{ji}$, $g_{ij} = -g_{ji}$, $k_{ij} = k_{ji}$,

$$\omega_0 \eta_{ji} = \pi N_i \frac{\left\langle \left(k_{ij} + \omega_0^2 m_{ij}\right)^2 + \omega_0^2 g_{ij}^2 \right\rangle_\psi}{2m_j m_i \omega_0^2 \Delta\omega} \tag{3.51}$$

By comparing eqns (3.50) and (3.51), we get

$$N_i \eta_{ij} = N_j \eta_{ji} \tag{3.52}$$

This is the so-called **reciprocity relationship**.

A second situation which may arise in practice is a set of resonators in weak interaction with a single resonator. In this case, we may consider that ω_i is a random variable with a uniform probability density function while ω_j remains deterministic. The relevant mean value is $\langle \beta_{ij}\rangle_{\omega_i}$ of eqn (3.44). Substituting eqn (3.44) into eqn (3.33) gives

$$\omega_0 \eta_{ij} = \pi N_j \frac{\left\langle \left(k_{ij} + \omega_j^2 m_{ij}\right)^2 + \omega_j^2 g_{ij}^2 \right\rangle_\psi}{2m_i m_j \omega_j^2 \Delta\omega} \tag{3.53}$$

For the reciprocal coupling loss factor, the relevant mean value is $\langle \beta_{ji}\rangle_{\omega_i}$ equal to $\langle \beta_{ij}\rangle_{\omega_i}$ by virtue of the symmetry $\beta_{ij} = \beta_{ji}$. We get

$$\omega_0 \eta_{ji} = \pi N_i \frac{\left\langle \left(k_{ij} + \omega_j^2 m_{ij}\right)^2 + \omega_j^2 g_{ij}^2 \right\rangle_\psi}{2m_i m_j \omega_j^2 \Delta\omega} \tag{3.54}$$

Again, we observe that reciprocity applies,

$$N_i \eta_{ij} = N_j \eta_{ji} \tag{3.55}$$

Validity of the approximations

Equation (3.50) results from eqn (3.33) after several approximations. To evaluate the total error induced by these approximations, we may consider the ratio of $\omega_0\eta_{ij}$ given in eqn (3.33) before the approximations to the one of eqn (3.50) after the approximations. This ratio is expected to be close to one. In the special case of an elastic coupling,

$$r_{\text{stiff}} = N_j\langle k_{ij}^2\rangle_\psi \frac{\langle Q_0\rangle}{m_i m_j} \times \frac{2m_i m_j \omega_0^2 \Delta\omega}{\pi N_j \langle k_{ij}^2\rangle_\psi} = \frac{2}{\pi}\omega_0^2\Delta\omega\langle Q_0\rangle \tag{3.56}$$

Similarly for a gyroscopic coupling alone,

$$r_{\text{gyr}} = \frac{2}{\pi}\Delta\omega\langle Q_1\rangle \tag{3.57}$$

while for an inertial coupling,

$$r_{\text{mass}} = \frac{2}{\pi}\frac{\Delta\omega}{\omega_0^2}(\langle Q_2\rangle + \Delta_i\Delta_j\langle Q_1\rangle) \tag{3.58}$$

These ratios neither depend on the masses m_i, m_j nor the coupling parameters k_{ij}, g_{ij}, m_{ij}. We are going to show that they depend on $\Delta\omega/\omega_0$ but not separately on ω_0 and $\Delta\omega$.

Let K_n be

$$K_n = \langle Q_n\rangle \frac{\Delta\omega^2}{\omega_0^{2n-1}} \tag{3.59}$$

By definition,

$$K_n = \frac{1}{\omega_0^{2n-1}} \int_{\omega_{\text{min}}}^{\omega_{\text{max}}} \int_{\omega_{\text{min}}}^{\omega_{\text{max}}} \frac{\Delta_i\omega_j^{2n} + \Delta_j\omega_i^{2n}}{\left(\omega_i^2 - \omega_j^2\right)^2 + (\Delta_i + \Delta_j)\left(\Delta_i\omega_j^2 + \Delta_j\omega_i^2\right)} \,\mathrm{d}\omega_i \mathrm{d}\omega_j \tag{3.60}$$

Let $\zeta_i = \Delta_i/2\omega_0$, $\zeta_j = \Delta_j/2\omega_0$ be the effective damping loss factors of the frequency band. By the change of variables $(\omega_i, \omega_j) \mapsto (\alpha_i\omega_0, \alpha_j\omega_0)$,

$$K_n = \int_{\omega_{\text{min}}/\omega_0}^{\omega_{\text{max}}/\omega_0} \int_{\omega_{\text{min}}/\omega_0}^{\omega_{\text{max}}/\omega_0} \frac{2\left(\zeta_i\alpha_j^{2n} + \zeta_j\alpha_i^{2n}\right)}{\left(\alpha_i^2 - \alpha_j^2\right)^2 + 4(\zeta_i + \zeta_j)\left(\zeta_i\alpha_j^2 + \zeta_j\alpha_i^2\right)} \,d\alpha_i d\alpha_j \tag{3.61}$$

where $\omega_{\text{min}}/\omega_0$ and $\omega_{\text{max}}/\omega_0$ are functions of $\Delta\omega/\omega_0$. The above integral only depends on the ratio $\Delta\omega/\omega_0$. Then by combining eqns (3.56), (3.57), (3.58), and (3.61),

$$r_{\text{stiff}} = \frac{2\omega_0}{\Delta\omega}\frac{K_0}{\pi}, \quad r_{\text{gyr}} = \frac{2\omega_0}{\Delta\omega}\frac{K_1}{\pi}, \quad r_{\text{mass}} = \frac{2\omega_0}{\Delta\omega}\frac{K_2 + 4\zeta_i\zeta_j K_1}{\pi} \tag{3.62}$$

We observe that the right-hand side only depends on the relative bandwidth $\Delta\omega/\omega_0$ and the effective damping factors ζ_i, ζ_j.

In Fig. 3.5 numerical estimations of the ratios r_{stiff}, r_{gyr}, and r_{mass} as defined above versus ζ_i and ζ_j for an octave band $\Delta\omega/\omega_0 = \sqrt{2}/4$ are plotted. It is apparent that the

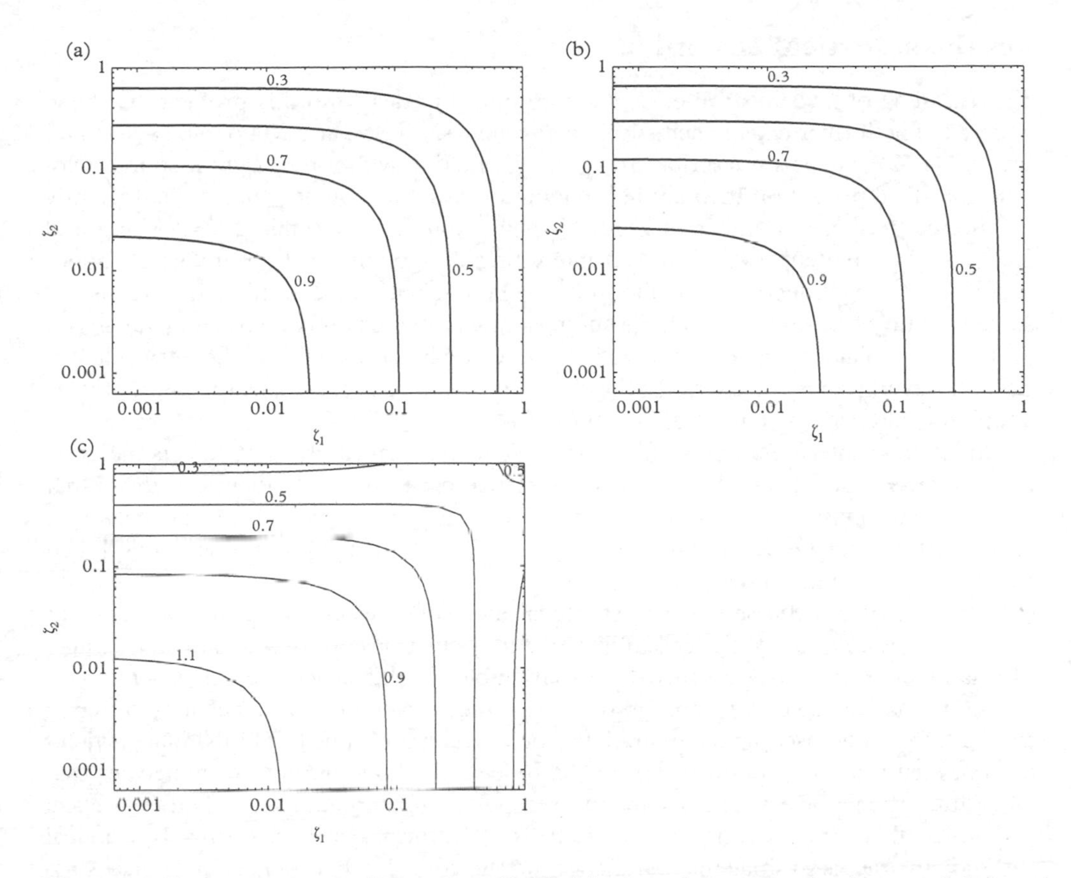

Figure 3.5 *Ratios r_{stiff}, r_{gyr}, and r_{inert} of eqn (3.62) of the theoretical coupling loss factors and their approximations of eqn (3.50) for an octave band. (a) elastic coupling; (b) gyroscopic coupling; (c) inertial coupling.*

approximation (3.50) gives the coupling loss factor within 10% whenever ζ_i, ζ_j < 0.03 for an elastic or a gyroscopic coupling. The case of an inertial coupling tends to underestimate by 20% the actual coupling loss factor for small damping. This is exactly the error induced by approximating by 1 the factor of eqn (3.47).

3.4 Comments

The fact that at high frequencies a mechanical structure can no longer be considered deterministic, in the sense that its natural frequencies become sensitive to the inevitable small variations in structural detail, has been the first motivation for the development of a statistical approach.

On the statistical ensemble

The meaning of the statistical ensemble introduced to deal with this problem has been debated. The former texts in statistical energy analysis (Lyon and Maidanik (1962) and Smith (1962)) aimed to describe structures interacting with a reverberant sound field. The statistical prediction then applies to a unique system provided that it is sufficiently disordered and/or its number of modes is high enough (see Renji (2004), Wang and Lai (2005), and Renji (2005) for an interesting discussion on the number of modes required). As mentioned by Woodhouse (1981) this is natural 'in architectural acoustics, since the number of modes of a large auditorium within the audible frequency range run into tens of millions. In such a situation one can expect statistics to be very reliable, while deterministic methods are obviously out of the question.' This is the point of view adopted when deriving eqns (3.19) and (3.20).

But another interpretation is often preferred. The statistical theory predicts the average response of an ensemble of dynamic systems to be nominally identical. From that point of view, statistical energy analysis may not predict the behaviour of a particular system but gives only the trend of the entire population. This is the point of view developed by Mace (1992) and Finnveden (1995).

This situation is the same as in statistical mechanics where the actual system, for instance a gas composed of a large number of molecules, is considered to be one sample of a large ensemble, the so-called Gibbs ensemble. In this ensemble, all systems have the same macroscopic state, that is pressure, volume, temperature, but they differ in their different microscopic configurations; the position and speed of individual particles are not identical throughout the ensemble. Indeed, the huge number of molecules that constitutes the gas is a strong argument in favour of a mean approach. The theoretical probability that a given configuration differs from the mean system described by classical thermodynamics is so small that in practice, we can consider that such a non-typical state does not exist. The actual system chosen at random from the ensemble is statistically very close to the mean system and we can consider that the *actual system behaves like the mean system*.

Nevertheless, the problems must handled carefully in vibroacoustics because, as remarked by Sestieri and Carcaterra (2013), 'the ensemble average of very simple systems . . . is completely different to any of the samples of the population'. These counter-examples stem from the non-ergodicity of linear systems in free vibration. See Carcaterra (2005) for more details.

On the probability density function

When defining the statistical ensemble of resonators, we have chosen to reduce the number of random variables to only one, the natural frequency ω_i, while m_i and Δ_i remained deterministic. Newland 1968 considered a tenth-order joint probability density function $p(\omega_i, \omega_j, \zeta_i, \zeta_j, m_i, m_j, U_i, U_j, k_{ij}, m_{ij})$ where m_{ij} is a coupling mass term. In Newland's notations, U_i and U_j designate the mean oscillator energies when $\epsilon = 0$, that is $U_r = S_r/4\zeta_r m_r \omega_r$. Newland assumed the statistical independence by setting

$$p(\omega_i, \omega_j, \zeta_i, \zeta_j, m_i, m_j, U_i, U_j, k_{ij}, m_{ij}) \\ = p(\omega_i)p(\omega_j)p(\zeta_i)p(\zeta_j)p(m_i)p(m_j)p(U_i)p(U_j)p(k_{ij})p(m_{ij}) \tag{3.63}$$

Newland did not specify the probabilistic behaviour of the variables ζ_i, ζ_j, m_i, m_j, U_i, U_j, k_{ij}, m_{ij}, but concerning ω_i and ω_j he assumed the uniform distribution $p(\omega_i) = p(\omega_j) = 1/\Delta\omega$.

Mace and Ji (2007) start from the most general joint probability density function $p(\omega_i, \omega_j, \Delta_i, \Delta_j, m_i, m_j, S_i, S_j, k_{ij})$. They assume that all parameters are known exactly except for ω_i and ω_j. In particular the choice is therefore to maintain constant the half-power bandwidth $\Delta_i = 2\zeta_i\omega_i$ instead of the damping ratio ζ_i. They also consider a uniform probability density function on $\omega_i - \omega_j$.

On equipartition of energy

The most common method to derive the coupling power proportionality between packets of resonators is based on the assumption of equipartition of modal energy. This is the method adopted in Lyon and DeJong (1995), Ungar (1966), Lyon (1973), Lesueur (1988), Fahy (1994), and Wijker (2009). Considering any pair $(i\alpha, j\beta)$ of resonators (one in each subsystem), the coupling power is

$$\langle P_{i\alpha,j\beta}\rangle = \beta_{i\alpha,j\beta}\left(\langle E_{i\alpha}\rangle - \langle E_{j\beta}\rangle\right) \tag{3.64}$$

where $\beta_{i\alpha,j\beta}$ is given by any relationship of Chapter 2 and $E_{i\alpha}$ the energy of mode i, α. The total exchanged power is simply the sum of all pairs $\langle P_{ij}\rangle = \sum_{\alpha,\beta}\langle P_{i\alpha,j\beta}\rangle$. So, if we introduce the critical assumption that each mode has the same energy within a subsystem $\langle E_{i\alpha}\rangle = \langle E_i\rangle/N_i$ and similarly $\langle E_{j\beta}\rangle = \langle E_j\rangle/N_j$, then

$$\langle P_{ij}\rangle = \left(\sum_{\alpha,\beta}\beta_{i\alpha,j\beta}\right)\left(\frac{\langle E_i\rangle}{N_i} - \frac{\langle E_j\rangle}{N_j}\right) \tag{3.65}$$

where $\langle E_i\rangle$ and $\langle E_j\rangle$ are the total energy of subsystems. This is the most direct way to obtain eqn (3.10) jointly with reciprocity. However, equipartition of modal energy is difficult to justify a priori and we have systematically avoided its usage in this chapter. Burroughs et al. (1997) argue that equipartition arises in multiresonant subsystems when all modes are submitted to the same broadband excitation and have the same damping loss factor.

Let us mention the so-called statistical modal energy distribution analysis (Maxit and Guyader (2003), Totaro and Guyader (2012), and Maxit et al. (2014)), a method introduced to extend the calculations in situations where equipartition is not verified. The principle of the method is based on a mode-by-mode energy balance with a direct application of eqn (3.64) where the $\beta_{i\alpha,j\beta}$ factors are calculated by equations of Chapter 2.

On the factor β

A question related to equipartition is the definition of the factor β between two groups of oscillators. The most common approach to define it is as the sum of factors β for all pairs of oscillators as done in eqn (3.65) as a result of energy equipartition. However, Mace (2007) underlined that taking the expected value from the coupling power of an individual pair,

$$\langle\langle P_{ij}\rangle\rangle = \langle\beta_{ij}\langle E_i\rangle\rangle_\omega - \langle\beta_{ji}\langle E_j\rangle\rangle_\omega \tag{3.66}$$

where as usual the innermost bracket denotes average for random forces and the outermost bracket average for random natural frequency, the term $\langle\beta_{ij}\langle E_i\rangle\rangle_\omega$ cannot be split generally into $\langle\beta_{ij}\rangle_\omega\langle\langle E_i\rangle\rangle$. The reason is that $\langle E_i\rangle - \langle E_j\rangle$ and β_{ij} are not statistically independent since if β_{ij} is large ($\omega_i \sim \omega_j$) then $\langle E_i\rangle - \langle E_j\rangle$ is small, and vice-versa. There is a strong correlation between $\langle E_i\rangle$ and β_{ij}. This difficulty is avoided in eqns (3.10) and (3.11) where the sums running over modes have been introduced separately on the numerator and the denominator.

4

Continuous Systems

In this chapter, we study continuous systems. The method consists in reducing the dynamics of continuous components to sets of discrete resonators by projecting the vibrating fields on a considered modal basis. We shall obtain *infinite* sets of discrete resonators in weak interaction. Nevertheless, the mathematical developments of Chapter 3 apply to *finite* sets of resonators; this will be the main difficulty of this chapter.

4.1 Resonant Modes

Normal modes

All finite structures admit a set of normal modes. Normal modes are defined as the eigenvectors of the stiffness operator. They form a complete set of orthogonal functions.

In a general manner, the governing equation of a vibrating structure has the form

$$m\frac{\partial^2 u}{\partial t^2} + c\frac{\partial u}{\partial t} + \mathcal{K}u = f(x, t) + \text{boundary conditions} \tag{4.1}$$

where m is the mass per unit length, area, or volume, $\mathcal{K}$ is the stiffness operator which acts on the spatial coordinate, and $f(x, t)$ the external force field. Boundary conditions are usually prescribed in terms of the field or its normal derivatives on the boundary. The normal modes $(\psi_\alpha)_{\alpha \geq 1}$ are the infinite sequence of eigenvectors of the stiffness operator which verify the homogeneous boundary conditions

$$\mathcal{K}\psi_\alpha = m\omega_\alpha^2 \psi_\alpha, \quad \alpha = 1, 2, \ldots \tag{4.2}$$

where $m\omega_\alpha^2$ are the corresponding eigenvalues, always non-negative. The sequence of eigenvalues is unbounded. Normal modes are orthogonal and conventionally normalized,

$$\int \psi_\alpha(x)\psi_\beta(x)\,\mathrm{d}x = \delta_{\alpha\beta} \tag{4.3}$$

Foundation of Statistical Energy Analysis in Vibroacoustics. First Edition. A. Le Bot.

where the integral is performed over the whole domain (beam length, plate area, enclosure volume, and so on). Finally, the normal modes verify the so-called closure relationship,

$$\sum_{\alpha=1}^{\infty} \psi_\alpha(x)\psi_\alpha(x') = \delta(x-x') \tag{4.4}$$

The closure property ensures that normal modes form a basis, that is all solutions u may be developed as a series of normal modes. If $u(x,t)$ is a solution of eqn (4.1) then at any time t,

$$\begin{aligned} u(x,t) &= \int u(x',t)\delta(x'-x)\,\mathrm{d}x' \\ &= \int u(x',t)\sum_\alpha \psi_\alpha(x)\psi_\alpha(x')\,\mathrm{d}x' \\ &= \sum_\alpha \int u(x',t)\psi_\alpha(x')\,\mathrm{d}x'\,\psi_\alpha(x) \\ &= \sum_\alpha U_\alpha(t)\psi_\alpha(x) \end{aligned} \tag{4.5}$$

where we have noted

$$U_\alpha(t) = \int u(x',t)\psi_\alpha(x')\,\mathrm{d}x' \tag{4.6}$$

the modal amplitude of u. Similarly, the external force may be decomposed as a series,

$$f(x,t) = \sum_\alpha F_\alpha(t)\psi_\alpha(x) \tag{4.7}$$

where the modal force is

$$F_\alpha(t) = \int f(x,t)\psi_\alpha(x)\,\mathrm{d}x \tag{4.8}$$

Rain-on-the-roof excitation

A rain-on-the-roof excitation is defined as a random force field $f(x,t)$ which is stationary, homogeneous, isotropic, and δ-correlated in space and time. The space–time correlation of a rain-on-the-roof force is, by definition,

$$R_{ff}(x,x',t,t') = \langle f(x,t)f(x',t')\rangle = S\delta(x'-x)\delta(t'-t) \tag{4.9}$$

The correlation function R_{ff} only depends on the distance $\chi = |x'-x|$ (homogeneity and isotropy) and the time delay $\tau = t'-t$ (stationarity). The name 'rain-on-the-roof' comes from an analogy with drops of rain falling on a flat surface. The intensity of rain depends on neither position nor time and all drops are events uncorrelated in both space and time.

Since f is a random process, the modal forces F_α are also random processes. Let us calculate the correlation of modal forces resulting from a rain-on-the-roof excitation,

$$\begin{aligned}\langle F_\alpha(t)F_\beta(t+\tau)\rangle &= \left\langle \int f(x,t)\psi_\alpha(x)\,\mathrm{d}x \times \int f(x',t+\tau)\psi_\beta(x')\,\mathrm{d}x' \right\rangle \\ &= \int\int \langle f(x,t)f(x',t+\tau)\rangle \psi_\alpha(x)\psi_\beta(x')\,\mathrm{d}x\mathrm{d}x'\end{aligned}$$

In the second equality, we have permuted integral and expectation symbols. We may recognize the correlation function of f. Therefore, by virtue of eqn (4.9),

$$\begin{aligned}\langle F_\alpha(t)F_\beta(t+\tau)\rangle &= \int\int S\delta(x'-x)\delta(\tau)\psi_\alpha(x)\psi_\beta(x')\,\mathrm{d}x\mathrm{d}x' \\ &= S\delta(\tau)\int \psi_\alpha(x)\psi_\beta(x)\,\mathrm{d}x \\ &= S\delta_{\alpha\beta}\delta(\tau)\end{aligned} \tag{4.10}$$

In the second equality, the double integral reduces to a single integral by the property of the δ-function and the last equality is obtained by orthonormality. Hence all modal forces are white noises with the same spectrum S and are mutually uncorrelated,

$$S_{F_\alpha F_\beta}(\omega) = S\delta_{\alpha\beta} \tag{4.11}$$

The condition of Assumption 2b is then fulfilled by a rain-on-the-roof excitation.

The converse is also true. If the modal forces are mutually uncorrelated white noises with the same power spectral density, i.e. eqn (4.10) holds, then the space–time correlation of f is

$$\begin{aligned}\langle f(x,t)f(x',t+\tau)\rangle &= \left\langle \sum_\alpha F_\alpha(t)\psi_\alpha(x)\sum_\beta F_\beta(t+\tau)\psi_\beta(x')\right\rangle \\ &= \sum_{\alpha,\beta} \langle F_\alpha(t)F_\beta(t+\tau)\rangle \psi_\alpha(x)\psi_\beta(x') \\ &= \sum_{\alpha,\beta} S\delta_{\alpha\beta}\delta(\tau)\psi_\alpha(x)\psi_\beta(x') \\ &= S\delta(\tau)\sum_\alpha \psi_\alpha(x)\psi_\beta(x') \\ &= S\delta(\tau)\delta(x'-x)\end{aligned}$$

where the closure property has been used in the last equality. The random force f is therefore a rain-on-the-roof field.

An example of a field with modal correlations is a point force even if the force is a white noise. A point force may be viewed as a force field having the form

$f(x,t) = \delta(x - x_0)f(t)$ where x_0 is the source position. The modal forces may readily be calculated by eqn (4.8),

$$F_\alpha(t) = \psi_\alpha(x_0)f(t) \tag{4.12}$$

Since $f(t)$ is assumed to be a white noise of power spectral density S,

$$R_{ff}(\tau) = \langle f(t)f(t+\tau)\rangle = S\delta(\tau) \tag{4.13}$$

The cross-correlations of modal forces are then

$$R_{F_\alpha F_\beta}(\tau) = \langle \psi_\alpha(x_0)f(t)\psi_\beta(x_0)f(t+\tau)\rangle = \psi_\alpha(x_0)\psi_\beta(x_0)S\delta(\tau) \tag{4.14}$$

And, by a Fourier transform, the cross-power spectral densities are

$$S_{F_\alpha F_\beta}(\omega) = \psi_\alpha(x_0)\psi_\beta(x_0)S \tag{4.15}$$

We can observe that, since the power spectral densities are independent of ω, the modal forces are also white noises. But, at the same time, cross-correlations exist when $\alpha \neq \beta$ and therefore *the modal forces induced by a point excitation are not mutually uncorrelated.* Equivalently, the power spectral densities of modal forces are not identical for all modes. One important assumption of statistical energy analysis is violated. The question raised is whether this kind of excitation is suitable in statistical energy analysis. From the point of view of what has been presented up to now the answer is clearly no, even if the force is a white noise. A short discussion of this problem is given by Fahy (1970).

Energy equipartition

Energy equipartition as defined in statistical physics is a state of complex systems in which the energy is equally shared in average by all degrees of freedom. Note that this does not mean that at a fixed time all degrees of freedom have the same energy, but only that their time-average energy is the same when observed over a long period. Equipartition may arrive when non-linear processes, such as shocks between molecules, ensure a perfect mixing of energy. But in the present context of linear systems, equipartition generally does not occur. For instance in a set of coupled linear oscillators free of external excitation, the initial repartition of energy over modes remains unchanged in time since modes are uncoupled and therefore do not exchange energy. However, when applying random external excitation, we may expect that in some particular conditions sources provide the same amount of energy to all modes. Let us explore this property.

By substituting eqn (4.5) into eqn (4.1),

$$m\sum_\alpha \ddot{U}_\alpha\psi_\alpha + c\sum_\alpha \dot{U}_\alpha\psi_\alpha + \sum_\alpha U_\alpha\mathcal{K}\psi_\alpha = f \tag{4.16}$$

But $\mathcal{K}\psi_\alpha$ is proportional to ψ_α by eqn (4.2),

$$m\sum_\alpha \left[\ddot{U}_\alpha + \Delta\dot{U}_\alpha + \omega_\alpha^2 U_\alpha\right]\psi_\alpha = f \tag{4.17}$$

where we have introduced the half-power bandwidth $\Delta = c/m$. By multiplying the above by ψ_α (after renaming the dummy subscript β), integrating, and re-ordering the sum and integral signs,

$$m \sum_\beta \left[\ddot{U}_\beta + \Delta \dot{U}_\beta + \omega_\beta^2 U_\beta \right] \int \psi_\beta \psi_\alpha \, \mathrm{d}x = \int f \psi_\alpha \, \mathrm{d}x \tag{4.18}$$

By orthonormality after eqn (4.3),

$$m \left[\ddot{U}_\alpha + \Delta \dot{U}_\alpha + \omega_\alpha^2 U_\alpha \right] = F_\alpha \tag{4.19}$$

where we have substituted the right-hand side with the modal force defined in eqn (4.8). Thus, each mode behaves like an independent linear oscillator excited by the modal force F_α. The modal energy is

$$E_\alpha = \frac{1}{2} m \dot{U}_\alpha^2 + \frac{1}{2} m \omega_\alpha^2 U_\alpha^2 \tag{4.20}$$

and we may check in all special cases of interest that the total vibrational energy is also the sum of modal energies. If f is a rain-on-the-roof field, all F_α are white noise with the same power spectral density S_0. The mean energy of an oscillator excited by a white noise has already been calculated at the beginning of Chapter 2,

$$\langle E_\alpha \rangle = \frac{S_0}{4\pi} \int_{-\infty}^{\infty} m\omega^2 \,|\, H_\alpha \,|^2 \, \mathrm{d}\omega = \frac{S_0}{2m\Delta} \tag{4.21}$$

The conditions for equipartition to hold are now apparent. The mean energy is the same for all modes when modal forces have the same power spectral density (rain-on-the-roof excitation) and the half-power bandwidth is constant. Conditions for equipartition in coupled subsystems are discussed by Ungar (1966). See also Magionesi and Carcaterra (2009) for a general discussion on energy equipartition in structural dynamics.

Broadband excitation

In practice a white noise never exists; not only because it is difficult to maintain a power spectral density that is perfectly constant over a wide frequency band, but also because a white noise is an idealized signal which contains infinite energy. In the best case, external excitations are random with a power spectral density constant over a broad but limited frequency band. Let us consider an idealized situation where the power spectral density is constant over the frequency band $[\omega_{\min}, \omega_{\max}]$ and zero elsewhere,

$$S(\omega) = \begin{cases} S_0 & \text{if } |\,\omega\,| \in [\omega_{\min}, \omega_{\max}] \\ 0 & \text{otherwise} \end{cases} \tag{4.22}$$

As in Chapter 3, we also note $\Delta\omega = \omega_{\max} - \omega_{\min}$ the frequency span and ω the centre frequency.

When calculating the modal energy, the integral over frequency reduces to

$$\langle E_\alpha \rangle = \frac{1}{4\pi} \int_{-\infty}^{\infty} m\omega^2 \,|\, H_\alpha \,|^2 S(\omega) \,\mathrm{d}\omega = \frac{S_0}{2\pi} \int_{\omega_{\min}}^{\omega_{\max}} m\omega^2 \,|\, H_\alpha \,|^2 \,\mathrm{d}\omega \tag{4.23}$$

In the second integral a factor 2 has been introduced since $\omega \mapsto \omega^2 \,|\, H_\alpha \,|^2$ is even. To apply all preceding results and in particular the coupling power proportionality, we must wonder if the second integral may be approximated by

$$\int_{\omega_{\min}}^{\omega_{\max}} \omega^2 \,|\, H_\alpha \,|^2 \,\mathrm{d}\omega \sim \int_0^{\infty} \omega^2 \,|\, H_\alpha \,|^2 \,\mathrm{d}\omega \tag{4.24}$$

where the limits of integration have been extended. Two different situations may arise. We shall speak of **resonant mode** when the natural frequency lies within the band $[\omega_{\min}, \omega_{\max}]$ and **non-resonant mode** otherwise. It is clear that most of the infinite integral is contained in the vicinity of the natural frequency and therefore the approximation will be satisfactory only for resonant modes. For non-resonant modes, the integral over $[\omega_{\min}, \omega_{\max}]$ is negligible,

$$\int_{\omega_{\min}}^{\omega_{\max}} \omega^2 \,|\, H_{\alpha'} \,|^2 \,\mathrm{d}\omega \ll \int_{\omega_{\min}}^{\omega_{\max}} \omega^2 \,|\, H_\alpha \,|^2 \,\mathrm{d}\omega \sim \int_0^{\infty} \omega^2 \,|\, H_\alpha \,|^2 \,\mathrm{d}\omega \tag{4.25}$$

where α' is the index of a non-resonant mode and α that of a resonant mode. We shall note $\alpha' \notin \Delta\omega$ and $\alpha \in \Delta\omega$. So, we introduce the further assumption of statistical energy analysis:

Assumption 6 *Contribution of non-resonant modes is neglected.*

So only resonant modes contribute significantly to the total energy.

The interpretation of the frequency band $\Delta\omega$ introduced in Section 3.1 is now more clear. In particular among all modes of a continuous structure only those whose natural frequency lies within $\Delta\omega$ are excited by external forces and therefore contribute to the vibrational response. Thus, the probability density function of resonant modes is confined to $\Delta\omega$, as claimed in Chapter 3.

4.2 Single Resonator Attached to a String

The first example of a continuous system that we study in this chapter is a resonator attached to a vibrating string. This is a simplified version of the resonator to plate studied in Lyon and Eichler (1964). It is chosen for its simplicity rather than for its practical interest.

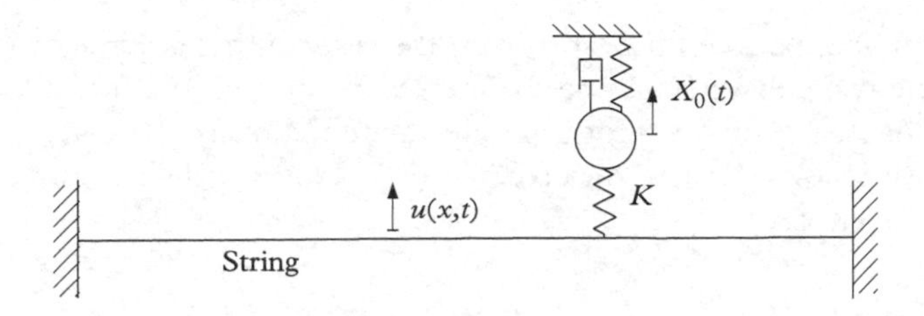

Figure 4.1 *Resonator attached to a string.*

Governing equation

The system is shown in Fig. 4.1. It is composed of a vibrating string of length L to which a resonator is attached at position x_0. The resonator has mass m_0, spring stiffness k_0, and damping coefficient c_0 connected in parallel to a rigid foundation. The resonator is coupled to the string by a spring of stiffness K assumed to be small (Assumption 3). No damper is placed in the coupling (Assumption 1). Furthermore the string is subjected to a random external force field $f(x, t)$ of type rain-on-the-roof, the resonator to a white noise external force $f_0(t)$, and f and f_0 are uncorrelated (Assumption 2).

The governing equation of the resonator is

$$m_0\ddot{X}_0 + c_0\dot{X}_0 + k_0X_0 = K(u_0 - X_0) + f_0(t) \tag{4.26}$$

The reaction of the coupling spring at position x_0 is $K(X_0 - u_0)$ where $u_0 = u(x_0, t)$ is the string deflection at x_0 and X_0 the moving mass position.

The governing equation of the string is

$$m\frac{\partial^2 u}{\partial t^2} - T\frac{\partial^2 u}{\partial x^2} = K(X_0 - u)\,\delta(x - x_0) + f(x, t) \tag{4.27}$$

where T is the tension and m the mass per unit length. The boundary conditions are

$$u(0, t) = u(L, t) = 0 \tag{4.28}$$

at the fixed points $x = 0$ and $x = L$.

Energy

The total vibrational energy E of the whole system is

$$E = \frac{1}{2}m_0\dot{X}_0^2 + \frac{1}{2}\int_0^L m\dot{u}^2\,\mathrm{d}x + \frac{1}{2}k_0X_0^2 + \frac{1}{2}\int_0^L Tu'^2\,\mathrm{d}x + \frac{1}{2}K(u_0 - X_0)^2 \tag{4.29}$$

where a prime mark denotes a space-derivative. The first two terms are respectively the kinetic energy of the moving mass and the string, and the next three terms are the elastic energy stored in respectively the spring k_0, the string, and the coupling spring K.

Since the coupling between the string and the resonator is assumed to be weak, it is natural to define two subsystems, one for the resonator and one for the string. But, as discussed in Chapter 2, one cannot separate unambiguously the elastic energy of the coupling. So, we follow the method of Chapter 2. And we define subsystem 0 as the subsystem having energy

$$E_0 = \frac{1}{2}m_0\dot{X}_0^2 + \frac{1}{2}k_0X_0^2 + \frac{1}{2}K\left(X_0^2 - X_0u_0\right) \tag{4.30}$$

Similarly, we define subsystem 1 by

$$E_1 = \frac{1}{2}\int_0^L m\dot{u}^2\,\mathrm{d}x + \frac{1}{2}\int_0^L Tu'^2\,\mathrm{d}x + \frac{1}{2}K\left(u_0^2 - X_0u_0\right) \tag{4.31}$$

By these definitions, we ensure that the sum of subsystem energies is equal to the energy of the entire system $E = E_0 + E_1$.

Power balance

The power balance of subsystem 0 is obtained by multiplying eqn (4.26) by $\dot{X}_0$,

$$m_0\ddot{X}_0\dot{X}_0 + c_0\dot{X}_0^2 + k_0X_0\dot{X}_0 = K\left(u_0 - X_0\right)\dot{X}_0 + f_0\dot{X}_0 \tag{4.32}$$

By time integration,

$$\frac{\mathrm{d}}{\mathrm{d}t}\left[\frac{1}{2}m_0\dot{X}_0^2 + \frac{1}{2}k_0X_0^2 + \frac{1}{2}K\left(X_0^2 - X_0u_0\right)\right] + c_0\dot{X}_0^2 + \frac{1}{2}K\left(X_0\dot{u}_0 - \dot{X}_0u_0\right) = f_0\dot{X}_0 \tag{4.33}$$

where $\dot{u}_0 = \dot{u}(x_0, t)$. Since the first term is the time derivative of E_0, we recognize successively the expressions of the power dissipated by the dashpot,

$$P_{\mathrm{diss},0} = c_0\dot{X}_0^2 \tag{4.34}$$

the net power exchanged between the string and the resonator,

$$P_{01} = \frac{1}{2}K\left(X_0\dot{u}_0 - \dot{X}_0u_0\right) \tag{4.35}$$

and the power injected in the resonator by the external force f_0,

$$P_0 = f_0\dot{X}_0 \tag{4.36}$$

The power balance of subsystem 0 finally reads

$$\frac{\mathrm{d}E_0}{\mathrm{d}t} + P_{\mathrm{diss},0} + P_{01} = P_0 \tag{4.37}$$

The power balance of subsystem 1 is obtained by multiplying eqn (4.27) by $\dot{u}$,

$$m\ddot{u}\dot{u} - Tu''\dot{u} = \delta_{x_0} K (X_0 - u_0)\, \dot{u} + f\dot{u} \tag{4.38}$$

Integrating the above equation over space gives

$$\int_0^L m\ddot{u}\dot{u}\, \mathrm{d}x - \int_0^L Tu''\dot{u}\, \mathrm{d}x = K (X_0 - u_0)\, \dot{u}_0 + \int_0^L f\dot{u}\, \mathrm{d}x \tag{4.39}$$

Integrating by parts the second term gives

$$\int_0^L m\ddot{u}\dot{u}\, \mathrm{d}x - \left[Tu'\dot{u} \right]_{x=0}^{x=L} + \int_0^L Tu'\dot{u}'\, \mathrm{d}x = K (X_0 - u_0)\, \dot{u}_0 + \int_0^L f\dot{u}\, \mathrm{d}x \tag{4.40}$$

By virtue of boundary conditions (4.28), the second term in brackets vanishes. The first and third terms are time derivatives. Some re-arrangement in the right-hand side leads to

$$\frac{\mathrm{d}}{\mathrm{d}t}\left[\frac{1}{2}\int_0^L m\dot{u}^2\, \mathrm{d}x + \frac{1}{2}\int_0^L Tu'^2\, \mathrm{d}x + \frac{1}{2}K (u_0^2 - X_0 u_0) \right] + \frac{1}{2}K \left(\dot{X}_0 u_0 - X_0 \dot{u}_0 \right) = \int_0^L f\dot{u}\, \mathrm{d}x \tag{4.41}$$

The exchanged power P_{10} is therefore

$$P_{10} = \frac{1}{2}K \left(\dot{X}_0 u_0 - X_0 \dot{u}_0 \right) \tag{4.42}$$

which is the opposite of the exchanged power P_{01}. The power balance of subsystem 1 becomes

$$\frac{\mathrm{d}E_1}{\mathrm{d}t} + P_{10} - P_1 \tag{4.43}$$

where $P_1 = \int f\dot{u}\, \mathrm{d}x$ is the power injected in the string by external forces.

Blocked modes

Let us consider the stiffness operator

$$-T\frac{\mathrm{d}^2}{\mathrm{d}x^2} \cdot + K\delta_{x_0} \cdot$$

applied to functions which verify boundary conditions (4.28). This linear operator admits eigenvalues and eigenvectors respectively noted $m\omega_\alpha^2$ and ψ_α, $\alpha = 1, 2, \ldots$ They verify

$$\left[-T\frac{\mathrm{d}^2}{\mathrm{d}x^2} \cdot + K\delta_{x_0} \cdot \right] \psi_\alpha(x) = m\omega_\alpha^2 \psi_\alpha(x) \tag{4.44}$$

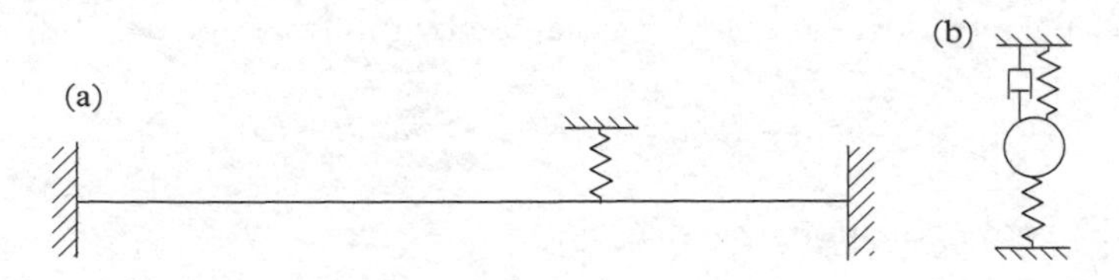

Figure 4.2 *Blocked modes of (a) the string and (b) the mechanical resonator.*

The eigenvectors ψ_α are the modes of the string of eqn (4.27) for which X_0 is 'blocked', i.e. has been set to zero. This is why these modes are referred to as 'blocked modes'. The corresponding structure is shown in Fig. 4.2. Blocked modes are different from free modes due to the presence of the coupling spring which stiffens the structure.

Reduction to sets of resonators

Let us substitute eqn (4.5) into eqn (4.27),

$$m\frac{\partial^2}{\partial t^2}\sum_{\alpha=1}^{\infty}U_\alpha\psi_\alpha - T\frac{\partial}{\partial x^2}\sum_{\alpha=1}^{\infty}U_\alpha\psi_\alpha = \delta_{x_0}K\left[X_0 - \sum_{\alpha=1}^{\infty}U_\alpha\psi_\alpha\right] + f \tag{4.45}$$

In the product $U_\alpha\psi_\alpha$, a time-derivation acts on U_α while a spatial-derivative acts on ψ_α. By permuting derivation and sum signs, it yields

$$\sum_{\alpha=1}^{\infty}\left[m\ddot{U}_\alpha\psi_\alpha + U_\alpha\left(-T\frac{\mathrm{d}^2}{\mathrm{d}x^2}\cdot +K\delta_{x_0}\cdot\right)\psi_\alpha\right] = \delta_{x_0}KX_0 + f \tag{4.46}$$

Applying the spectral property (4.44) of modes gives

$$\sum_{\alpha=1}^{\infty}\left[m\ddot{U}_\alpha + m\omega_\alpha^2U_\alpha\right]\psi_\alpha = \delta_{x_0}KX_0 + f \tag{4.47}$$

By multiplying by ψ_β, integrating over $[0,\ L]$, and applying orthonormality (4.3), we obtain

$$m\ddot{U}_\beta + m\omega_\beta^2U_\beta = K\psi_\beta(x_0)X_0 + F_\beta \tag{4.48}$$

where

$$F_\beta = \int_0^L f(t)\psi_\beta(x)\,\mathrm{d}x \tag{4.49}$$

is the modal force. This equation shows that the governing equation of a string is reducible to a set of undamped resonators U_β having modal mass m and modal stiffness $m\omega_\beta^2$. They are mutually uncoupled but coupled to resonator 0 by the stiffness

$$k_{\beta 0} = K\psi_\beta(x_0) \tag{4.50}$$

Similarly, the governing equation of resonator X_0 is found by introducing eqn (4.5) into eqn (4.26),

$$m_0\ddot{X}_0 + c_0\dot{X}_0 + (k_0 + K)X_0 = \sum_{\beta=1}^{\infty} K\psi_\beta(x_0)U_\beta + f_0 \tag{4.51}$$

So that resonator 0 is coupled to all resonators β. The coupling stiffness between resonator 0 and mode β is

$$k_{0\beta} = K\psi_\beta(x_0) \tag{4.52}$$

It can be observed from eqns (4.50) and (4.52) that the symmetry $k_{0\beta} = k_{\beta 0}$ holds.

The second step for the reduction to a set of resonators is to express the energies in terms of resonator amplitudes. The elastic energy of resonator 0 is straightforward. Substituting eqn (4.5) into eqn (4.30),

$$E_0 = \frac{1}{2}m_0\dot{X}_0^2 + \frac{1}{2}(k_0 + K)X_0^2 - \frac{1}{2}K\sum_{\beta=1}^{\infty} X_0U_\beta\psi_\beta(x_0) \tag{4.53}$$

Again, we may recognize the inter-modal stiffness $k_{0\beta}$.

For the string energy, the calculation is a little more complicated. Two integrals appear in eqn (4.31). By substituting eqn (4.5) into the first one,

$$\int_0^L \dot{u}^2\,\mathrm{d}x = \int_0^L \left(\sum_{\beta=1}^{\infty}\dot{U}_\beta\psi_\beta\right)^2 \mathrm{d}x = \sum_{\alpha,\beta}\dot{U}_\alpha\dot{U}_\beta\int_0^L \psi_\alpha(x)\psi_\beta(x)\,\mathrm{d}x$$

Orthonormality of modes leads to

$$\int_0^L \dot{u}^2\,\mathrm{d}x = \sum_{\beta=1}^{\infty}\dot{U}_\beta^2 \tag{4.54}$$

For the second integral, a similar calculation gives

$$\int_0^L u'^2\,\mathrm{d}x = \sum_{\alpha,\beta}U_\alpha U_\beta\int_0^L \psi'_\alpha\psi'_\beta\,\mathrm{d}x$$

But an integration by parts gives

$$\int_0^L \psi'_\alpha \psi'_\beta \, dx = [\psi_\alpha \psi'_\beta]_{x=0}^{x=L} - \int_0^L \psi_\alpha \psi''_\beta \, dx = -\int_0^L \psi_\alpha \psi''_\beta \, dx \tag{4.55}$$

The term in brackets vanishes by application of the boundary conditions (4.28) to ψ_α. The vibrational energy of the string becomes

$$E_1 = \frac{1}{2}\sum_\beta m\dot{U}_\beta^2 - \frac{1}{2}\sum_{\alpha,\beta} TU_\alpha U_\beta \int_0^L \psi_\alpha \psi''_\beta \, dx + \frac{1}{2}K\left(\sum_{\alpha,\beta} U_\alpha U_\beta \psi_\alpha(x_0)\psi_\beta(x_0) - X_0 \sum_\beta U_\beta \psi_\beta(x_0)\right)$$

Hence,

$$E_1 = \frac{1}{2}\sum_\beta m\dot{U}_\beta^2 + \frac{1}{2}\sum_{\alpha,\beta} U_\alpha U_\beta \int_0^L \psi_\alpha \left(-T\psi''_\beta + K\delta_{x_0}\psi_\beta\right) dx - \sum_{\beta=1}^{\infty} \frac{1}{2} K\psi_\beta(x_0) X_0 U_\beta$$

We may now apply the spectral property in the integral,

$$E_1 = \frac{1}{2}\sum_\beta m\dot{U}_\beta^2 + \frac{1}{2}\sum_{\alpha,\beta} m\omega_\beta^2 U_\alpha U_\beta \int_0^L \psi_\alpha \psi_\beta \, dx - \sum_{\beta=1}^{\infty} \frac{1}{2} K\psi_\beta(x_0) X_0 U_\beta$$

And by orthonormality,

$$E_1 = \sum_{\beta=1}^{\infty} \frac{1}{2} m\dot{U}_\beta^2 + \frac{1}{2} m\omega_\beta^2 U_\beta^2 - \frac{1}{2} K\psi_\beta(x_0) X_0 U_\beta \tag{4.56}$$

The last step in the reduction process to a set of resonators is to find the coupling power. Substituting eqn (4.5) into eqn (4.42) leads to

$$P_{10} = \sum_{\beta=1}^{\infty} \frac{1}{2} k_{0\beta} \left(\dot{X}_0 U_\beta - X_0 \dot{U}_\beta\right) \tag{4.57}$$

Canonical problem

Let us summarize all these results. The governing equations of subsystems are

$$m_0\ddot{X}_0 + c_0\dot{X}_0 + (k_0 + K)X_0 = \sum_{\beta \in \Delta\omega} k_{0\beta} U_\beta + f_0 \tag{4.58}$$

$$m\ddot{U}_\beta + m\omega_\beta^2 U_\beta = k_{0\beta} X_0 + F_\beta \tag{4.59}$$

where $f_0(t)$ and $F_\beta(t)$ are white noise random processes mutually uncorrelated. The vibrational energies are

$$E_0 = \frac{1}{2}m_0\dot{X}_0^2 + \frac{1}{2}(k_0 + K)X_0^2 - \sum_{\beta\in\Delta\omega} \frac{1}{2}k_{0\beta}X_0U_\beta \tag{4.60}$$

$$E_1 = \sum_{\beta\in\Delta\omega} \frac{1}{2}m\dot{U}_\beta^2 + \frac{1}{2}m\omega_\beta^2U_\beta^2 - \frac{1}{2}k_{0\beta}X_0U_\beta \tag{4.61}$$

The coupling power is

$$P_{10} = -P_{01} = \sum_{\beta\in\Delta\omega} \frac{1}{2}k_{0\beta}\left(\dot{X}_0U_\beta - X_0\dot{U}_\beta\right) \tag{4.62}$$

where the coupling stiffness is $k_{0\beta} = k_{\beta 0} = K\psi_\beta(x_0)$. In the above equations the sums have been limited to resonant modes $\beta \in \Delta\omega$ whose number is noted N_1.

The set of equations (4.58–4.62) constitutes the reduction of the string–oscillator system to a packet of resonators. The string is reduced to a set of resonators whose amplitudes are U_β, $\beta = 1, 2, \ldots$. According to eqn (4.58) resonator 0 is coupled to all other resonators $\beta \geq 1$ through the elastic force $k_{0\beta}U_\beta$. And by eqn (4.59), all resonators $\beta \geq 1$ are coupled to resonator 0 through the term $k_{0\beta}X_0$ but not to other resonators $\alpha \geq 1$. Resonators $\beta \geq 1$ are therefore mutually uncoupled. Furthermore, all coupling stiffnesses $k_{0\beta} = K\psi_\beta(x_0)$ between resonator 0 and the other ones are proportional to K which has been assumed to be small.

The situation is summarized in Fig. 4.3. Subsystem 1 (the string) is composed of N_1 resonators mutually uncoupled but coupled to subsystem 0 (the mechanical oscillator).

Coupling loss factor

Formally, the set of equations (4.58–4.62) is equivalent to eqns (3.1–3.5) and therefore all the conclusions of Chapter 3 apply. In particular, the whole system can be modelled

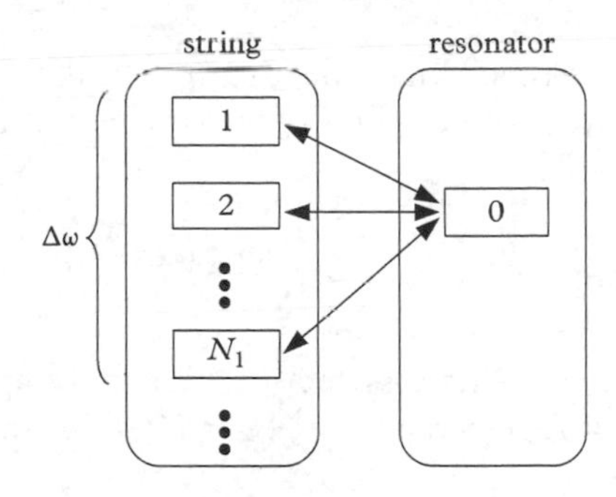

Figure 4.3 *Interaction of modes in a string–resonator system.*

by random resonators (the string) lightly coupled to a deterministic resonator (the mechanical oscillator). In particular ω_0 has a fixed value while ω_1 is a random variable ranging in $[\omega_{\min}, \omega_{\max}]$ with centre frequency $\omega = (\omega_{\min}\omega_{\max})^{1/2}$.

To determine the probability density function p_ω of ω_1, we must examine the actual sequence of natural frequencies of the blocked modes. But since the coupling is light, this sequence is not very different from the sequence of free modes and in particular we may expect that the modal densities are equal. The natural frequencies of a simple string are

$$\omega_\alpha = \pi c \frac{\alpha}{L} \tag{4.63}$$

where $c = (T/m)^{1/2}$ is the wave speed in a string. The natural frequencies are equally spaced with a frequency span $\delta\omega = \pi c/L$. In particular, the number of modes in the frequency band $\Delta\omega$ is $N_1 = L\Delta\omega/\pi c$. Therefore, the best probability density function is the uniform probability as in Section 3.3 and the relevant approximations of the coupling loss factors are those of eqns (3.53) and (3.54).

Likewise, the modes of the simple string ($K = 0$) are

$$\psi_\alpha(x) = \sqrt{\frac{2}{L}} \sin\left(\alpha\pi \frac{x}{L}\right) \tag{4.64}$$

In order to evaluate the bracket $\langle k_{0\alpha}^2 \rangle_\psi$, a technical result is required. For all $u \in]0, \pi[$,

$$\lim_{N\to\infty} \frac{1}{N} \sum_{\alpha=1}^{N} \sin^2 \alpha u = \frac{1}{2} \tag{4.65}$$

The bracket $\langle k_{0\alpha}^2 \rangle_\psi$ is the mean value of $k_{0\alpha}^2$ where $\alpha \in \Delta\omega$,

$$\frac{1}{N_1} \sum_{\alpha\in\Delta\omega} K^2 \psi_\alpha^2(x_0) \tag{4.66}$$

where the sum contains N_1 terms. Under the assumption that the number of modes N_1 is large enough, the limit $N_1 \to \infty$ is a good approximation of the finite sum,

$$\langle k_{0\alpha}^2 \rangle_\psi = \lim_{N_1\to\infty} \frac{K^2}{N_1} \sum_{\alpha=1}^{N_1} \frac{2}{L} \sin^2\left(\alpha\pi \frac{x_0}{L}\right) = \frac{K^2}{L} \tag{4.67}$$

We can now obtain the coupling loss factors by simply applying eqns (3.53) and (3.54). We set $N_0 = 1$, $N_1 = L\Delta\omega/\pi c$, and $\langle k_{0\alpha}^2 \rangle_\psi = K^2/L$ in eqn (3.53),

$$\omega\eta_{10} = \frac{\pi K^2}{2m_0 m L \omega_0^2 \Delta\omega} \tag{4.68}$$

where mL is the total mass of the string. Similarly, applying eqn (3.54) leads to the reciprocal coupling loss factor,

$$\omega\eta_{01} = \frac{K^2}{2m_0 mc\omega_0^2} \tag{4.69}$$

These are the coupling loss factors of a string–resonator system.

4.3 Beams Connected by a Torsional Spring

The second example studied in this section was introduced by Crandall and Lotz (1971). The system is composed by two Euler–Bernoulli beams subjected to rain-on-the-roof random forces. The beams of length L_i are simply supported at their ends and are connected by a small torsional spring. As shown in Fig. 4.4, a local x-axis is used for each beam. They are chosen in order to ensure the symmetry of the problem.

Governing equation

The equation of motion of beam i is

$$m_i \frac{\partial^2 u_i}{\partial t^2} + c_i \frac{\partial u_i}{\partial t} + E_i I_i \frac{\partial^4 u_i}{\partial x^4} = f_i \tag{4.70}$$

where m_i is the mass per unit length, c_i the viscous damping coefficient, E_i the Young modulus, I_i the moment of inertia, and $E_i I_i$ the bending stiffness. The transverse deflection is noted $u_i(x, t)$ while $f_i(x, t)$ is a random force field applied to beams and assumed to be rain-on-the-roof.

The beams are simply supported at the ends. This imposes that both deflection and moment are zero at the right end,

$$u_i(L_i, t) = \frac{\partial^2 u_i}{\partial x^2}(L_i, t) = 0 \tag{4.71}$$

at any time t.

At the connection point the deflections are also set to zero,

$$u_i(0, t) = 0 \tag{4.72}$$

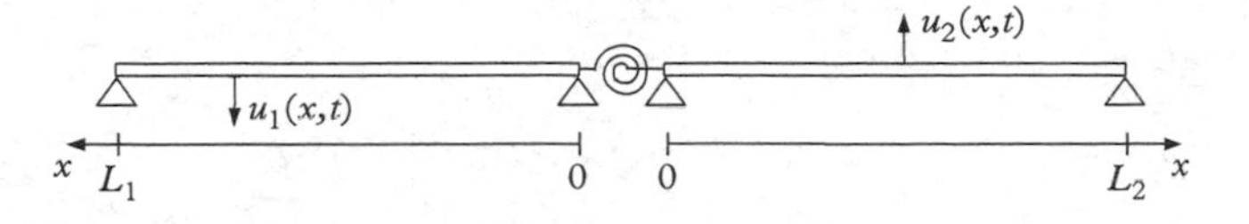

Figure 4.4 *Beams coupled by a torsional spring.*

but the moment has a value imposed by the presence of a torsional spring of stiffness K,

$$E_i I_i \frac{\partial^2 u_i}{\partial x^2}(0,t) = K\left[\frac{\partial u_i}{\partial x}(0,t) - \frac{\partial u_j}{\partial x}(0,t)\right] \tag{4.73}$$

where $j \neq i$.

Energy

The vibrational energy of beam i is

$$E_i = \frac{1}{2}\int_0^{L_i} m_i \dot{u}_i^2(x,t)\,\mathrm{d}x + \frac{1}{2}\int_0^{L_i} E_i I_i u_i''^2(x,t)\,\mathrm{d}x + \frac{K}{2}\left(u_i'^2(0,t) - u_1'(0,t)u_2'(0,t)\right) \tag{4.74}$$

where the first term is the kinetic energy, the second term the elastic energy, and the third term the contribution of the coupling following the method introduced in Chapter 2. The elastic energy stored in the coupling is $1/2 \times K\left(u_2' - u_1'\right)^2$ at $x = 0$ which has been separated in the sum of $1/2 \times K\left(u_1'^2 - u_1'u_2'\right)$ and $1/2 \times K\left(u_2'^2 - u_1'u_2'\right)$.

Power balance

The power exchanged between the two beams is

$$P_{ij} = \frac{1}{2}K\left(u_i'(0,t)\dot{u}_j'(0,t) - \dot{u}_i'(0,t)u_j'(0,t)\right) \tag{4.75}$$

We observe that $P_{ji} = -P_{ij}$.

To derive this expression we must write the power balance. Let us multiply eqn (4.70) by $\dot{u}_i$,

$$m_i \ddot{u}_i \dot{u}_i + c_i \dot{u}_i^2 + E_i I_i u_i^{(iv)} \dot{u}_i = f_i \dot{u}_i$$

where $u_i^{(iv)}$ designates the fourth spatial derivative of u_i. After integrating over x and performing a double integration by parts, it yields

$$\frac{\mathrm{d}}{\mathrm{d}t}\left[\frac{1}{2}\int_0^{L_i} m_i \dot{u}_i^2 + E_i I_i u_i''^2\,\mathrm{d}x\right] + \int_0^{L_i} c_i \dot{u}_i^2\,\mathrm{d}x + E_i I_i\left[u_i''' \dot{u}_i - u_i'' \dot{u}_i'\right]_{x=0}^{x=L_i} = \int_0^{L_i} f_i \dot{u}_i\,\mathrm{d}x$$

By applying boundary conditions (4.72), (4.71), we see $u_i''' \dot{u}_i$ is zero at $x = 0$ and $x = L_i$. Therefore, by the boundary condition (4.72) and the coupling condition (4.73), the last term of the left-hand side becomes

$$\begin{aligned} E_i I_i\left[u_i''' \dot{u}_i - u_i'' \dot{u}_i'\right]_{x=0}^{x=L_i} &= K\left[u_i'(0,t) - u_j'(0,t)\right]\dot{u}_i'(0,t) \\ &= \frac{\mathrm{d}}{\mathrm{d}t}\left[\frac{1}{2}K\left(u_i'^2(0,t) - u_i'(0,t)u_j'(0,t)\right)\right] + P_{ij} \end{aligned}$$

where $j \neq i$. The power balance of subsystem i follows,

$$\frac{\mathrm{d}E_i}{\mathrm{d}t} + P_{\mathrm{diss},i} + P_{ij} = P_i \tag{4.76}$$

where the dissipated power is

$$P_{\mathrm{diss},i} = \int_0^{L_i} c_i \dot{u}_i^2 \,\mathrm{d}x \tag{4.77}$$

and the injected power is

$$P_i = \int_0^{L_i} f_i \dot{u}_i \,\mathrm{d}x \tag{4.78}$$

This completes the proof of eqn (4.75).

Blocked modes

The blocked modes are found by considering the time-independent solutions to the homogeneous problem of eqns (4.70–4.73)

$$E_i I_i \psi_{i\alpha}^{(iv)} = m_i \omega_{i\alpha}^2 \psi_{i\alpha} \tag{4.79}$$

with the homogeneous boundary conditions

$$\psi_{i\alpha}(0) = E_i I_i \psi_{i\alpha}''(0) - K\psi_{i\alpha}'(0) = \psi_{i\alpha}(L_i) = \psi_{i\alpha}''(L_i) = 0 \tag{4.80}$$

These are the eigenmodes of the system shown in Fig. 4.5 where the adjacent beam has been blocked.

Reduction to sets of resonators

To reduce the equation of motion (4.70) to sets of resonators, as done in Section 4.2 for the string–resonator system, we immediately encounter a difficulty. Projection of eqn (4.70) on a modal basis does not lead to any coupling between modes of beam 1 and modes of beam 2. The reason is that eqns (4.70) for $i = 1, 2$ are uncoupled. The coupling appears only through the coupling condition (4.73). To manage such a situation, we

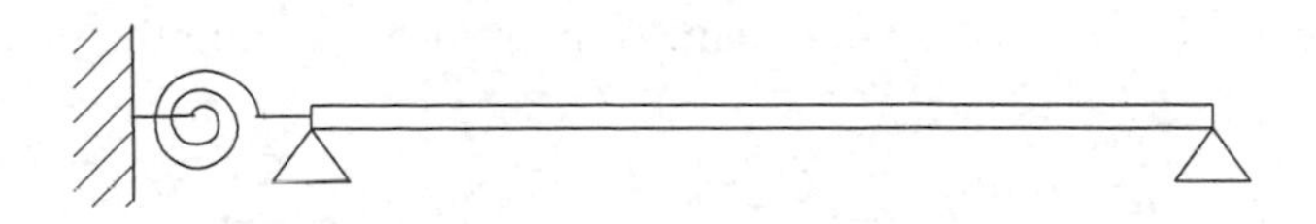

Figure 4.5 *Blocked modes of a beam with a torsional spring at one of its ends.*

follow the method introduced by Karnopp (1966) and we re-formulate eqns (4.70) and (4.73) into a single equation.

The most general equation of motion of beam with an external force density f_i and moment density g_i is

$$m_i\ddot{u}_i + c_i\dot{u}_i + E_iI_iu_i^{(iv)} = f_i - g_i' \tag{4.81}$$

where g_i' is the space-derivative of g_i. In eqn (4.73), the term $-Ku_j'(0,t)$ may be viewed as a concentrated moment applied at the origin. Such a discontinuity is equivalent to an external moment field $g_i(x) = -Ku_j'(0,t)\delta(x)$. Equation (4.70) with boundary conditions (4.71), (4.72), (4.73) is therefore equivalent to

$$m_i\ddot{u}_i + c_i\dot{u}_i + E_iI_iu_i^{(iv)} = f_i + Ku_j'(0,t)\delta' \tag{4.82}$$

with the homogeneous boundary conditions,

$$u_i(0,t) = E_iI_iu_i''(0,t) - Ku_i'(0,t) = u_i(L_i,t) = u_1''(L_i,t) = 0 \tag{4.83}$$

The problem is now to find the solution as a series (4.5). Since blocked modes verify the homogeneous boundary conditions (4.80), any series of type (4.5) automatically verifies the boundary conditions (4.83).

We are now going to reduce the governing equation of coupled beams to sets of resonators. After substituting the series (4.5) into (4.82),

$$\sum_{\alpha}\left[m_i\ddot{U}_{i\alpha}\psi_{i\alpha} + c_i\dot{U}_{i\alpha}\psi_{i\alpha} + E_iI_iU_{i\alpha}\psi_{i\alpha}^{(iv)}\right] = f_i + \sum_{\beta}KU_{j\beta}\psi_{j\beta}'(0)\delta' \tag{4.84}$$

By multiplying by $\psi_{i\alpha}$, integrating over x, and applying orthonormality (4.5), it yields

$$m_i\ddot{U}_{i\alpha} + c_i\dot{U}_{i\alpha} + m_i\omega_{i\alpha}^2U_{i\alpha} = F_{i\alpha} + \sum_{\beta}K\psi_{i\alpha}'(0)\psi_{j\beta}'(0)U_{j\beta} \tag{4.85}$$

where $F_{i\alpha}$ is the modal force,

$$F_{i\alpha}(t) = \int_0^{L_i} f_i(x,t)\psi_{i\alpha}(x)\,\mathrm{d}x \tag{4.86}$$

Formally, eqn (4.85) is analogous to eqn (3.1) of the canonical problem provided that we denote the modal stiffness by $k_{i\alpha} = m_i\omega_{i\alpha}^2$ and the inter-modal coupling stiffness by $k_{1\alpha,2\beta} = K\psi_{1\alpha}'(0)\psi_{2\beta}'(0)$.

To calculate the elastic energy of modes, one must project eqn (4.74) on the modal basis. The integral $\int \dot{u}_i^2\,\mathrm{d}x$ is easy to obtain. A direct projection in the modal basis gives

$$\int_0^{L_i}\dot{u}_i^2\,\mathrm{d}x = \sum_{\alpha,\beta}\dot{U}_{i\alpha}\dot{U}_{i\beta}\int_0^{L_i}\psi_{i\alpha}\psi_{i\beta}\,\mathrm{d}x = \sum_{\alpha}\dot{U}_{i\alpha}^2 \tag{4.87}$$

by the orthonormality property.

Let us now examine the second term of (4.74),

$$\int_0^{L_i} u_i''^2 \, dx = \sum_{\alpha,\beta} U_{i\alpha} U_{i\beta} \int_0^{L_i} \psi_{i\alpha}'' \psi_{i\beta}'' \, dx$$

where eqn (4.5) has been introduced. With a double integration by parts, one obtains

$$\int_0^{L_i} u_i''^2 \, dx = \sum_{\alpha,\beta} U_{i\alpha} U_{i\beta} \left(\left[\psi_{i\alpha}' \psi_{i\beta}'' - \psi_{i\alpha} \psi_{i\beta}''' \right]_{x=0}^{x=L_i} + \int_0^{L_i} \psi_{i\alpha} \psi_{i\beta}^{(iv)} \, dx \right)$$

Applying the boundary conditions of modes and orthonormality leads to

$$E_i I_i \int_0^{L_i} u_i''^2 \, dx = \sum_{\alpha,\beta} U_{i\alpha} U_{i\beta} \left(-K \psi_{i\alpha}'(0) \psi_{i\beta}'(0) + m_i \omega_{i\alpha}^2 \delta_{\alpha\beta} \right) \tag{4.88}$$

where $\delta_{\alpha\beta}$ is the Kronecker symbol.

The last term of (4.74) is more direct to project,

$$u_i'^2(0,t) - u_1'(0,t) u_2'(0,t) = \sum_{\alpha,\beta} U_{i\alpha} U_{i\beta} \psi_{i\alpha}'(0) \psi_{i\beta}'(0) - U_{1\alpha} U_{2\beta} \psi_{1\alpha}'(0) \psi_{2\beta}'(0) \tag{4.89}$$

Introducing eqns (4.87), (4.88), and (4.89) in eqn (4.74) and simplifying,

$$E_i = \sum_{\alpha} \left(\frac{1}{2} m_i \dot{U}_{i\alpha}^2 + \frac{1}{2} m_i \omega_{i\alpha}^2 U_{i\alpha}^2 - \sum_{\beta} \frac{1}{2} K \psi_{1\alpha}'(L_1) \psi_{2\beta}'(0) U_{1\alpha} U_{2\beta} \right) \tag{4.90}$$

Equation (4.90) is formally equivalent to eqns (3.2) and (3.3).

The last step is to reduce eqn (4.75) in the modal basis. Following the same method as for previous calculations,

$$u_i'(0,t) \dot{u}_j'(0,t) = \sum_{\alpha,\beta} U_{i\alpha} \dot{U}_{j\beta} \psi_{i\alpha}'(0) \psi_{j\beta}'(0) \tag{4.91}$$

and the coupling power of eqn (4.75) becomes

$$P_{ij} = \sum_{\alpha,\beta} \frac{1}{2} K \psi_{i\alpha}'(L_1) \psi_{j\beta}'(0) \left[U_{i\alpha} \dot{U}_{j\beta} - \dot{U}_{i\alpha} U_{j\beta} \right] \tag{4.92}$$

Equation (4.92) is formally equivalent to eqns (3.3) and (3.4).

Equations (4.85), (4.90), and (4.92) constitute the reduction to the canonical problem. Each beam is a packet of resonators coupled to all resonators of the other beam (see Fig. 4.6). The coupling stiffness of modes α and β is found to be

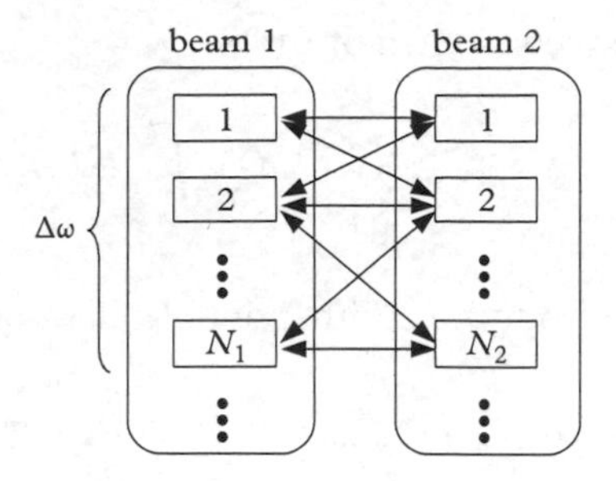

Figure 4.6 *Interaction of modes in a beam–beam system.*

$$k_{i\alpha,j\beta} = K\psi'_{i\alpha}(0)\psi'_{j\beta}(0) \tag{4.93}$$

The coupling stiffness is symmetric $k_{i\alpha,j\beta} = k_{j\beta,i\alpha}$.

Coupling loss factor

Since we have reduced the problem to its canonical form, we can immediately apply the conclusions of Chapter 3. We have two sets of resonators, one for each beam, coupled by elastic elements. We may therefore consider that ω_i, $i = 1, 2$ are random variables. The coupling power proportionality holds and the relevant approximation of the coupling loss factor is given by eqn (3.50),

$$\omega\eta_{ij} = \frac{N_j \pi k_{ij}^2}{2m_i m_j \omega^2 \Delta\omega} \tag{4.94}$$

where $N_j/\Delta\omega$ is the mean modal density and k_{ij}^2 is the mean square of the coupling stiffnesses $k_{i\alpha,j\beta}$, $\alpha = 1, 2, \ldots$, $\beta = 1, 2, \ldots$.

As in Section 4.2, we don't take the exact sequence of blocked modes but just its limits when $K \to 0$.

$$\omega_{i\alpha} = \frac{\alpha^2\pi^2}{L_i^2}\sqrt{\frac{E_i I_i}{m_i}} \tag{4.95}$$

In view of forthcoming approximations, considering the exact values of blocked modes would not fundamentally modify the final result. The corresponding mode shapes are

$$\psi_{i\alpha}(x) = \sqrt{\frac{2}{L_i}}\sin\left(\alpha\pi\frac{x}{L_i}\right) \tag{4.96}$$

The number of modes in the bandwidth $\Delta\omega$ is the exact number of integers whose corresponding natural frequency $\omega_{j\beta}$ is greater than $\omega_{\min}$ and lower than $\omega_{\max}$. When this number is large enough, a rough approximation is obtained by considering that β is real-valued and taking the derivative with respect to $\omega_{j\beta}$. It yields

$$\frac{N_j}{\Delta\omega} = \frac{d\beta}{d\omega_{j\beta}} = \frac{L_j}{2\pi\sqrt{\omega}}\left(\frac{m_j}{E_jI_j}\right)^{1/4} \tag{4.97}$$

The second statistical term of eqn (4.94) is k_{ij}. From eqn (4.93) and the natural frequencies and mode shapes given in eqns (4.95) and (4.96),

$$k_{i\alpha,j\beta} = \frac{2K}{\sqrt{L_iL_j}} \times \frac{\alpha\pi}{L_i} \times \frac{\beta\pi}{L_j} = \frac{2K}{\sqrt{L_iL_j}}\sqrt{\omega_{i\alpha}}\left(\frac{m_i}{E_iI_i}\right)^{1/4}\sqrt{\omega_{j\beta}}\left(\frac{m_j}{E_jI_j}\right)^{1/4} \tag{4.98}$$

So, by considering that ω_i and ω_j are random variables and taking their mean values ω,

$$k_{ij}^2 = \langle k_{i\alpha,j\beta}^2 \rangle = \frac{4K^2\omega^2}{L_iL_j}\left(\frac{m_im_j}{E_iI_iE_jI_j}\right)^{1/2} \tag{4.99}$$

Finally, substituting eqns (4.97) and (4.99) into (4.94) gives

$$\eta_{ij} = \frac{K^2}{L_i\,(m_iE_iI_i)^{1/2}\,m_j^{1/4}\,(E_jI_j)^{3/4}\,\omega^{3/2}} \tag{4.100}$$

This is the expression of the coupling loss factor between two beams lightly coupled by a rotational spring (Crandall and Lotz, 1971).

4.4 Fluid-Loaded Plate

We now outline a vibroacoustical problem involving a coupling between structural vibrations and acoustics. A vibrating structure and the surrounding fluid exchange energy by two processes: sound radiation and structural response. In this example, a plate having simply supported edges is coupled to a parallelepipedic room. For the sake of simplicity, the example is limited to the two-dimensional case. The plate has length L_x and the room has dimensions $L_x \times L_z$ as shown in Fig. 4.7. The room walls are assumed to be perfectly rigid (Neumann's condition).

Acoustic potential

In linear acoustics, the acoustical pressure p is a small local fluctuation of fluid pressure about the static value p_0. In a similar fashion, the velocity of fluid particle $\mathbf{v}$ is a variation around zero if the fluid is assumed to be at rest. Since thermodynamics and fluid mechanics show that an acoustical velocity field is irrotational, we can define a velocity potential Φ, also called acoustic potential, such as

$$\mathbf{v} = \nabla\Phi \tag{4.101}$$

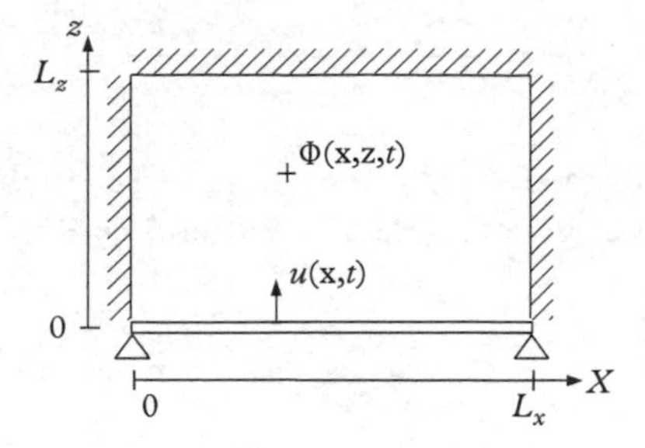

Figure 4.7 *Plate coupled to an acoustical cavity.*

The acoustic potential and acoustic pressure are then related by

$$p = -\rho_0 \frac{\partial \Phi}{\partial t} \tag{4.102}$$

where ρ_0 is the density of fluid at rest (1.2 kg/m^3 at 20 °C and 1 atm for air). The acoustic potential verifies the wave equation

$$\frac{1}{c_0^2} \frac{\partial^2 \Phi}{\partial t^2} - \Delta \Phi = \frac{f_0}{\rho_0} \tag{4.103}$$

where c_0 is the speed of sound (343 m/s at 20 °C for air) and f_0 a source term interpreted as a mass source.

Governing equation

The system shown in Fig. 4.7 can be reduced to the following set of equations. In fluid, the acoustical potential is governed by the wave equation

$$\frac{1}{c_0^2} \frac{\partial^2 \Phi}{\partial t^2} - \left(\frac{\partial^2 \Phi}{\partial x^2} + \frac{\partial^2 \Phi}{\partial z^2} \right) = \frac{f_0}{\rho_0} \tag{4.104}$$

where f_0 is a rain-on-the-roof stochastic process. On the three rigid walls $x = 0$, $x = L_x$, and $z = L_z$, the normal component of the velocity **v** is set to zero. Hence,

$$\frac{\partial \Phi}{\partial x}(0, z, t) = \frac{\partial \Phi}{\partial x}(L_x, z, t) = \frac{\partial \Phi}{\partial z}(x, L_z, t) = 0 \tag{4.105}$$

This is the Neumann condition.

The equation of motion of a plate is

$$m \frac{\partial^2 u}{\partial t^2} + c \frac{\partial u}{\partial t} + D \frac{\partial^4 u}{\partial x^4} = f_1 + \rho_0 \frac{\partial \Phi}{\partial t} \tag{4.106}$$

where f_1 is a force density and the second term in the right-hand side is the contribution of the acoustic pressure. The source term f_1 is also assumed to be a rain-on-the-roof stochastic process. As usual m is the mass per unit area, c a viscous damping coefficient, and D the bending stiffness. On simply supported edges, the conditions are

$$u(0,t) = \frac{\partial^2 u}{\partial x^2}(0,t) = u(L_x,t) = \frac{\partial^2 u}{\partial x^2}(L_x,t) = 0 \tag{4.107}$$

for the edges $x = 0$ and $x = L_x$.

Finally, a coupling condition between structural vibration and fluid must be specified. We write

$$\frac{\partial \Phi}{\partial z}(x,0,t) = \frac{\partial u}{\partial t} \tag{4.108}$$

for the continuity of normal velocities.

In eqn (4.106), the term $\rho_0\dot{\Phi}$ reveals the influence of fluid on structural vibration. However, no structural term appears in eqn (4.104). The influence of structural vibration on acoustics is visible through the coupling condition (4.108). In order to highlight the reciprocal link between fluid and structural degrees of freedom, it is more convenient to replace eqns (4.104) and (4.108) with the modified wave equation

$$\frac{1}{c_0^2}\frac{\partial^2 \Phi}{\partial t^2} - \left(\frac{\partial^2 \Phi}{\partial x^2} + \frac{\partial^2 \Phi}{\partial z^2}\right) = \frac{f_0}{\rho_0} - \frac{\partial u}{\partial t}(x,t)\delta(z) \tag{4.109}$$

and the homogeneous coupling condition

$$\frac{\partial \Phi}{\partial z}(x,0,t) = 0 \tag{4.110}$$

Although this is not obvious, eqns (4.104), (4.108) and eqns (4.109), (4.110) are equivalent. They give the same field Φ inside the domain $z > 0$ but not on the boundary $z = 0$.

Energy

The acoustical energy density is composed of two terms. The kinetic energy is $1/2 \times \rho_0|\mathbf{v}|^2$ and the elastic energy stored in the fluid is $1/2 \times p^2/\rho_0 c_0^2$. The total vibrational energy contained in the cavity is therefore

$$E_0 = \frac{\rho_0}{2}\int_0^{L_z}\int_0^{L_x}\left(\frac{\partial \Phi}{\partial x}\right)^2 + \left(\frac{\partial \Phi}{\partial z}\right)^2 \mathrm{d}x\mathrm{d}z + \frac{\rho_0}{2c_0^2}\int_0^{L_z}\int_0^{L_x}\left(\frac{\partial \Phi}{\partial t}\right)^2 \mathrm{d}x\mathrm{d}z \tag{4.111}$$

The structural energy is also composed of two terms: the kinetic energy $1/2 \times m\dot{u}^2$ and elastic energy $1/2 \times Du''^2$. The total vibrational energy of the plate is

$$E_1 = \frac{m}{2}\int_0^{L_x}\left(\frac{\partial u}{\partial t}\right)^2 \mathrm{d}x + \frac{D}{2}\int_0^{L_x}\left(\frac{\partial^2 u}{\partial x^2}\right)^2 \mathrm{d}x \tag{4.112}$$

Power balance

The power exchanged between plate and fluid is the flux of normal intensity $p\dot{u}$,

$$P_{01} = \rho_0 \int_0^{L_x} \frac{\partial \Phi}{\partial t}\frac{\partial u}{\partial t}\,\mathrm{d}x \tag{4.113}$$

Naturally, the reciprocal power is $P_{10} = -P_{01}$.

Blocked modes

Acoustic blocked modes are obtained by setting $u = 0$ in eqn (4.109) and studying the related spectral problem. They verify

$$-\left(\frac{\partial^2 \phi_\alpha}{\partial x^2} + \frac{\partial^2 \phi_\alpha}{\partial z^2}\right) = \frac{\omega_{0\alpha}^2}{c_0^2}\phi_\alpha \tag{4.114}$$

and the Neumann boundary condition on the walls,

$$\frac{\partial \phi_\alpha}{\partial x}(0,z) = \frac{\partial \phi_\alpha}{\partial x}(L_x,z) = \frac{\partial \phi_\alpha}{\partial z}(x,0) = \frac{\partial \phi_\alpha}{\partial z}(x,L_z) = 0 \tag{4.115}$$

Blocked modes are sketched in Fig. 4.8. They are found to be

$$\phi_\alpha(x,z) = \sqrt{\frac{4}{L_x L_z}}\cos\left(p\pi\frac{x}{L_x}\right)\cos\left(q\pi\frac{z}{L_z}\right) \tag{4.116}$$

where we have introduced the double subscript $\alpha = p, q$. The natural frequencies are

$$\omega_{0\alpha} = c_0\sqrt{\left(\frac{p\pi}{L_x}\right)^2 + \left(\frac{q\pi}{L_z}\right)^2} \tag{4.117}$$

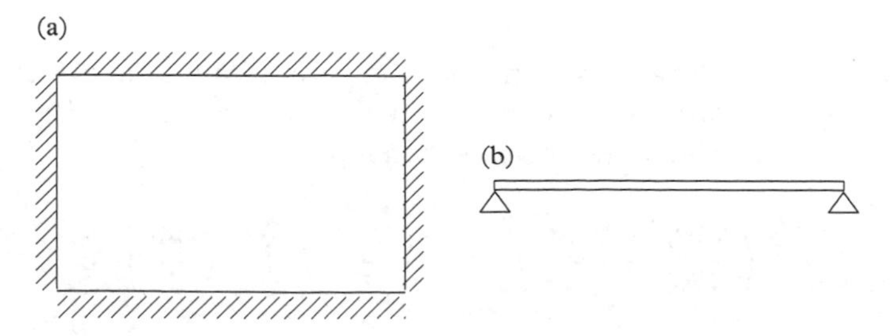

Figure 4.8 *Blocked modes of the acoustical cavity with (a) Neumann's condition, and (b) a simply supported plate.*

Acoustic blocked modes form an orthonormal basis. Orthonormality reads

$$\int_0^{L_z}\int_0^{L_x} \phi_\alpha(x,z)\phi_{\alpha'}(x,z)\,\mathrm{d}x\mathrm{d}z = \delta_{\alpha\alpha'} \tag{4.118}$$

and completeness

$$\Phi(x,z,t) = \sum_{\alpha=0}^{\infty} P_\alpha(t)\phi_\alpha(x,z) \tag{4.119}$$

for any acoustical potential Φ verifying a Neumann's boundary condition.

Blocked modes of plate are found by imposing $\Phi = 0$ in eqn (4.106). They verify

$$D\psi_\beta^{(iv)}(x) = m\omega_{1\beta}^2\psi_\beta(x) \tag{4.120}$$

with homogeneous simply supported boundary conditions at the edges,

$$\psi_\beta(0) = \psi_\beta''(0) = \psi_\beta(L_x) = \psi_\beta''(L_x) = 0 \tag{4.121}$$

These are the modes of a simply supported beam (Blevins, 1979),

$$\psi_\beta(x) = \sqrt{\frac{2}{L_x}}\sin\left(\beta\pi\frac{x}{L_x}\right) \tag{4.122}$$

with the natural frequencies

$$\omega_{1\beta} = \frac{\beta^2\pi^2}{L_x^2}\sqrt{\frac{D}{m}} \tag{4.123}$$

Orthonormality of modes reads

$$\int_0^{L_x} \psi_\beta(x)\psi_{\beta'}(x)\,\mathrm{d}x = \delta_{\beta\beta'} \tag{4.124}$$

and completeness

$$u(x,t) = \sum_{\beta=1}^{\infty} U_\beta(t)\psi_\beta(x) \tag{4.125}$$

Reduction to sets of resonators

Let us substitute eqns (4.125) and (4.119) into eqn (4.109),

$$\sum_{\alpha'} \frac{1}{c_0^2}\ddot{P}_{\alpha'}\phi_{\alpha'} - P_{\alpha'}\left(\frac{\partial^2\phi_{\alpha'}}{\partial x^2} + \frac{\partial^2\phi_{\alpha'}}{\partial z^2}\right) = \frac{f_0}{\rho_0} - \sum_\beta \dot{U}_\beta\psi_\beta\delta(z) \tag{4.126}$$

In the now familiar way, using eqn (4.114), multiplying by ϕ_α, integrating over $[0, L_x] \times [0, L_z]$, and applying orthonormality (4.118) gives

$$\frac{\rho_0}{c_0^2}\left(\ddot{P}_\alpha + \omega_{0\alpha}^2 P_\alpha\right) = F_{0\alpha} - \sum_\beta g_{0\alpha,1\beta}\dot{U}_\beta \tag{4.127}$$

where $F_{0\alpha} = \int\int f_0\phi_\alpha \,\mathrm{d}x\mathrm{d}z$ is the modal force and the gyroscopic constants are

$$g_{0\alpha,1\beta} = \rho_0 \int_0^{L_x} \psi_\beta(x)\phi_\alpha(x,0)\,\mathrm{d}x \tag{4.128}$$

This constitutes the reduction of the wave equation to a set of undamped resonators of magnitude P_α, mass ρ_0/c_0^2, and stiffness $\rho_0\omega_{0\alpha}^2/c_0^2$, randomly excited by modal forces $F_{0\alpha}$, and gyroscopically coupled to resonators U_β.

In a similar fashion, introducing eqns (4.125) and (4.119) into eqn (4.106) yields

$$\sum_\beta \left[m\ddot{U}_\beta\psi_\beta + c\dot{U}_\beta\psi_\beta + DU_\beta\psi_\beta^{(iv)}\right] = f_1 + \rho_0\sum_\alpha \dot{P}_\alpha\phi_\alpha \tag{4.129}$$

And, after a calculation similar to that previously done, we find

$$m\ddot{U}_\beta + c\dot{U}_\beta + m\omega_{1\beta}^2 U_\beta = F_{1\beta} - \sum_\alpha g_{1\beta,0\alpha}\dot{P}_\alpha \tag{4.130}$$

where $F_{1\beta} = \int f_1\psi_\beta \,\mathrm{d}x$ and

$$g_{1\beta,0\alpha} = -\rho_0 \int_0^{L_x} \psi_\beta(x)\phi_\alpha(x,0)\,\mathrm{d}x \tag{4.131}$$

The plate is therefore reduced to a set of resonators of magnitude U_β, mass m, damping coefficient c, stiffness $m\omega_{1\beta}^2$, randomly excited by modal forces $F_{1\beta}$, and gyroscopically coupled to resonators P_α. We naturally observe that $g_{1\beta,0\alpha} = -g_{0\alpha,1\beta}$.

For the energies, we also introduce eqns (4.125) and (4.119) into respectively eqns (4.112) and (4.111). For fluid, we get

$$E_0 = \int_0^{L_z}\int_0^{L_x} \frac{\rho_0}{2}\left[\left(\sum_\alpha P_\alpha\frac{\partial\phi_\alpha}{\partial x}\right)^2 + \left(\sum_\alpha P_\alpha\frac{\partial\phi_\alpha}{\partial z}\right)^2\right] + \frac{\rho_0}{2c_0^2}\left(\sum_\alpha \dot{P}_\alpha\phi_\alpha\right)^2 \mathrm{d}x\mathrm{d}z$$

Let us first compute the last integral,

$$\int_0^{L_z}\int_0^{L_x}\left(\sum_\alpha \dot{P}_\alpha\phi_\alpha\right)^2 \mathrm{d}x\mathrm{d}z = \sum_{\alpha,\alpha'}\dot{P}_\alpha\dot{P}_{\alpha'}\int_0^{L_z}\int_0^{L_x}\phi_\alpha\phi_{\alpha'}\,\mathrm{d}x\mathrm{d}z = \sum_\alpha \dot{P}_\alpha^2$$

where the last equality stems from orthonormality. For the first integral, we get

$$\int_0^{L_z}\int_0^{L_x}\left(\sum_\alpha P_\alpha \frac{\partial \phi_\alpha}{\partial x}\right)^2 \mathrm{d}x\mathrm{d}z = \sum_{\alpha,\alpha'} P_\alpha P_{\alpha'} \int_0^{L_z}\int_0^{L_x} \frac{\partial \phi_\alpha}{\partial x}\frac{\partial \phi_{\alpha'}}{\partial x}\,\mathrm{d}x\mathrm{d}z$$

$$= \sum_{\alpha,\alpha'} P_\alpha P_{\alpha'} \int_0^{L_z}\left(\left[\phi_\alpha \frac{\partial \phi_{\alpha'}}{\partial x}\right]_{x=0}^{x=L_x} - \int_0^{L_x} \phi_\alpha \frac{\partial^2 \phi_{\alpha'}}{\partial x^2}\,\mathrm{d}x\right)\mathrm{d}z$$

$$= -\sum_{\alpha,\alpha'} P_\alpha P_{\alpha'} \int_0^{L_z}\int_0^{L_x} \phi_\alpha \frac{\partial^2 \phi_{\alpha'}}{\partial x^2}\,\mathrm{d}x\mathrm{d}z.$$

In the second line, an integration by parts has been done and the boundary conditions (4.115) have been substituted into the third line. By reason of symmetry,

$$\int_0^{L_z}\int_0^{L_x}\left(\sum_\alpha P_\alpha \frac{\partial \phi_\alpha}{\partial x}\right)^2 + \left(\sum_\alpha P_\alpha \frac{\partial \phi_\alpha}{\partial z}\right)^2 \mathrm{d}x\mathrm{d}z$$

$$= \sum_{\alpha,\alpha'} P_\alpha P_{\alpha'} \int_0^{L_z}\int_0^{L_x} \phi_\alpha \left(-\frac{\partial^2 \phi_{\alpha'}}{\partial x^2} - \frac{\partial^2 \phi_{\alpha'}}{\partial z^2}\right)\mathrm{d}x\mathrm{d}z$$

$$= \sum_{\alpha,\alpha'} P_\alpha P_{\alpha'} \frac{\omega_{0\alpha'}^2}{c_0^2} \int_0^{L_z}\int_0^{L_x} \phi_\alpha \phi_{\alpha'}\,\mathrm{d}x\mathrm{d}z$$

$$= \sum_\alpha \frac{\omega_{0\alpha}^2}{c_0^2} P_\alpha^2$$

The result is again obtained by applying orthonormality of modes. Thus, the acoustic energy of the whole cavity is

$$E_0 = \sum_\alpha \left(\frac{\rho_0}{2c_0^2}\dot{P}_\alpha^2 + \frac{\rho_0 \omega_{0\alpha}^2}{2c_0^2} P_\alpha^2\right) \tag{4.132}$$

We observe that the total energy is the sum of individual energies of modes. The kinetic energy of modes is the first term where the modal mass is ρ_0/c_0^2 and the elastic energy of modes is the second term where $\rho_0\omega_{0\alpha}^2/c_0^2$ is the modal stiffness. It can also be observed that no energy is stored in the coupling as is expected for a gyroscopic coupling.

For the plate, we get

$$E_1 = \frac{m}{2}\int_0^{L_x}\left(\sum_\beta \dot{U}_\beta \psi_\beta\right)^2 \mathrm{d}x + \frac{D}{2}\int_0^{L_x}\left(\sum_\beta U_\beta \psi_\beta''\right)^2 \mathrm{d}x$$

The calculation of the above expression was carried out in Section 4.3. The result is given in eqn (4.90) as the special case $K = 0$,

$$E_1 = \sum_\beta \left(\frac{1}{2} m \dot{U}_\beta^2 + \frac{1}{2} m \omega_{1\beta}^2 U_\beta^2 \right) \tag{4.133}$$

Again, we find the total energy of the plate to be a sum of modal energies, kinetic energy for the first term with modal mass m, and elastic energy for the second term with modal stiffness $m\omega_{1\beta}^2$. No energy is stored in the coupling.

Finally, the coupling power is reduced by introducing eqns (4.125) and (4.119) in (4.113),

$$P_{01} = \rho_0 \int_0^{L_x} \sum_\alpha \dot{P}_\alpha \phi_\alpha \times \sum_\beta \dot{U}_\beta \psi_\beta \, \mathrm{d}x \tag{4.134}$$

which, after re-arrangement, leads to

$$P_{01} = -\sum_{\alpha,\beta} g_{0\alpha,1\beta} \dot{P}_\alpha \dot{U}_\beta \tag{4.135}$$

Equations (4.127), (4.130) and (4.132), (4.133), and (4.135) constitute the reduction of the vibroacoustic problem to the canonical problem. The coupling is gyroscopic. There is no elastic or inertial coupling between resonators.

Coupling loss factor

The formal equivalence of the above equations with eqns (3.1–3.5) leads to the following analogy: $m_0 \leftrightarrow \rho_0/c_0^2$, $m_1 \leftrightarrow m$. The coupling loss factor given in eqn (3.50) then reads

$$\omega\eta_{10} = \frac{\pi N_0 c_0^2 \langle g_{10}^2 \rangle}{2 m \rho_0 \Delta\omega} \tag{4.136}$$

where N_0 is the number of acoustical modes within the frequency band $\Delta\omega$ and $\langle g_{10}^2 \rangle$ the mean square value of $g_{1\beta,0\alpha}$ for all modes in the same frequency band. By a direct calculation of eqn (4.131) where ψ_β and ϕ_α are respectively given in eqns (4.122) and (4.116),

$$g_{1\beta,0\alpha} = \begin{cases} 0 & \text{if } p = \beta \\ -\frac{\rho_0}{\pi}\sqrt{\frac{8}{L_z}} \left[1 - (-1)^{\beta+p}\right] \frac{\beta}{\beta^2 - p^2} & \text{elsewhere} \end{cases} \tag{4.137}$$

where $\alpha = p, q$ is a double subscript. For a given β and varying p, q, the largest values of $g_{1\beta,0\alpha}^2$ are reached when $p = \beta \pm 1$. For those modes, $g_{1\beta,0\alpha}^2 = 8\rho_0^2/\pi^2 L_z (1 \pm 1/2\beta)^2$. When p is greater than $\beta + 1$ or lower than $\beta - 1$, $g_{1\beta,0\alpha}^2$ rapidly decreases and becomes negligible. Since furthermore β is large by virtue of Assumption 4, the term $1 \pm 1/2\beta$ may be conveniently approximated by 1 and

$$\langle g_{10}^2 \rangle = \frac{8\rho_0^2}{\pi^2 L_z} \times \frac{N_p}{N_0} \tag{4.138}$$

where N_p is the number of acoustical modes p, q in the frequency band such that $p = \beta \pm 1$. These modes have maximum proximate mode coupling to the structural mode β for two reasons. First, they are resonant and therefore their natural frequencies are close to the natural frequency of the structural mode. Second, they have an x-component of their wavenumbers $p\pi/L_x$ close to the structural wavenumber $\beta\pi/L_x$. The problem therefore reduces to counting the proximate acoustical modes.

Two situations must be distinguished according to whether the structural wavenumber $k_1 = (m\omega^2/D)^{1/4}$ is greater or lower than the acoustical wavenumber ω/c_0 where ω is the centre frequency of the band $\Delta\omega$. The situation is represented in Fig. 4.9 where the acoustical modes are the grid points in the $p\pi/L_x$, $q\pi/L_z$-plane. Resonant acoustical modes are localized in a shell of thickness $\Delta\omega/c_0$ and equation $(p\pi/L_x)^2 + (q\pi/L_z)^2 = \omega^2/c_0^2$.

When $k_1(\omega) > k_0(\omega)$, the resonant structural modes are subsonic. For a resonant structural mode β, the wavenumber $\beta\pi/L_x$ is outside the shell of resonant acoustical modes (Fig. 4.9(a)). It is then clear that no resonant acoustical mode verifies $p = \beta \pm 1$ and the number of proximate modes is zero. The sound radiation is low.

When $k_1(\omega) < k_0(\omega)$, the resonant structural modes are supersonic. A resonant structural mode $\beta\pi/L_z$ is inside the shell and the acoustical proximate modes are grid points whose x component of wavenumber is $(\beta \pm 1)\pi/L_x$ (Fig. 4.8(b)). Their number is twice the number of grid points on the intersection of the shell with the vertical line passing through $\beta\pi/L_x$. Let θ be the angle between k_0 and k_1. We get

$$N_p = \frac{2\Delta\omega}{c_0 \sin\theta} \times \frac{L_z}{\pi} \tag{4.139}$$

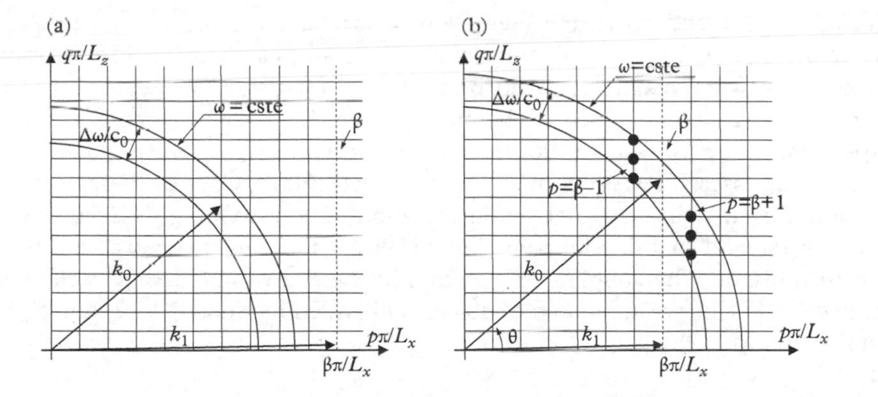

Figure 4.9 *Count of proximate modes. (a) no proximate mode in the subsonic case; (b) six proximate modes (dark dots) in the supersonic case.*

But the angle θ verifies $\cos\theta = k_1/k_0$. And since $k_1 = (m\omega^2/D)^{1/4}$ and $k_0 = \omega/c_0$, it yields

$$\sin\theta = \sqrt{1-\cos^2\theta} = \sqrt{1-\frac{\omega_c}{\omega}} \tag{4.140}$$

where $\omega_c = c_0^2(m/D)^{1/2}$ is the so-called coincidence frequency. By substituting eqns (4.140), (4.139), and (4.138) into eqn (4.136), it yields

$$\omega\eta_{10} = \frac{2\rho_0 c_0}{\pi m\sqrt{1-\frac{\omega_c}{\omega}}} \tag{4.141}$$

This result will be confirmed in Chapter 9 except that the factor $2/\pi$ will turn out to be 1. This slight underestimation is due to the neglected terms when estimating $\langle g_{10}^2\rangle$.

4.5 Comments

On the damping

The damping models in the three examples studied in this chapter have not been chosen arbitrarily. They all verify the condition of **proportional damping**. This condition is important to allow a development of the damped solution on undamped normal modes. A point damping with, for instance, a localized viscous dashpot, would give rise to coupling between all the modes, leading to a more complicated solution. Nevertheless, Keane and Price (1991) have shown that both damping models, point damping and proportional damping, lead to sensibly identical frequency averaged power flow provided that the damping level remains light.

On sound radiation

The problem of sound–structure interaction is among the most studied in statistical energy analysis due to its practical importance. The statistical theories of sound radiation are numerous and are mentioned in the former texts on statistical energy analysis (Smith, 1962; Lyon and Maidanik, 1962). The most popular formulas for sound radiation are perhaps obtained by Maidanik (1962), then Crocker and Price (1969) in both subsonic and supersonic cases. The reader may refer to Wijker (2009, Chapter 4) for a summary of available formulas in statistical sound radiation. The reasoning of Section 4.4 is a simplified version of that developed by Fahy (1969). The full reasoning leads to better explicit formulas for the coupling loss factors in the supersonic case as well as in the subsonic case. Further reading on the subject is provided by Soize (1995) and Fahy and Gardonio (2007).

5

Wave Propagation

So far, we have only considered solutions to the governing equations based on normal mode development. Normal modes are standing waves. By projecting the solution onto the set of normal modes, we solved the governing equation which served as a starting point for statistical considerations. This method is also known as the modal approach of statistical energy analysis. But another point of view is possible. Instead of standing waves, we may consider propagating waves. A vibrational field may be decomposed as a superposition of propagating waves and, if these waves are sufficiently disordered, statistical reasonings become possible and lead to new results. This alternative way to proceed is known as the wave approach to statistical energy analysis.

This chapter outlines the main concepts and features of waves, with a special emphasis on energy. For a more complete review the reader may refer to the numerous standard books on wave motion (Graff, 1991; Billinham and King, 2000; Pierce, 2006)

5.1 Propagating Wave

In a one-dimensional space, a propagating wave is a particular solution to the governing equation having the form $\psi(x-ct)$ where c is the propagation speed of the signal. Indeed, it will turn out that it is not always possible to find such a solution, or at least, when one of them has been found, some constraints may restrict the form of ψ.

D'Alembert's solution

The simplest case to highlight the existence of propagating waves is the so-called D'Alembert's solution to the wave equation. Let us consider the one-dimensional wave equation,

$$\frac{1}{c^2}\frac{\partial^2\psi}{\partial t^2} - \frac{\partial^2\psi}{\partial x^2} = 0 \tag{5.1}$$

where c is the sound speed and let us perform the change of variables,

$$\xi = x - ct, \quad \zeta = x + ct \tag{5.2}$$

Foundation of Statistical Energy Analysis in Vibroacoustics. First Edition. A. Le Bot.

By the chain rule, we have

$$\frac{\partial \psi}{\partial x} = \frac{\partial \psi}{\partial \xi}\frac{\partial \xi}{\partial x} + \frac{\partial \psi}{\partial \zeta}\frac{\partial \zeta}{\partial x} = \frac{\partial \psi}{\partial \xi} + \frac{\partial \psi}{\partial \zeta} \tag{5.3}$$

$$\frac{\partial \psi}{\partial t} = \frac{\partial \psi}{\partial \xi}\frac{\partial \xi}{\partial t} + \frac{\partial \psi}{\partial \zeta}\frac{\partial \zeta}{\partial t} = c\left(-\frac{\partial \psi}{\partial \xi} + \frac{\partial \psi}{\partial \zeta}\right) \tag{5.4}$$

and for the second derivatives,

$$\frac{\partial^2 \psi}{\partial x^2} = \frac{\partial^2 \psi}{\partial \xi^2} + 2\frac{\partial^2 \psi}{\partial \xi \partial \zeta} + \frac{\partial^2 \psi}{\partial \zeta^2}, \quad \frac{\partial^2 \psi}{\partial t^2} = c^2\left(\frac{\partial^2 \psi}{\partial \xi^2} - 2\frac{\partial^2 \psi}{\partial \xi \partial \zeta} + \frac{\partial^2 \psi}{\partial \zeta^2}\right) \tag{5.5}$$

Substituting the above derivatives into the wave equations gives

$$\frac{\partial^2 \psi}{\partial \xi \partial \zeta} = 0 \tag{5.6}$$

This equation may be integrated directly. Since its first derivative with respect to ξ is zero, $\partial \psi / \partial \zeta$ is a constant which may depend on the second variable ζ,

$$\frac{\partial \psi}{\partial \zeta} = G(\zeta) \tag{5.7}$$

However, a similar reasoning gives

$$\psi(\xi, \zeta) = f(\xi) + g(\zeta) \tag{5.8}$$

where g is any function whose derivative is G. By changing back to x, t variables, we finally obtain

$$\psi(x, t) = f(x - ct) + g(x + ct) \tag{5.9}$$

A term like $f(x-ct)$ may be interpreted as follows. In a frame moving from left to right at speed c the local abscissa is $\xi = x - ct$ and the disturbance $f(\xi)$ appears to be stationary since it remains unchanged in time. In the reference frame, the term $f(x-ct)$ is therefore a **right-travelling wave**. Similarly, $g(x+ct)$ is a **left-travelling wave**. Thus, all solutions of the wave equation may be written as a sum of a right-travelling wave and a left-travelling wave whose propagation speed is exactly the constant c appearing in the wave equation (see Fig. 5.1).

Figure 5.1 *D'Alembert's solution to the wave equation.*

Two features of D'Alembert's solution may be underlined. The first point is that f and g are arbitrary functions. They are determined by the initial conditions. The second point to emphasize is that the shapes of f and g remain unchanged during propagation. So, the initial disturbance may have any shape and will propagate as two waves towards left and right without distortion. However, these two characteristics, arbitrary shape and permanence of the disturbance, are not representative of all systems.

Euler–Bernoulli beam

Most often, the shape of f and g must be imposed. For instance, let us consider the governing equation of transverse disturbance in beams,

$$m\frac{\partial^2 u}{\partial t^2} + EI\frac{\partial^4 u}{\partial x^4} = 0 \tag{5.10}$$

A right-travelling wave $u(x, t) = f(x - c_p t)$ verifies

$$mc_p^2 f'' + EI\,f^{(iv)} = 0 \tag{5.11}$$

We have changed the notation of speed for a reason that will be clarified later. Since m and EI are positive numbers, $f(\xi)$ is of the form $A\cos\left(\xi\,(mc_p^2/EI)^{1/2} - \varphi\right) + B\xi + C$ where A, B, C, and φ are four constants. But $B\xi + C$ corresponds to a rotation and vertical translation of the beam in the moving frame ξ, t. In the reference frame, this is a rigid body movement and is therefore not associated with a vibrational state. Hence the vibrating part of the solution,

$$u(x, t) = A\cos\left(\sqrt{\frac{mc_p^2}{EI}}\,(x - c_p t) - \varphi\right) \tag{5.12}$$

may be written as

$$u(x, t) = A\cos(\omega t - \kappa x + \varphi) \tag{5.13}$$

where we have set $\omega = c_p^2(m/EI)^{1/2}$ and $\kappa = c_p(m/EI)^{1/2}$. The constant ω is called the **angular frequency** and κ the **wavenumber**. The speed c_p is related to ω and κ by $c_p = \omega/\kappa$ (a summary of sound speed for various waves is shown in Table 5.1).

In a similar fashion, a right-travelling solution is found by substituting $x+c_p t$ to $x-c_p t$.

In opposition with the wave equation, a propagating wave may travel in a beam but only under the condition that its shape is a cosine function (with an arbitrary phase). As a counterpart, all propagation speeds are possible $c_p = \omega/\kappa$ but the angular frequency and the wavenumber are constrained by the relationship,

$$m\omega^2 = EI\kappa^4 \tag{5.14}$$

This is the so-called **dispersion equation**.

Table 5.1 *Waves and phase speed in various systems with E Young's modulus, ν Poisson's ratio, $G = E/2(1-\nu)$ shear modulus, $\lambda = E\nu/(1+\nu)(1-2\nu)$ and $\mu = E/2(1+\nu)$ Lame's parameters, h thickness, $D = Eh^3/12(1-\nu^2)$ bending stiffness, b width, $I = bh^3/12$ inertia, T tension, ρ density, and m mass per unit length or area. $\mathbf{u} = (u, v, w)$ deformation along x-, y-, and z- direction, p acoustical pressure, $\mathbf{v}$ particle velocity.*

System	Wave	Motion	Governing equation	Dispersion equation	Phase speed
String	transverse	deflection w	$T\partial_x^2 w - m\partial_t^2 w = 0$	$T\kappa^2 - m\omega^2 = 0$	$\sqrt{T/m}$
Beam	longitudinal	elongation u	$E\partial_x^2 u - \rho\partial_t^2 u = 0$	$E\kappa^2 - \rho\omega^2 = 0$	$\sqrt{E/\rho}$
Beam	torsional	rotation angle θ	$G\partial_x^2\theta - \rho\partial_t^2\theta = 0$	$G\kappa^2 - \rho\omega^2 = 0$	$\sqrt{G/\rho}$
Beam	bending	deflection w	$EI\partial_x^4 w + m\partial_t^2 w = 0$	$EI\kappa^4 - m\omega^2 = 0$	$(EI/m)^{1/4}\sqrt{\omega}$
Membrane	transverse	deflection w	$T\Delta w - m\partial_t^2 w = 0$	$T\kappa^2 - m\omega^2 = 0$	$\sqrt{T/m}$
Plate	longitudinal	$\left\{ u = \partial_x\varphi + \partial_y\psi \right.$	$\Delta\varphi - \frac{1}{c_L^2}\partial_t^2\varphi = 0$	$\kappa = \omega/c_L$	$c_L = \sqrt{E/\rho(1-\nu^2)}$
Plate	shear	$\left. v = \partial_y\varphi - \partial_x\psi \right.$	$\Delta\psi - \frac{1}{c_T^2}\partial_t^2\psi = 0$	$\kappa = \omega/c_T$	$c_T = \sqrt{G/\rho}$
Plate	bending	deflection w	$D\Delta^2 w + m\partial_t^2 w = 0$	$D\kappa^4 - m\omega^2 = 0$	$c_B = (D/m)^{1/4}\sqrt{\omega}$
Solid	P-wave	$\left\{ \mathbf{u} = \nabla\Phi + \nabla\wedge\mathbf{H} \right.$	$(\lambda+2\mu)\Delta\Phi - \rho\partial_t^2\Phi = 0$	$(\lambda+2\mu)\kappa^2 - \rho\omega^2 = 0$	$c_1 = \sqrt{(\lambda+2\mu)/\rho}$
Solid	S-wave	$\nabla\cdot\mathbf{H} = 0$	$\mu\Delta\mathbf{H} - \rho\partial_t^2\mathbf{H} = 0$	$\mu\kappa^2 - \rho\omega^2 = 0$	$c_2 = \sqrt{\mu/\rho}$
Fluid	pressure	$p = -\rho\partial_t\Phi$, $\mathbf{v} = \nabla\Phi$	$\Delta\Phi - \frac{1}{c_0^2}\partial_t^2\Phi = 0$	$\kappa = \omega/c_0$	343 m/s at 20 °C

Multidimensional propagation

In a multidimensional medium, a propagating wave may have various shapes. We shall distinguish these waves by the shape of the line (two dimensions) or surface (three dimensions) of constant perturbation. Thus, a plane wave is a solution whose perturbation at a fixed time is constant in all planes normal to a given direction. In other words, the solution does not depend on the position in a fixed normal plane but depends on the position of this plane along the propagation axis. A cylindrical wave depends on the radius r but not on φ and z where r, φ, and z are cylindrical coordinates (see Fig. 5.2). A spherical wave depends on r but not on θ or φ. Let us remark that when we say constant on a surface we mean constant at a fixed time, but when time passes the disturbance varies at a fixed point.

Plane waves

Let us detail what a plane wave is in a three-dimensional space. The wave equation reads

$$\frac{1}{c^2}\frac{\partial^2\psi}{\partial t^2} - \left(\frac{\partial^2\psi}{\partial x^2} + \frac{\partial^2\psi}{\partial y^2} + \frac{\partial^2\psi}{\partial y^2}\right) = 0 \tag{5.15}$$

Considering a direction given by the unit vector $n = (n_x, n_y, n_z)$, we are looking for solutions propagating along that direction and constant along the other directions. So, we choose a frame in which the first coordinate is $x' = n_x x + n_y y + n_z z$, the distance to the plane normal to n. The two other coordinates y' and z' are arbitrarily chosen in the plane normal to n. The expected solutions depend on time and x' but not on the y' and z' so that we can write $\psi(x', t)$. By the chain rule,

$$\frac{\partial\psi}{\partial x} = \frac{\partial\psi}{\partial x'}\frac{\partial x'}{\partial x} + \frac{\partial\psi}{\partial y'}\frac{\partial y'}{\partial x} + \frac{\partial\psi}{\partial z'}\frac{\partial z'}{\partial x} = n_x\frac{\partial\psi}{\partial x'} \tag{5.16}$$

since $\partial\psi/\partial y' = 0$ and $\partial\psi/\partial z' = 0$. For the second derivative,

$$\frac{\partial^2\psi}{\partial x^2} = n_x^2\frac{\partial^2\psi}{\partial x'^2} \tag{5.17}$$

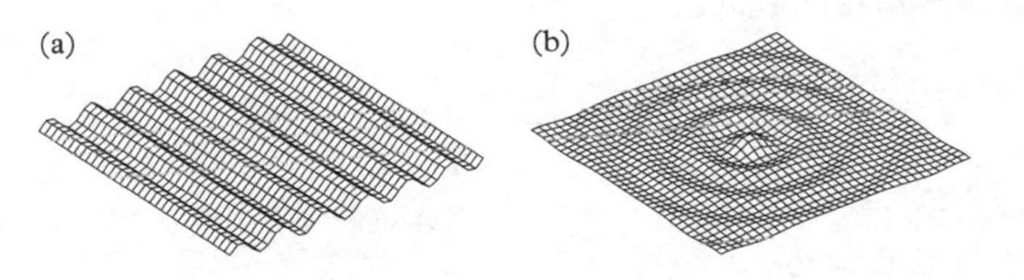

Figure 5.2 *Wave travelling on a surface. (a) plane wave; (b) cylindrical wave.*

Indeed, similar relationships hold for partial derivatives with respect to y and z. Thus, by substituting $\psi(x', t)$ into the wave equation and remarking that $n_x^2 + n_y^2 + n_z^2 = 1$, the wave equation reduces to the one-dimensional wave equation,

$$\frac{1}{c^2}\frac{\partial^2 \psi'}{\partial t^2} - \frac{\partial^2 \psi'}{\partial x'^2} = 0 \tag{5.18}$$

But D'Alembert's solution gives $\psi(x', t) = f(x' - ct) + g(x' + ct)$ and therefore in the original frame,

$$\psi(x, y, z, t) = f(n_x x + n_y y + n_z z - ct) + g(n_x x + n_y y + n_z z + ct) \tag{5.19}$$

f is a plane wave travelling in the positive x'-direction while g travels in the negative x'-direction.

Spherical waves

For spherical waves, we must look for solutions $\psi(r, t)$ to the wave equation verifying a spherical symmetry about the origin. So, by developing the Laplacian operator in spherical coordinates, it yields

$$\frac{1}{c^2}\frac{\partial^2 \psi}{\partial t^2} - \frac{1}{r^2}\frac{\partial}{\partial r}\left(r^2 \frac{\partial \psi}{\partial r}\right) = 0 \tag{5.20}$$

By substituting $\psi(r, t) = \phi(r, t)/r$, we get

$$\frac{1}{c^2}\frac{\partial^2 \phi}{\partial t^2} - \frac{\partial^2 \phi}{\partial r^2} = 0 \tag{5.21}$$

Again by D'Alembert's solution $\phi(r, t) = f(r - ct) + g(r + ct)$ and therefore

$$\psi(r, t) = \frac{1}{r}\left[f(r - ct) + g(r + ct)\right] \tag{5.22}$$

The field is found to be the sum of an outgoing wave f/r and an incoming wave g/r. Their magnitudes depend on the radius r.

As an example of a spherical wave, the impulse response of the wave equation in dimension three is the solution of

$$\frac{1}{c^2}\frac{\partial^2 \psi}{\partial t^2} - \Delta\psi = \delta(r)\delta(t) \tag{5.23}$$

where δ is the Dirac function. It is well known (Barton, 1989, p. 267) that the solution in infinite medium at rest for $t < 0$ is

$$\psi(r, t) = \frac{c}{4\pi r}\delta(r - ct) \tag{5.24}$$

where r is the distance to the origin. Thus, the perturbation is an infinitely short pulse, the magnitude of which decreases like $1/r$. This feature explains why speech is possible. If a pulse were to spread in time, then phonemes clearly separated by the speaker would overlap at the receiver point and the words would rapidly become impossible to understand.

5.2 Harmonic Wave

A harmonic plane wave may be sought as a particular solution to the governing equation of the form $A_0 \cos(\omega t - \kappa x + \varphi)$ where κ is the wavenumber, ω the angular frequency, φ the phase, and x the distance along the axis of propagation. Generally, ω and κ are constrained by the dispersion equation that we shall introduce in this section.

Initial-value problem

In order to highlight how harmonic waves may be used to solve differential problems and what the link between ω and κ is, let us consider the following example. The differential problem is

$$\begin{cases} m\dfrac{\partial^2 u}{\partial t^2} + c\dfrac{\partial u}{\partial t} - T\dfrac{\partial^2 u}{\partial x^2} = 0 \\ u(x,0) = f(x), \quad \dot{u}(x,0) = 0 \end{cases} \tag{5.25}$$

This is the governing equation of a stretched string with a viscous damping force. The initial shape is imposed $f(x)$ while the initial speed is zero. Let $U(\kappa, t)$ be the Fourier transform of $u(x,t)$ with respect to x, and $F(\kappa)$ the Fourier transform of $f(x)$. The problem becomes

$$\begin{cases} m\dfrac{\mathrm{d}^2 U}{\mathrm{d}t^2} + c\dfrac{\mathrm{d}U}{\mathrm{d}t} + T\kappa^2 U = 0 \\ U(\kappa,0) = F(\kappa), \quad \dot{U}(\kappa,0) = 0 \end{cases} \tag{5.26}$$

We have employed the ordinary derivation symbol in order to point out that only the derivation with respect to time appears in the problem (5.26). The variable κ can therefore be considered as a parameter. Thus, the above problem is an ordinary differential equation of order two with initial conditions. This is a Cauchy problem and since $F(\kappa)$ is complex-valued, the solution must be found in the space of complex-valued functions.

It is well known that the solutions to the homogeneous differential equation lie in a linear space of dimension two whose basis is composed of well-chosen functions $\mathrm{e}^{\lambda t}$. If λ is written as $\lambda = \imath\omega$ then ω is the solution to

$$-m\omega^2 + \imath c\omega + T\kappa^2 = 0 \tag{5.27}$$

Since κ is real, this polynomial has two complex roots ω_0 and $-\overline{\omega}_0$ and the solution U has the form

$$U(\kappa, t) = a_1(\kappa)\mathrm{e}^{\imath\omega_0 t} + a_2(\kappa)\mathrm{e}^{-\imath\overline{\omega}_0 t} \tag{5.28}$$

Applying the initial conditions,

$$\begin{pmatrix} 1 & 1 \\ \omega_0 & -\overline{\omega}_0 \end{pmatrix} \begin{pmatrix} a_1 \\ a_2 \end{pmatrix} = \begin{pmatrix} F(\kappa) \\ 0 \end{pmatrix} \tag{5.29}$$

Solving this set of linear equations, substituting the solutions into eqn (5.28), and coming back to $u(x, t)$ gives

$$u(x, t) = \int_{-\infty}^{\infty} F(\kappa) \left[\frac{\overline{\omega}_0}{\overline{\omega}_0 + \omega_0} \mathrm{e}^{\imath(\omega_0 t + \kappa x)} + \frac{\omega_0}{\overline{\omega}_0 + \omega_0} \mathrm{e}^{\imath(-\overline{\omega}_0 t + \kappa x)} \right] \frac{\mathrm{d}\kappa}{2\pi} \tag{5.30}$$

where $\omega_0(\kappa)$ is a function of κ. This is the solution to the original problem.

Let us discuss this result. The solution is constructed by superimposing harmonic waves $\mathrm{e}^{\imath(\omega t + \kappa x)}$ where ω is a complex solution to eqn (5.27). Although the waves $\mathrm{e}^{\imath(\omega t + \kappa x)}$ are complex-valued, the integral is real-valued and u remains real in time. In eqn (5.27), κ is a real-valued parameter and ω is naturally found as a complex function of κ. But this is not always the case.

Boundary-value problem

Let us consider a second example.

$$\begin{cases} m\dfrac{\partial^2 u}{\partial t^2} + c\dfrac{\partial u}{\partial t} - T\dfrac{\partial^2 u}{\partial x^2} = 0 \\ u(0, t) = f(t), \quad u(L, t) = 0 \end{cases} \tag{5.31}$$

This is a boundary-value problem. This time, we introduce $U(x, \omega)$ and $F(\omega)$, the Fourier transforms of respectively $u(x, t)$ and $f(t)$ with respect to time. The problem becomes

$$\begin{cases} (-m\omega^2 + \imath c\omega)U - T\dfrac{\mathrm{d}^2 U}{\mathrm{d}x^2} = 0 \\ U(0, \omega) = F(\omega), \quad U(L, \omega) = 0 \end{cases} \tag{5.32}$$

This is an ordinary differential equation of order two whose basis solutions must be found with the form $\mathrm{e}^{\lambda x}$. Writing $\lambda = \imath\kappa$, we get

$$-m\omega^2 + \imath c\omega + T\kappa^2 = 0 \tag{5.33}$$

In this equation, ω is real-valued. This equation admits two complex solutions κ_0 and $-\kappa_0$ which verify $\kappa_0(-\omega) = \overline{\kappa}_0(\omega)$. The solution of the original problem therefore has the form

$$U(x, \omega) = a_1(\omega)e^{\imath\kappa_0 x} + a_2(\omega)e^{-\imath\kappa_0 x} \tag{5.34}$$

By applying the boundary conditions,

$$\begin{pmatrix} 1 & 1 \\ e^{\imath\kappa_0 L} & e^{-\imath\kappa_0 L} \end{pmatrix} \begin{pmatrix} a_1 \\ a_2 \end{pmatrix} = \begin{pmatrix} F(\omega) \\ 0 \end{pmatrix} \tag{5.35}$$

Again, solving this set of equations and going back to $u(x, t)$ yields

$$u(x, t) = \int_{-\infty}^{\infty} F(\omega) \left[\frac{e^{-\imath\kappa_0 L}}{e^{-\imath\kappa_0 L} - e^{\imath\kappa_0 L}} e^{\imath(\omega t + \kappa_0 x)} - \frac{e^{\imath\kappa_0 L}}{e^{-\imath\kappa_0 L} - e^{\imath\kappa_0 L}} e^{\imath(\omega t - \kappa_0 x)} \right] \frac{d\omega}{2\pi} \tag{5.36}$$

This is the solution of the boundary-value problem constructed as a superposition of harmonic waves $e^{\imath(\omega t + \kappa x)}$. The wavenumber κ is generally complex but the parameter ω is real. The integral is real by the properties $\kappa_0(-\omega) = \overline{\kappa}_0(\omega)$ and $F(-\omega) = \overline{F}(\omega)$.

Dispersion equation

We have underlined the importance of harmonic waves $Ae^{\imath(\omega t + \kappa x)}$ for solving linear partial differential equations. In both initial-value and boundary-value problems, we have found the solution to be a linear superposition of harmonic waves. This highlights why harmonic waves are so fundamental. However, ω and κ cannot take any value. In the above example, they are constrained by the condition $-m\omega^2 + \imath c\omega + T\kappa^2 = 0$. But there exists a strong difference between eqns (5.27) and (5.33). In the first case, κ is a real-valued parameter and ω is a complex root of the polynomial equation while this is the converse in the second case. So, if we introduce the polynomial $P(\omega, \kappa) = -m\omega^2 + \imath c\omega + T\kappa^2$ in two variables, the **dispersion equation** is

$$-m\omega^2 + \imath c\omega + T\kappa^2 = 0 \tag{5.37}$$

where ω and κ may now both be complex-valued. Regarding all solutions of $P(\omega, \kappa) = 0$, not all of them are useful. It depends on the type of problem we want to solve. In the first example, we were only interested in solutions $\omega(\kappa)$ where $\kappa \in \mathbb{R}$, while in the second example we found $\kappa(\omega)$ where $\omega \in \mathbb{R}$.

We are now in a position to introduce the dispersion equation in a more general case. Assume that $\psi_k(x, t)$ are p functions depending on space and time and that they are governed by a set of coupled linear partial differential equations. We wonder if harmonic waves of the form $\psi_k = A_k e^{\imath(\omega t + \kappa x)}$ may freely propagate. In a precise manner, we look for solutions to the homogeneous differential equations (without source) in an infinite medium (no boundary) and without worrying about the initial state (no prescribed initial conditions). By deriving with respect to space and time such a solution, we have the correspondence,

$$\frac{\partial}{\partial t} \leftrightarrow \imath\omega, \qquad \frac{\partial}{\partial x} \leftrightarrow \imath\kappa \tag{5.38}$$

Therefore, by substituting $A_k e^{\imath(\omega t+\kappa x)}$ into the set of linear partial differential equations, we get

$$\sum_{l=1}^{p} D_{kl}(\kappa,\omega)A_l = 0, \quad k = 1,\ldots,p \tag{5.39}$$

where the coefficients $D_{kl}(\kappa,\omega)$ are polynomials of κ and ω. We get a system of linear equations on the unknowns A_l. A non-trivial solution (i.e. not all $A_l = 0$) only exists under the condition that the determinant is zero,

$$\det \mathbf{D}(\kappa,\omega) = 0 \tag{5.40}$$

where $\mathbf{D}$ is the matrix whose entries are D_{kl}. This is a polynomial equation called the **dispersion equation**. The coefficients of the polynomial are complex-valued.

The dispersion equation is also called the **characteristic equation** of a linear differential equation. The solution $\omega(\kappa)$ or alternatively $\kappa(\omega)$ is called the dispersion relationship. All information on waves is contained in the dispersion equation, such as the number and type of waves, the cutoff frequencies, and the speed of propagation, as we shall see in forthcoming examples.

Complex notation

The link between real-valued solutions to the governing equations and complex-valued solutions is narrow. We have shown that an initial-value problem and a boundary-value problem may be solved by a Fourier transform which leads us to construct the real-valued solution as a sum of complex-valued solutions.

The second point is that the governing equation is a linear partial differential equation with real-valued constant coefficients. In such conditions, two principles hold. The first one is the linear superposition principle which states that any linear combination of solutions is also a solution. The second one states that if ψ is a complex solution, then the conjugate $\overline{\psi}$ is also a solution. This is a simple consequence of

$$\overline{a\frac{\partial^{n+m}\psi}{\partial x^n \partial^m t}} = a\frac{\partial^{n+m}\overline{\psi}}{\partial x^n \partial t^m} \tag{5.41}$$

if $a \in \mathbb{R}$. By combining both principles, $\mathrm{Re}(\psi)$ and $\mathrm{Im}(\psi)$ are always valid real-valued solutions.

So, any real-valued solution is a sum of conjugate complex-valued solutions and, conversely, one can construct a real-valued solution by taking the real or imaginary part of any complex-valued solution.

From now on, we adopt the complex notation. A real harmonic wave $A_0 \cos(\omega t + \kappa x + \varphi)$ may be written $\mathrm{Re}\left[A\mathrm{e}^{\imath(\omega t+\kappa x)}\right]$ where the phase φ is absorbed in the complex magnitude $A = A_0\mathrm{e}^{\imath\varphi}$. If ω and κ are solutions of the dispersion equation, the latter is a complex solution and the former is a real solution. We keep in mind that when constructing a complex solution of a given problem by superimposing complex harmonic waves, the actual waves (real-valued) are the real part of the final result.

Phase speed

Let us consider a right-travelling harmonic wave $A\mathrm{e}^{\imath(\omega t-\kappa x)}$ where $\kappa \in \mathbb{R}$ and $\omega \in \mathbb{C}$. Let $A = A_0\mathrm{e}^{\imath\varphi}$ be the polar form of A with A_0 the magnitude and φ the argument. And let $\omega = \omega_r + \imath\omega_i$ be the Cartesian form of ω. The wave field becomes

$$A_0\mathrm{e}^{-\omega_i t}\mathrm{e}^{\imath(\omega_r t-\kappa x+\varphi)}$$

in the polar representation. The magnitude $A_0\mathrm{e}^{-\omega_i t}$ is a decreasing function in time while the phase $\omega_r t - \kappa x + \varphi$ is a function of both space and time. An example is plotted in Fig. 5.3. We observe that the wave shape is a sine function which propagates to the right but with a magnitude that decreases. The disturbance will vanish in the limit of infinite time. The speed at which the wave propagates is the **phase speed**.

At time t, the phase is $\omega_r t - \kappa x + \varphi$. A point located at a crest of a wave has phase $0 \pm 2k\pi$ where $k \in \mathbb{N}$. For instance, the crest $k = 0$ has abscissa $x_0(t) = (\omega_r t + \varphi)/\kappa$. This is the square dot in Fig. 5.3. At a later time $t = t + \mathrm{d}t$, the phase is $\omega_r(t + \mathrm{d}t) - \kappa x + \varphi$ and therefore the abscissa of *the same crest* is $x_0(t + \mathrm{d}t) = (\omega_r t + \omega_r \mathrm{d}t + \varphi)/\kappa$. This point x_0 goes on with speed $c_\mathrm{p} = (x_0(t + \mathrm{d}t) - x_0(t))/\mathrm{d}t$. This is the so-called phase speed of a wave,

$$c_\mathrm{p} = \frac{\omega_r}{\kappa} = \mathrm{Re}\left(\frac{\omega}{\kappa}\right), \quad \kappa \in \mathbb{R} \tag{5.42}$$

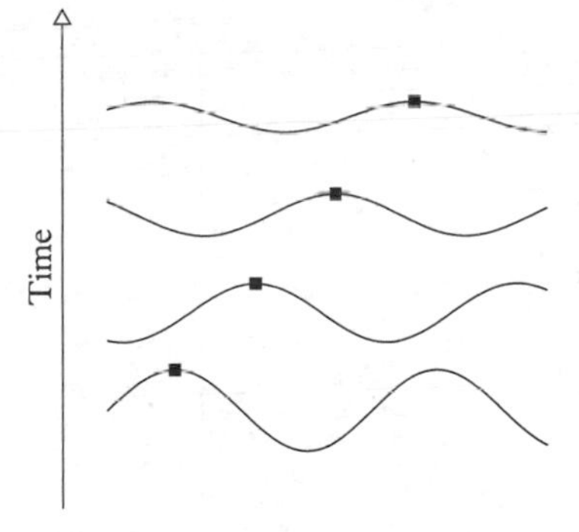

Figure 5.3 *Phase speed. A point of constant phase (square dot) moves at phase speed.*

The phase speed is therefore the speed of any point of constant phase. This is the speed of propagation of a crest of the wave. When ω is real, the above definition reduces to $c_p = \omega/\kappa$.

Dispersion of waves

As we have seen in Section 5.1, ω and κ are linked by the dispersion relation so that we may write $\omega(\kappa)$. Sinusoidal waves of different frequencies may have different wavenumbers κ and consequently different phase speeds ω/κ. If the dispersion relation is linear, $\omega = c\kappa$ where c is constant, then all sinusoidal waves travel with the same speed. This is the case for the wave equation. Such a medium is called **nondispersive**. But in general, the plot $\omega(\kappa)$ is a curved line and the phase speed does depend on the frequency. The medium is then called **dispersive**.

We can consider two different types of dispersion. In the first case, the phase speed decreases with the frequency. In other words since $\kappa = 2\pi/\lambda$, the phase speed increases with the wavelength λ. This is called **normal dispersion**. An example of normal dispersion is the well-known experiment of refraction of light by a glass prism. The refraction angle depends on the light colour. A red beam ($\lambda \approx 0.8\ \mu$m) is bent less than a violet beam ($\lambda \approx 0.4\ \mu$m) which shows that red beams travel faster than violet beams. In the second case the phase speed increases with the frequency and decreases with the wavelength. Next we shall talk about **anomalous dispersion**. In optics, anomalous dispersion is rarer than normal dispersion. Examples of normal dispersion and anomalous dispersion are shown in Fig. 5.4. In this figure, two plane waves travel with different phase speeds. The main wave has a large magnitude and a large wavelength. A small wave is superimposed on this main wave with a smaller wavelength. In the case of normal dispersion, the crest of the secondary wave travels less rapidly than the main wave crest. Therefore, as time passes the secondary crest (square dot), initially in front of the

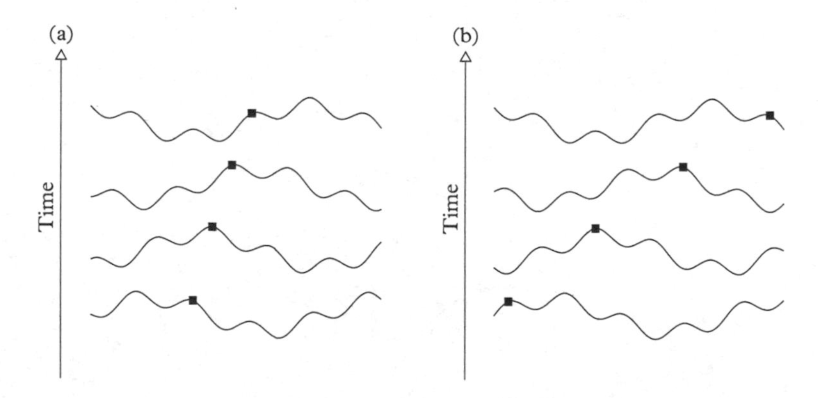

Figure 5.4 *(a) normal dispersion: the small wavelength wave (high frequency) is passed by the main wave crest (low frequency). (b) anomalous dispersion: the small wavelength wave is getting over the crest of the main wave.*

main crest, is caught again by the main crest and finally ends up behind the main crest. The converse phenomenon is observed for anomalous dispersion.

String on elastic foundation

As an illustrative example, let us consider a string resting on an elastic foundation. The governing equation is (Graff, 1991, p. 51)

$$\frac{1}{c^2}\frac{\partial^2 u}{\partial t^2} + \alpha^2 u - \frac{\partial^2 u}{\partial x^2} = 0 \tag{5.43}$$

where $\alpha = k/T$ with k the stiffness per unit length of the foundation and T the tension of the string, while $c = (T/m)^{1/2}$ is the free wave speed, i.e. the wave speed when $k = 0$. In the field of quantum mechanics, this equation is known as the Klein–Gordon equation, with $\alpha = mc/\hbar$, m being the mass of the particle, $\hbar$ the Planck's constant, and c the speed of light. We wonder if a wave can propagate in such a medium. So, we are looking for a solution of the form $u(x, t) = Ae^{\imath(\omega t+\kappa x)}$. By substituting this solution into the Klein–Gordon equation, we get

$$\left(-\frac{\omega^2}{c^2} + \alpha^2 + \kappa^2\right) Ae^{\imath(\omega t+\kappa x)} = 0 \tag{5.44}$$

so that a non-trivial solution $A \neq 0$ exists if and only if the following condition is verified:

$$-\frac{\omega^2}{c^2} + \alpha^2 + \kappa^2 = 0 \tag{5.45}$$

This is the dispersion equation of the Klein–Gordon equation. The dispersion equation shows that wavenumber κ and frequency ω are linked. The solution to eqn (5.45) is

$$\omega(\kappa) = \pm c\sqrt{\kappa^2 + \alpha^2} \tag{5.46}$$

This is the dispersion relationship. The relationship $\omega(\kappa)$ is plotted in Fig. 5.5. This relationship highlights that for any real wavenumber κ, two propagating waves exist $e^{\imath(\omega t+\kappa x)}$ and $e^{\imath(-\omega t+\kappa x)}$. The former is left-travelling and the latter is right-travelling when $\kappa > 0$. In both cases, the circular frequency ω is always greater than $c\alpha$.

An alternative form of the dispersion relationship is found by imposing that ω is real. We obtain

$$\kappa(\omega) = \pm\sqrt{\frac{\omega^2}{c^2} - \alpha^2} \tag{5.47}$$

Again, two waves are found. But depending on ω, the wavenumber κ may be real or imaginary. When $|\omega| \geq c\alpha$, the two waves are propagating. But when $|\omega| \leq c\alpha$, κ

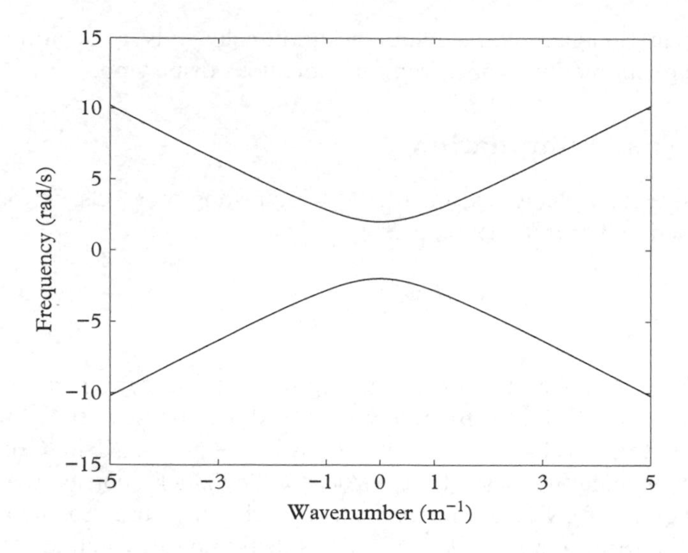

Figure 5.5 *Dispersion diagram for Klein–Gordon equation.*

is imaginary. So, if we introduce $\gamma = (\alpha^2 - \omega^2/c^2)^{1/2}$, the waves are $e^{\iota\omega t+\gamma x}$ and $e^{\iota\omega t-\gamma x}$. These waves do not propagate. Their magnitude rapidly decreases when going away to the left for the former and to the right for the latter. They are called **evanescent waves**. The transition frequency

$$\omega_c = c\alpha \tag{5.48}$$

between non-propagating and propagating behaviour is called **cutoff frequency**.

As a final remark, let us calculate the phase speed $c_\mathrm{p} = \omega/\kappa$,

$$c_\mathrm{p} = c\sqrt{1 + \frac{\alpha^2}{\kappa^2}} \tag{5.49}$$

We may observe that the phase speed depends on the wavenumber and therefore the Klein–Gordon equation is dispersive. Since the phase speed is a decreasing function of κ, we can even say that this is a normal dispersion. In the limit $\kappa \to \infty$, the phase speed is c.

Timoshenko beam

A more sophisticated example is the so-called Timoshenko beam. For such a beam cross-sections are no longer assumed to be orthogonal to the centroidal axis as was the case for Euler–Bernoulli beams. As a consequence, two state variables are required to describe the position of a cross-section; the deflection u and the absolute rotation angle ψ. The fact that the cross-section may not be orthogonal to the centroidal axis implies that

$\psi \neq \partial u/\partial x$ in general. Both state variables must therefore be considered as independent. The governing equations are (Graff, 1991, p. 183)

$$GS\alpha\left(\frac{\partial \psi}{\partial x} - \frac{\partial^2 u}{\partial x^2}\right) + \rho S\frac{\partial^2 u}{\partial t^2} = 0 \tag{5.50}$$

$$GS\alpha\left(\frac{\partial u}{\partial x} - \psi\right) + EI\frac{\partial^2 \psi}{\partial x^2} - \rho I\frac{\partial^2 \psi}{\partial t^2} = 0 \tag{5.51}$$

In these expressions, ρ denotes the mass density of the material, S the cross-sectional area, I the flexural inertia, E the Young modulus, and G the shear modulus. The parameter α is called Timoshenko shear coefficient and depends on the shape of the cross-section. It is usually determined by stress analysis for each cross-section but its exact value is of no importance for our purpose. To derive the dispersion equation, we assume a solution of a propagating wave of the form $u(x,t) = A_1 e^{\imath(\omega t + \kappa x)}$ and $\psi(x,t) = A_2 e^{\imath(\omega t + \kappa x)}$. Substituting this form of solution into the governing equations gives

$$(GS\alpha\kappa^2 - \rho S\omega^2)A_1 + \imath GS\alpha\kappa A_2 = 0 \tag{5.52}$$

$$\imath GS\alpha\kappa A_1 - (GS\alpha + EI\kappa^2 - \rho I\omega^2)A_2 = 0 \tag{5.53}$$

A non-trivial solution exists only under the condition that the determinant is not zero,

$$G^2S^2\alpha^2\kappa^2 - (GS\alpha\kappa^2 - \rho S\omega^2)(GS\alpha + EI\kappa^2 - \rho I\omega^2) = 0 \tag{5.54}$$

which after simplification gives

$$\frac{EI}{\rho S}\kappa^4 - \frac{I}{S}\left(1 + \frac{E}{G\alpha}\right)\kappa^2\omega^2 - \omega^2 + \frac{\rho I}{GS\alpha}\omega^4 = 0 \tag{5.55}$$

This is the dispersion equation of the Timoshenko beam whose solutions are shown in Fig. 5.6. Remember that the dispersion equation of the Euler–Bernoulli beam was $\rho S\omega^2 - EI\kappa^4 = 0$.

To investigate the propagating behaviour, two asymptotic cases may be considered. In the case of large wavenumbers, we may factorize κ^4 from the equation and let $\kappa \to \infty$. The equation reduces to

$$\frac{E}{\rho} - \left(1 + \frac{E}{G\alpha}\right)c_p^2 + \frac{\rho}{G\alpha}c_p^4 = 0 \tag{5.56}$$

where we have introduced the phase speed $c_p = \omega/\kappa$. This equation has the two roots,

$$c_p = \sqrt{\frac{G\alpha}{\rho}} \quad \text{and} \quad c_p = \sqrt{\frac{E}{\rho}} \tag{5.57}$$

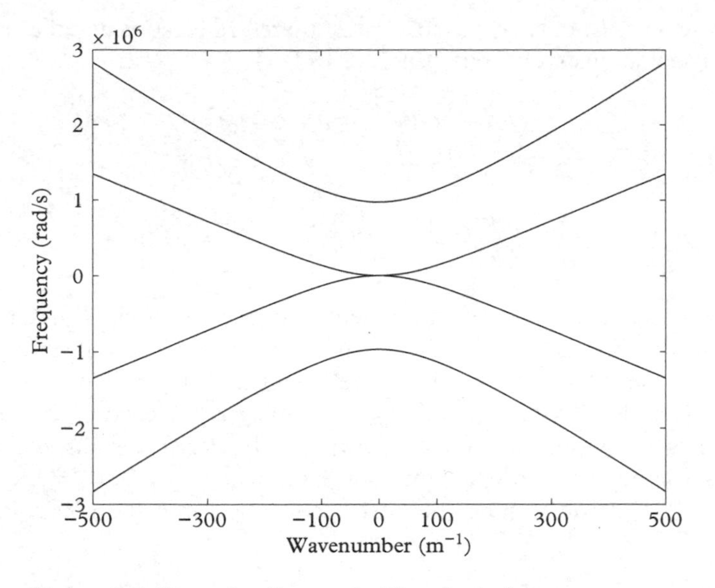

Figure 5.6 *Dispersion diagram for Timoshenko beam.*

Thus, two waves propagate in a Timoshenko beam as opposed to a single wave in a Euler–Bernoulli beam. Furthermore, the velocities are bounded at large wavenumbers. This corrects an important shortcoming of Euler–Bernoulli theory which predicts an infinite velocity $c_p = \kappa^2 (EI/\rho S)^{1/2}$ when $\kappa \to \infty$, a result which is of course not physical.

The second asymptotic behaviour of interest is when $\kappa \to 0$. The dispersion equation becomes

$$-\omega^2 + \frac{\rho I}{GS\alpha}\omega^4 = 0 \tag{5.58}$$

The two roots are 0 and

$$\omega_c = \sqrt{\frac{GS\alpha}{\rho I}} \tag{5.59}$$

A zero means that the corresponding root of the dispersion equation always exists and is real. But ω_c is a cutoff frequency for the second wave. For $\omega \geq \omega_c$, κ is real and the wave propagates. For $\omega \leq \omega_c$, κ is imaginary and the wave is evanescent.

5.3 Wave Packet

The harmonic waves $e^{\imath(\omega t + \kappa x)}$ are solutions to the governing equations provided that ω and κ are linked by the dispersion relationship. Since these equations are linear, a superposition of harmonic waves is also a solution. This is the way that we solved initial-value and boundary-value problems. We now turn to the concept of wave packet by considering a continuous sum of harmonic waves of the form

$$\psi(x,t) = \frac{1}{2\pi}\int_{-\infty}^{\infty} \Psi_0(\kappa)\mathrm{e}^{\imath(\omega(\kappa)t+\kappa x)}\,\mathrm{d}\kappa \tag{5.60}$$

where $\omega(\kappa)$ verifies the dispersion relationship and $\kappa \mapsto \Psi_0(\kappa)$ is an arbitrary complex-valued function.

The general form given in eqn (5.60) is always a valid solution to the homogeneous governing equation but complex-valued in general. By taking the real part of eqn (5.60), we get a real-valued solution.

The most interesting case is when the disturbance is confined to a bounded spatial region with all harmonic components propagating in the same direction. It is under these circumstances that we speak of **wave packet**. The concept of wave packet is useful in many fields of physics, such as acoustics, optics, seismology, and quantum mechanics (Cohen-Tannoudji et al., 1977). Wave packets have many interesting properties. We shall describe only a few of them here.

Initial state

The function $\Psi_0(\kappa)$ is determined by the initial state of the field. To prove this fact, let us remark that by fixing the time t and separating the exponential $\mathrm{e}^{\imath\kappa x}$ in eqn (5.60) we may recognize an inverse Fourier transform. So,

$$\Psi(\kappa,t) = \int_{-\infty}^{\infty} \psi(x,t)\mathrm{e}^{-\imath\kappa x}\,\mathrm{d}x \tag{5.61}$$

where $\Psi(\kappa,t) = \Psi_0(\kappa)\mathrm{e}^{\imath\omega(\kappa)t}$. In particular at $t = 0$, $\Psi_0(\kappa) = \Psi(\kappa,0)$ is merely the Fourier transform of the initial shape $\psi(x,0)$.

Reality condition

In physics, the field ψ is generally real-valued (excepted in quantum mechanics) but the integral eqn (5.60) is complex-valued. However, eqn (5.60) may remain real provided that $\Psi_0(\kappa)$ and $\omega(\kappa)$ verify two conditions.

Let us assume that ψ is real and let us substitute $\overline{\psi}(x,t) = \psi(x,t)$ into the conjugate of eqn (5.61).

$$\overline{\Psi}(\kappa,t) = \int_{-\infty}^{\infty} \psi(x,t)\mathrm{e}^{\imath\kappa x}\,\mathrm{d}x = \Psi(-\kappa,t)$$

Thus $\kappa \mapsto \Psi(\kappa,t)$ is a Hermitian function. In particular,

$$\overline{\Psi}_0(\kappa)\mathrm{e}^{-\imath\overline{\omega}(\kappa)t} = \Psi_0(-\kappa)\mathrm{e}^{\imath\omega(-\kappa)t}$$

At $t = 0$,

$$\overline{\Psi}_0(\kappa) = \Psi_0(-\kappa) \tag{5.62}$$

At any later time,

$$-\overline{\omega}(\kappa) = \omega(-\kappa) \tag{5.63}$$

Thus ψ is real if and only if Ψ_0 is Hermitian and ω anti-Hermitian.

Time evolution

All information on time evolution of a wave packet is contained in the dispersion relationship $\omega(\kappa)$. If the field is assumed to be known at $t = 0$, then the field is known at any later time $t > 0$ by eqn (5.60). More precisely, we know that the spectrum $\Psi(\kappa, t)$ at $t > 0$ is obtained by multiplying this initial spectrum $\Psi(\kappa, 0)$ by $e^{\imath\omega(\kappa)t}$. The phase of each spectral component is therefore shifted by $\omega(\kappa)t$.

Let us examine the case of a nondispersive medium. The frequency is proportional to the wavenumber $\omega(\kappa) = c\kappa$ and $\Psi(x, t)$ becomes

$$\Psi(\kappa, t) = \Psi(\kappa, 0)e^{\imath c\kappa t}$$

By taking the inverse Fourier transform,

$$\psi(x, t) = \frac{1}{2\pi}\int_{-\infty}^{\infty} \Psi(\kappa, 0)e^{\imath\kappa(ct+x)}\, d\kappa = \psi(x + ct, 0) \tag{5.64}$$

Thus a wave packet propagates without distortion in a nondispersive medium. The field at time t is simply the initial state translated by ct (to the left in the present case).

In the general case of a dispersive medium, the phase shift may evolve in a more complicated way. The spectrum is warped, the distortion being more important in time. In the space domain, the shape $\psi(x, t)$ is also deformed and this deformation is entirely driven by the time-evolution of the phase shift $\omega(\kappa)t$.

Group speed

We have mentioned above that a wave packet is a perturbation of limited spatial extension. The movement is localized and consequently the vibrational energy is also localized. As time passes the disturbance zone moves and we may naturally raise the question of the propagation speed of this perturbed zone.

Let us decompose $\Psi_0 = Ae^{\imath\varphi}$ and $\omega = \omega_r + \imath\omega_i$ where A, φ, ω_r, and ω_i are real and let $g(\kappa) = Ae^{-\omega_i t}$ and $\Phi(\kappa) = \omega_r t + \kappa x + \varphi$ where t is considered as a parameter. Equation (5.60) becomes

$$\psi(x,t) = \frac{1}{2\pi}\int_{-\infty}^{\infty} g(\kappa)e^{\imath\Phi(\kappa)}\,d\kappa \tag{5.65}$$

The wave packet is represented as a sum of vectors $ge^{\imath\Phi}$ in the complex plane.

This type of integral is evaluated by the stationary phase method (Bleistein and Handelsman, 1975, p. 219). One compares the rate of evolution of $\kappa \mapsto g(\kappa)$ with that of $\kappa \mapsto \Phi(\kappa)$ in the neighbourhood of a point κ_0. Two extreme cases are to be considered.

In the first case $\Phi(\kappa)$ varies rapidly. More precisely, $\kappa \mapsto g(\kappa)$ varies slowly when $\Phi(\kappa)$ changes by 2π. The vectors $ge^{\imath\Phi}$ have almost the same magnitude but their direction is indifferent as shown in Fig. 5.7(a). The integral becomes

$$g(\kappa_0)\int e^{\imath\Phi(\kappa)}\,d\kappa \sim 0 \tag{5.66}$$

and the sum of all vectors is zero or very small.

In the second case $\Phi(\kappa)$ is stationary at κ_0 and therefore $\Phi(\kappa)$ varies slowly with κ. Since $\kappa \mapsto \Phi(\kappa)$ is almost constant, say Φ_0, all vectors $ge^{\imath\Phi_0}$ have same direction. Their sum

$$e^{\imath\Phi_0}\int g(\kappa)\,d\kappa \neq 0 \tag{5.67}$$

is a large vector which contributes non-trivially to the integral value. The situation is shown in Fig. 5.7(b).

So, the only values of κ which contribute to the integral $\int ge^{\imath\Phi}\,d\kappa$ are the stationary points of $\kappa \mapsto \Phi(\kappa)$. The condition of stationary phase reads

$$\frac{d\Phi}{d\kappa} = \frac{d}{d\kappa}\left[\omega_r(\kappa)t - \kappa x + \varphi(\kappa)\right] = \frac{d\omega_r}{d\kappa}t - x + \frac{d\varphi}{d\kappa} = 0 \tag{5.68}$$

Since ω_r and φ do not depend on x and t, this gives a linear equation on x and t. The position x for which the phase Φ is stationary is

$$x(t) = \omega_r'(\kappa)t + \varphi'(\kappa) \tag{5.69}$$

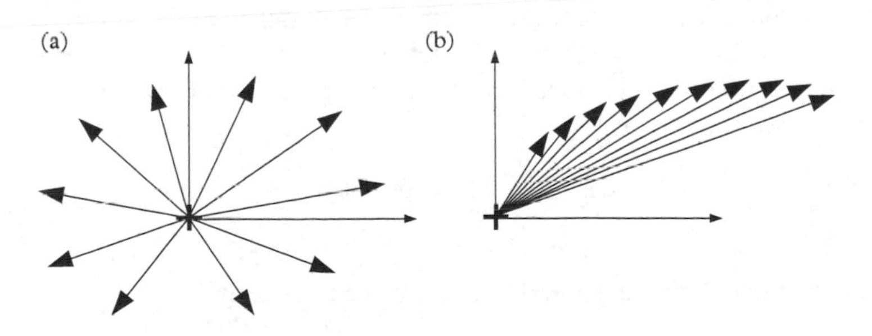

Figure 5.7 *Stationary phase method. (a) non-stationary phase; (b) stationary phase.*

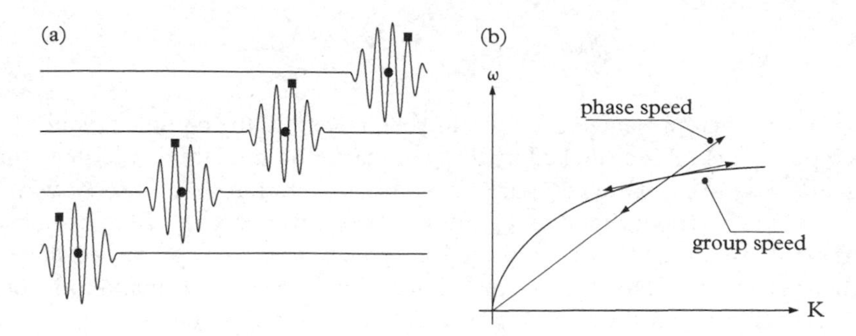

Figure 5.8 *(a) wave packet: the black circle moves at the group speed; the black square moves at the phase speed. (b) dispersion curve: the phase speed is the slope from origin; the group speed is the slope of the tangent.*

This point moves at a constant speed, called the **group speed**, defined by

$$c_g = \frac{\mathrm{d}\omega_r}{\mathrm{d}\kappa} = \mathrm{Re}\left(\frac{\mathrm{d}\omega}{\mathrm{d}\kappa}\right) \tag{5.70}$$

This is the speed of propagation of a wave packet whose energy is localized in the neighbourhood of a point. In quantum mechanics, the group speed is also the speed of the particle associated with the wave.

The difference between the group speed and the phase speed may be illustrated as follows. In Fig. 5.8(a) the time evolution of a wave packet is plotted. The phase speed is the speed of a wave crest (black square) while the group speed is the speed of the main disturbance (black circle). The corresponding relation $\omega(\kappa)$ is plotted in Fig. 5.8(b). At a given point on the curve, the phase speed is simply the slope ω/κ of a straight line from the origin to the point (κ, ω). The group speed is the slope of the tangent $\mathrm{d}\omega/\mathrm{d}\kappa$ at the same point. In the case of Fig. 5.8(a), the phase speed is greater than the group speed.

Uncertainty principle

In quantum mechanics $|\psi|^2$ corresponds to the probability of the presence of the particle at x. In classical mechanics, we may interpret $|\psi|^2$ as the local energy of the field. So, if we want to quantify the localization of energy, it is natural to define the mean position of the wave packet by

$$\langle x \rangle = \frac{\int x|\psi|^2\,\mathrm{d}x}{\int |\psi|^2\,\mathrm{d}x} \tag{5.71}$$

We may also define the spatial spread Δx of a wave packet by

$$\Delta x^2 = \frac{\int (x - \langle x \rangle)^2 |\psi|^2\,\mathrm{d}x}{\int |\psi|^2\,\mathrm{d}x} \tag{5.72}$$

In quantum mechanics Δx is the standard deviation of the particle position. In classical mechanics, Δx is the standard deviation of the localization of energy. If the wave packet has a limited spatial extension—that is the field ψ has a bounded support—then Δx is half the wave packet width.

Similarly, by taking the Fourier transform Ψ of ψ, we may introduce the mean wavenumber,

$$\langle\kappa\rangle = \frac{\int \kappa |\Psi|^2 \, d\kappa}{\int |\Psi|^2 \, d\kappa} \tag{5.73}$$

When ψ is real Ψ is Hermitian and therefore $\kappa \mapsto |\Psi|^2$ is even, hence $\langle\kappa\rangle = 0$. The spectral spread $\Delta\kappa$ of the wave packet is

$$\Delta\kappa^2 = \frac{\int (\kappa - \langle\kappa\rangle)^2 |\Psi|^2 \, d\kappa}{\int |\Psi|^2 \, d\kappa} \tag{5.74}$$

Then spatial spread and spectral spread are related by the following inequality,

$$\Delta x \times \Delta\kappa \geq \frac{1}{2} \tag{5.75}$$

This is the so-called **uncertainty principle**.

The uncertainty principle is a general limitation imposed on any Fourier pair. The more spread the signal is, the more concentrated is its spectral component and vice versa.

In quantum mechanics, Δx is the standard deviation of the particle position. And the wavenumber is related to the impulse by de Broglie's relation $p = \hbar\kappa$. The uncertainty inequality is then known as Heisenberg's principle $\Delta x \times \Delta p \geq \hbar/2$ which states the impossibility of measuring simultaneously the position and momentum of a particle with an arbitrary precision.

If an observer is at a fixed position x, the passage of a wave packet has a certain duration $\Delta t = \Delta x/c_g$ where c_g is the group speed. In addition if $\Delta\kappa$ is small enough, a first-order development gives $\Delta\omega = \omega'(\kappa)\Delta\kappa$ or $\Delta\omega = c_g\Delta\kappa$. The uncertainty inequality then takes the alternative form,

$$\Delta t \times \Delta\omega \geq \frac{1}{2} \tag{5.76}$$

The shorter the duration of a wave packet, the larger its frequency width and vice versa (Fig. 5.9). In quantum mechanics the energy of a particle is related to the frequency of the associated wave by $E = \hbar\omega$. The uncertainty principle then reads

$$\Delta t \times \Delta E \geq \frac{\hbar}{2} \tag{5.77}$$

It is also impossible to measure duration and energy accurately and simultaneously.

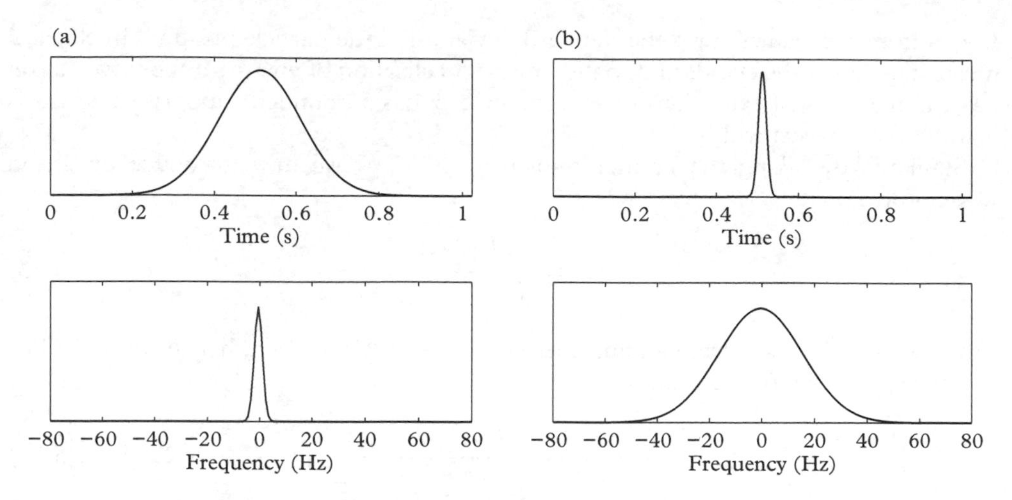

Figure 5.9 *The shorter the pulse, the wider the spectrum and vice versa. (a) long pulse and narrow spectrum. (b) short pulse and wide spectrum.*

We are going to prove a more general inequality in which $\langle x \rangle$ and $\langle \kappa \rangle$ may be replaced by any arbitrary real numbers a and b (Folland, 1983). The inequality to prove is

$$\frac{\int (x-a)^2 |\psi|^2 \, \mathrm{d}x}{\int |\psi|^2 \, \mathrm{d}x} \times \frac{\int (\kappa - b)^2 |\Psi|^2 \, \mathrm{d}\kappa}{\int |\Psi|^2 \, \mathrm{d}\kappa} \geq \frac{1}{4} \tag{5.78}$$

for any real a and b and any Fourier pair ψ and Ψ. We start from Schwarz's inequality,

$$\left| \int f\bar{g} \, \mathrm{d}x \right|^2 \leq \int |f|^2 \, \mathrm{d}x \times \int |g|^2 \, \mathrm{d}x \tag{5.79}$$

for any complex-valued functions f and g whose square is integrable. By setting $f = (x-a)\psi$ and $g = \partial_x \psi - \imath b \psi$, we get

$$\left| \int (x-a)\psi \left(\partial_x \overline{\psi} + \imath b \overline{\psi} \right) \mathrm{d}x \right|^2 \leq \int (x-a)^2 |\psi|^2 \, \mathrm{d}x \times \int |\partial_x \psi - \imath b \psi|^2 \, \mathrm{d}x \tag{5.80}$$

Furthermore, by integrating by parts,

$$\int (x-a) \left[\overline{\psi}(\partial_x \psi - \imath b \psi) + \psi(\partial_x \overline{\psi} + \imath b \overline{\psi}) \right] \mathrm{d}x = \left[(x-a)|\psi|^2 \right] - \int |\psi|^2 \, \mathrm{d}x \tag{5.81}$$

But for a concentrated wave packet, we may assume that $(x-a)|\psi|^2 \to 0$ when $|x| \to \infty$. Therefore the bracket vanishes and eqn (5.81) reduces to

$$\mathrm{Re} \int (x-a)\psi(\partial_x \overline{\psi} + \imath b \overline{\psi}) \, \mathrm{d}x = -\frac{1}{2} \int |\psi|^2 \, \mathrm{d}x \tag{5.82}$$

By squaring, remarking that $[\mathrm{Re}(z)]^2 \leq |z|^2$ for any complex number z, and combining with eqn (5.80),

$$\frac{1}{4}\left(\int |\psi|^2 \,\mathrm{d}x\right)^2 \leq \int (x-a)^2 |\psi|^2 \,\mathrm{d}x \times \int |\partial_x \psi - \imath b\psi|^2 \,\mathrm{d}x \tag{5.83}$$

Furthermore, by the Parseval equality,

$$\int |\psi|^2 \,\mathrm{d}x = \frac{1}{2\pi}\int |\Psi|^2 \,\mathrm{d}\kappa \tag{5.84}$$

Since the Fourier transform of $\partial_x \psi - \imath b\psi$ is $\imath(\kappa - b)\Psi$, again by the Parseval equality,

$$\int |\partial_x \psi - \imath b\psi|^2 \,\mathrm{d}x = \frac{1}{2\pi}\int (\kappa - b)^2 |\Psi|^2 \,\mathrm{d}\kappa \tag{5.85}$$

Substituting eqns (5.84) and (5.85) into eqn (5.83) leads to the expected inequality. This completes the proof.

Gaussian wave packet

A real Gaussian wave packet has the initial form

$$\psi(x,0) = \mathrm{e}^{\frac{-x^2}{4\sigma_0^2}} \times \cos(\kappa_0 x) \tag{5.86}$$

where κ_0, σ_0 are two constants. We may also introduce the complex Gaussian wave packet whose initial form is

$$\psi_c(x,0) = \mathrm{e}^{\frac{-x^2}{4\sigma_0^2}} \times \mathrm{e}^{\imath\kappa_0 x} \tag{5.87}$$

By applying eqn (5.60), we shall calculate the full time evolution of the complex Gaussian wave packet whose initial state is $\psi_c(x,0)$. It is clear that by taking the real part of this complex solution, we shall obtain a real solution to the governing equation whose initial state is $\mathrm{Re}\,[\psi_c(x,0)] = \psi(x,0)$.

The Fourier transform of the initial state is

$$\Psi_0(\kappa) = \int_{-\infty}^{\infty} \mathrm{e}^{\frac{-x^2}{4\sigma_0^2}} \times \mathrm{e}^{\imath\kappa_0 x} \times \mathrm{e}^{-\imath\kappa x} \,\mathrm{d}x = \int_{-\infty}^{\infty} \mathrm{e}^{\frac{-1}{4\sigma_0^2}\left[x^2 + 4\imath\sigma_0^2(\kappa-\kappa_0)x\right]} \,\mathrm{d}x \tag{5.88}$$

But,

$$x^2 + 4\imath\sigma_0^2(\kappa - \kappa_0)x = \left[x + 2\imath\sigma_0^2(\kappa - \kappa_0)\right]^2 + 4\sigma_0^4(\kappa - \kappa_0)^2 \tag{5.89}$$

and the integral becomes

$$\Psi_0(\kappa) = \int_{-\infty}^{\infty} \mathrm{e}^{-\frac{1}{4\sigma_0^2}\left[x+2\imath\sigma_0^2(\kappa-\kappa_0)\right]^2} \mathrm{d}x \times \mathrm{e}^{-\sigma_0^2(\kappa-\kappa_0)^2} \tag{5.90}$$

In Appendix C, we calculate the integral,

$$\int_{-\infty}^{\infty} \mathrm{e}^{-\alpha^2(x+\beta)^2}\,\mathrm{d}x = \frac{\sqrt{\pi}}{\alpha} \tag{5.91}$$

if $\mathrm{Re}\left(\alpha^2\right) > 0$. So, letting $\alpha = 1/2\sigma_0 > 0$ and $\beta = 2\imath\sigma_0^2(\kappa - \kappa_0)$ yields

$$\Psi_0(\kappa) = 2\sigma_0\sqrt{\pi}\,\mathrm{e}^{-\sigma_0^2(\kappa-\kappa_0)^2} \tag{5.92}$$

At a later time $t > 0$, the field $\psi_c(x, t)$ is given by eqn (5.60). We have

$$\psi_c(x,t) = \frac{\sigma_0}{\sqrt{\pi}} \int_{-\infty}^{\infty} \mathrm{e}^{-\sigma_0^2(\kappa-\kappa_0)^2}\,\mathrm{e}^{\imath(\omega(\kappa)t+\kappa x)}\,\mathrm{d}\kappa \tag{5.93}$$

The integral (5.93) cannot be calculated in so general a form since it depends on the dispersion relation $\omega(\kappa)$, which we have not yet specified. However, it is of interest to consider the special case where the dispersion relation takes the form

$$\omega(\kappa) = \omega_0 + \omega_0'(\kappa - \kappa_0) + \frac{1}{2}\omega_0''(\kappa - \kappa_0)^2 \tag{5.94}$$

Substituting the latter into eqn (5.93) and re-arranging the terms gives

$$\psi_c(x,t) = \frac{\sigma_0}{\sqrt{\pi}}\mathrm{e}^{\imath(\omega_0 t+\kappa_0 x)} \int_{-\infty}^{\infty} \mathrm{e}^{-\sigma_0^2(\kappa-\kappa_0)^2}\,\mathrm{e}^{\imath\omega_0' t(\kappa-\kappa_0)+\frac{\imath}{2}\omega_0'' t(\kappa-\kappa_0)^2+\imath(\kappa-\kappa_0)x}\,\mathrm{d}\kappa \tag{5.95}$$

But

$$\left(\tfrac{\imath}{2}\omega_0'' t - \sigma_0^2\right)(\kappa-\kappa_0)^2 + \imath(\omega_0' t + x)(\kappa-\kappa_0) = \left(\tfrac{\imath}{2}\omega_0'' t - \sigma_0^2\right)\left[(\kappa-\kappa_0) + \frac{\imath(\omega_0' t+x)}{\imath\omega_0'' t - 2\sigma_0^2}\right]^2 - \frac{(\omega_0' t+x)^2}{4\sigma_0^2 - 2\imath\omega_0'' t} \tag{5.96}$$

so that we can get the argument that does not depend on κ out of the integral

$$\psi_c(x,t) = \frac{\sigma_0}{\sqrt{\pi}}\mathrm{e}^{\imath(\omega_0 t+\kappa_0 x)} \int_{-\infty}^{\infty} \mathrm{e}^{-\left(\sigma_0^2-\frac{\imath}{2}\omega_0'' t\right)\left[(\kappa-\kappa_0)+\frac{\imath(\omega_0' t+x)}{\imath\omega_0'' t-2\sigma_0^2}\right]^2}\,\mathrm{d}\kappa \times \mathrm{e}^{-\frac{(\omega_0' t+x)^2}{4\sigma_0^2-2\imath\omega_0'' t}} \tag{5.97}$$

The integral now takes the standard form (5.91) with

$$\alpha^2 = \sigma_0^2 - \frac{\imath}{2}\omega_0'' t, \quad \beta = -\kappa_0 + \frac{\imath(\omega_0' t + x)}{\imath\omega_0'' t - 2\sigma_0^2} \tag{5.98}$$

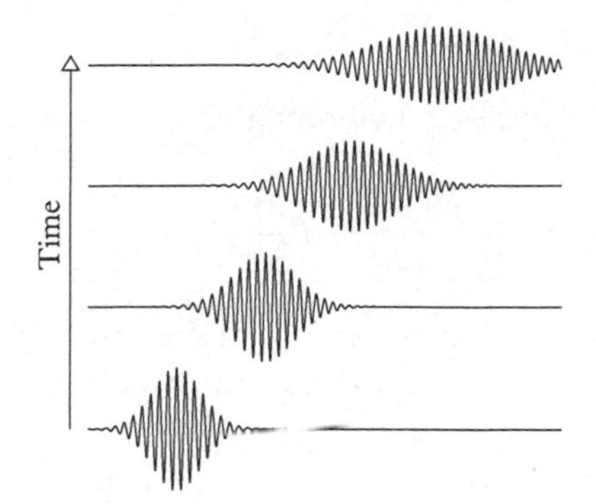

Figure 5.10 *Time evolution of a Gaussian wave packet.*

Let

$$\sigma(t) = \sqrt{\sigma_0^2 - \imath\omega_0'' t/2} \tag{5.99}$$

be the square root whose real part is positive. By applying eqn (5.91), we get

$$\psi_c(x,t) = \frac{\sigma_0}{\sigma(t)} e^{\imath(\omega_0 t + \kappa_0 x)} \times e^{-\frac{(\omega_0' t + x)^2}{4\sigma(t)^2}} \tag{5.100}$$

and subsequently,

$$\psi(x,t) = \mathrm{Re}\left(\frac{\sigma_0}{\sigma(t)} e^{\imath(\omega_0 t + \kappa_0 x)} \times e^{-\frac{(\omega_0' t + x)^2}{4\sigma(t)^2}}\right) \tag{5.101}$$

These equations give the time evolution of complex and real wave packets in a system whose dispersion relationship has the form (5.94). In Fig. 5.10 a typical time-evolution of a wave packet is shown. The rapid oscillation is controlled by the wavenumber κ_0 while the spatial extent is of an order of σ_0 at initial state. As time passes, the disturbance goes to the right and the spatial extent increases.

5.4 Energy Transport

We now turn to the question of energy involved in wave propagation. Two important features of waves are introduced in this section. The first one is the equality of local kinetic and elastic energies, a result encountered in Chapter 2 in another form. The second feature is the proportionality of energy flow and local vibrational energy. The proportionality constant will be found to be the group speed.

Travelling wave in strings

As a simple example, let us consider a vibrating string whose governing equation is

$$m\frac{\partial^2 u}{\partial t^2} - T\frac{\partial^2 u}{\partial x^2} = 0 \tag{5.102}$$

where m is the mass per unit length and T the tension. This is the wave equation with speed $c = (T/m)^{1/2}$. The kinetic energy density $K(x, t)$ and elastic energy density $V(x, t)$ are

$$K(x,t) = \frac{1}{2}m\dot{u}^2, \quad V(x,t) = \frac{1}{2}Tu'^2 \tag{5.103}$$

and the energy density $W(x, t) = K(x, t) + V(x, t)$. Furthermore, the intensity defined as the rate of energy travelling in a string is

$$I(x,t) = -Tu'\dot{u} \tag{5.104}$$

Both energy density and intensity characterize the local exchange of energy. The local power balance is

$$\frac{\partial I}{\partial x} + \frac{\partial W}{\partial t} = 0 \tag{5.105}$$

This relationship may be checked by substituting the definitions of W and I and applying the governing equation.

We now consider a single wave $u(x, t) = f(x-ct)$ where f is arbitrary. The local kinetic energy is

$$K(x,t) = \frac{1}{2}m\left[\frac{\partial}{\partial t}f(x-ct)\right]^2 = \frac{1}{2}mc^2 f'^2(x-ct) \tag{5.106}$$

and the local elastic energy is

$$V(x,t) = \frac{1}{2}T\left[\frac{\partial}{\partial x}f(x-ct)\right]^2 = \frac{1}{2}Tf'^2(x-ct) \tag{5.107}$$

From $mc^2 = T$, we get

$$K(x,t) = V(x,t) \tag{5.108}$$

So, a travelling wave in a string verifies the equality of local kinetic and elastic energies. This equality holds at any time t and any position x.

The second property concerns the intensity. It yields

$$I(x,t) = -T\frac{\partial}{\partial x}f(x-ct)\frac{\partial}{\partial t}f(x-ct) = Tcf'^2(x-ct) \quad (5.109)$$

Comparing this with the previous expressions of local energies yields

$$I(x,t) = cW(x,t) \quad (5.110)$$

For a travelling wave in a string, the intensity is proportional to the vibrational energy density. The proportionality constant is the sound speed. Again, this equality holds at any time t and any position x.

Time-average quadratic quantity

Before discussing a more general form of the proportionality $I = cW$, we need to introduce the complex notation associated with quadratic quantities such as energy and power.

If a harmonic wave propagates in a medium, the field of disturbance has the form $\mathrm{Re}\,[Ae^{\iota\omega t}]$ and $Ae^{\iota\omega t}$ is the associated complex wave. For instance $Ve^{\iota\omega t}$ denotes the velocity of a point of the medium and $Fe^{\iota\omega t}$ the internal force at that point. The instantaneous power is

$$P(t) = \mathrm{Re}\left[Fe^{\iota\omega t}\right].\mathrm{Re}\left[Ve^{\iota\omega t}\right] \quad (5.111)$$

With this time-dependent quantity we associate the complex number

$$\Pi = \frac{1}{2}F\overline{V} \quad (5.112)$$

More generally, if A and B are two complex numbers denoting respectively a harmonic force and velocity or a stress and strain rate, we associate the complex number $1/2 \times A\overline{B}$ with the time-dependent quantity $\mathrm{Re}\,[Ae^{\iota\omega t}].\mathrm{Re}\,[Be^{\iota\omega t}]$. The special case $A = B$ applies for instance to the kinetic energy $1/4 \times mV\overline{V}$ or other forms of local energies. The real part $\mathrm{Re}\,[\Pi]$ of the complex power is called **active power** while the imaginary part $\mathrm{Im}\,[\Pi]$ is called **reactive power**.

Let us write $F = |F|e^{\iota\varphi}$ and $V = |V|e^{\iota\psi}$ in polar coordinates; one gets

$$\begin{aligned} P(t) &= |F|\cos(\omega t + \varphi).|V|\cos(\omega t + \psi) \\ &= \frac{1}{2}|F|.|V|\,[\cos(\varphi - \psi) + \cos(2\omega t + \varphi + \psi)] \end{aligned}$$

Thus, the instantaneous power is the sum of a constant value and a periodic function whose frequency is twice the harmonic frequency. This is also the case for all quadratic quantities. The complex power Π may therefore be interpreted as follows. The real part or active power $P = \text{Re}(\Pi)$ is the time-averaged instantaneous power,

$$P = \frac{1}{2}|F|.|V|\cos(\varphi - \psi) \tag{5.113}$$

More generally, the real part of a product $1/2 \times A\overline{B}$ may always be interpreted as the time average of the underlying instantaneous quantity. While the time fluctuations of the instantaneous power about the mean value have magnitude,

$$\sqrt{P^2 + Q^2} = \frac{1}{2}|F|.|V| \tag{5.114}$$

where $Q = \text{Im}(\Pi)$ is the reactive power (Fig. 5.11). So, the minimal value of the instantaneous power is $P - (P^2 + Q^2)^{1/2}$. If the reactive power is non-null then the power alternately flows in both directions but with a preferred direction, like the ebb and flow of sea waves during the rising of the tide.

Harmonic wave in beams

As a second example, consider a transversely vibrating beam whose governing equation is

$$m\frac{\partial^2 u}{\partial t^2} + EI\frac{\partial^4 u}{\partial x^4} = 0 \tag{5.115}$$

where m is the mass per unit length and EI the bending stiffness. The kinetic and elastic energy densities are

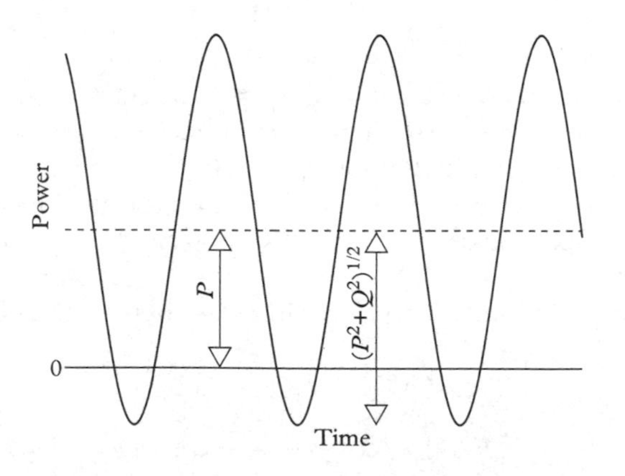

Figure 5.11 *Time evolution of power. The mean value is the active power P and the magnitude is* $(P^2 + Q^2)^{1/2}$ *where Q is the reactive power.*

$$K(x,t) = \frac{1}{2}m\dot{u}^2, \quad V(x,t) = \frac{1}{2}EIu''^2 \tag{5.116}$$

and of course, the vibrational energy density is the sum $W(x,t) = K(x,t) + V(x,t)$. The intensity defined as the rate of energy crossing a sectional area of the beam is

$$I(x,t) = EI\left(u'''\dot{u} - u''\dot{u}'\right) \tag{5.117}$$

Energy density and intensity are linked by the local power balance,

$$\frac{\partial I}{\partial x} + \frac{\partial W}{\partial t} = 0 \tag{5.118}$$

A harmonic propagating wave in beams has the form $u(x,t) = u_0 e^{\iota(\omega t - \kappa x)}$. When applied to the above equations, we obtain successively the time-averaged kinetic energy density,

$$K = \frac{1}{4}m\dot{u}\bar{\dot{u}} = \frac{1}{4}m\omega^2|u_0|^2 \tag{5.119}$$

the time-averaged elastic energy density,

$$V = \frac{1}{4}EIu''\overline{u''} = \frac{1}{4}EI\kappa^4|u_0|^2 \tag{5.120}$$

and the active intensity,

$$I = \frac{1}{2}EI \times \mathrm{Re}\left[u'''\bar{\dot{u}} - u''\overline{\dot{u}'}\right] = EI\kappa^3\omega|u_0|^2 \tag{5.121}$$

We may observe from the preceding three equations that time-average energies and intensity are constant in space. The energy is uniformly spread and flows with a constant rate. The dispersion equation $m\omega^2 = EI\kappa^4$ shows that kinetic and elastic energies are equal,

$$K = V \tag{5.122}$$

Furthermore, it is clear by combining eqns (5.119), (5.120), and (5.121) that I and $W = K + V$ are proportional,

$$I = c_g W \tag{5.123}$$

where $c_g = 2\omega/\kappa$. Thus, in accordance with eqn (5.110), time-average intensity and energy density are proportional at any position but the constant proportionality is shown to be the **group speed**. The group speed is therefore the speed of propagation of energy.

Wave packet in beams

The results of the previous subsection may be straightforwardly generalized to wave packet propagation.

We found in eqn (5.101) the time evolution of a Gaussian wave packet,

$$u(x,t) = \frac{\sigma_0}{\sigma(t)} e^{\imath(\omega_0 t + \kappa_0 x)} \times e^{-\frac{(\omega_0' t + x)^2}{4\sigma(t)^2}} \tag{5.124}$$

where $\sigma(t) = (\sigma_0^2 - \imath\omega_0'' t/2)^{1/2}$ and ω_0, ω_0', and ω_0'' are derivatives of Taylor's expansion of the dispersion relationship $\omega(\kappa)$ in the neighbourhood of κ_0. A Gaussian wave packet may be decomposed as a product $u(x,t) = u_0(x,t)e^{\imath(\omega_0 t - \kappa_0 x)}$ where $u_0(x,t)$ is the shape function and $e^{\imath(\omega_0 t - \kappa_0 x)}$ the harmonic component. Equation (5.124) shows that the shape function is

$$u_0(x,t) = \frac{\sigma_0}{\sigma(t)} e^{-\frac{(\omega_0' t + x)^2}{4\sigma(t)^2}} \tag{5.125}$$

Let us assume that the shape function varies slowly compared with fluctuations of the harmonic component,

$$\left|\frac{\partial u_0}{\partial t}\right| \ll \omega_0 |u_0|, \quad \left|\frac{\partial u_0}{\partial x}\right| \ll \kappa_0 |u_0| \tag{5.126}$$

In view of the shape function, this condition is satisfied if $\sigma_0 \gg 2\pi/\kappa_0$. The harmonic component wavelength is much smaller than the spatial extent of the wave packet. Since the spatial extent of a wave packet must be smaller than the typical size of the medium in which it propagates, this condition may be formulated as a high-frequency condition which will constitute the first assumption of the wave approach of statistical energy analysis:

Assumption 7 *The frequency is high.*

In other words, we adopt the geometrical optics framework.

According to eqn (5.126), the spatial and time derivatives of u_0 may always be neglected compared with derivatives of the harmonic component. We then arrive at the following rule. When deriving u, the derivative only applies to u_0. For instance if we calculate the time-average energies of a wave packet in a beam by eqns (5.119) and (5.120),

$$K(x,t) = \frac{1}{4} m\omega_0^2 |u_0(x,t)|^2 \tag{5.127}$$

and

$$V(x,t) = \frac{1}{4} EI\kappa_0^4 \, |u_0(x,t)|^2 \tag{5.128}$$

Again, by the dispersion equation $m\omega_0^2 = EI\kappa_0^4$ we obtain the equality of time-average kinetic and elastic energies,

$$K(x,t) = V(x,t) \tag{5.129}$$

Similarly, the active intensity calculated by eqn (5.121) gives

$$I(x,t) = EI\kappa_0^3 \omega_0 \, |u_0(x,t)|^2 \tag{5.130}$$

which compared to the total energy density $W(x,t) = K(x,t) + V(x,t)$ leads to the following proportionality,

$$I(x,t) = c_g W(x,t) \tag{5.131}$$

where $c_g = 2\omega_0/\kappa_0$ is the group velocity.

Remember that the two equalities $K = V$ and $I = c_g W$ are now valid locally but subject to the high-frequency assumption.

The proportionality of the energy flow with the energy density has been introduced here in the special case of waves in beams and strings. However, the result is quite general, applying for instance in electromagnetism and elastodynamics. The fact that the group velocity is also the speed of energy has been recognized for a long time. Interested readers may consult Biot (1957) and Brillouin (1960).

6

Modal Densities

The modal density is a statistical property of systems which can be determined by calculating the exact sequence of eigenfrequencies. This sequence depends on the wave type, geometry, and boundary conditions, and is usually a complex problem. Nevertheless, when the number of modes is sufficiently high and/or when the complexity of geometry induces a disorder in the eigenfrequency sequence, some asymptotic relationships are available. Surprisingly, the asymptotic modal density depends on relatively few parameters. This is what we shall explore in this chapter.

6.1 Definition

Modal density is defined as the number of modes per unit frequency. More precisely, if we consider a frequency band $[\omega_{\min}, \omega_{\max}]$ of span $\Delta\omega = \omega_{\max} - \omega_{\min}$ containing ΔN modes, the modal density, denoted by $n(\omega)$, is

$$n(\omega) = \frac{\Delta N}{\Delta\omega} \tag{6.1}$$

Since the angular frequency ω has unit rad/s, the physical unit of modal density is second. Other authors have adopted the alternative definition

$$n(f) = \frac{\Delta N}{\Delta f} \tag{6.2}$$

where f is the cyclic frequency in Hz. Obviously, $n(f) = 2\pi\, n(\omega)$.

The modal density is a statistical property of systems which has a meaning only when the number of modes is high enough in the frequency band of interest. The inverse of the modal density $\delta\bar{\omega} = 1/n$ is the mean frequency span between successive modes. It plays the role of a representative elementary volume for the calculation of 'effective' characteristics in homogenization theory. In statistical energy analysis, $\delta\bar{\omega}$ is the smallest frequency resolution for which the theory makes sense.

6.2 Analytic Modal Densities

In this section, we study some examples of systems for which the sequence of natural frequencies is exactly known. Except for strings for which we give all details of the proof, the sequences of natural frequencies are discussed in outline. The proofs may be found in all standard books (Meirovitch, 1967; Graff, 1991; Soedel, 2004). A large number of formulae is summarized by Blevins (1979). This approach of deriving mode count and modal densities from exact natural frequency sequences has been followed by Xie et al. (2002a, 2002b).

String

The normal modes of a string are the eigenvectors of the operator $d^2./dx^2$ in the linear space of functions $x \mapsto \psi(x)$ null at $x = 0$ and $x = L$. They solve the problem

$$-T\frac{d^2\psi}{dx^2} = \lambda\psi \tag{6.3}$$

under the prescribed boundary conditions

$$\psi(0) = \psi(L) = 0 \tag{6.4}$$

The constant λ is the eigenvalue attached to the eigenvector ψ.

It is easy to see that the case $\lambda \leq 0$ leads to the only trivial solution $\psi(x) = 0$. When $\lambda = 0$, then $\psi = Ax + B$ where the constants A and B are determined by the boundary conditions. But $\psi(0) = 0$ imposes $B = 0$ and $\psi(L) = 0$ gives $A = 0$. Hence $\psi = 0$. Likewise when $\lambda < 0$, then $\psi = Ae^{-\kappa x} + Be^{\kappa x}$ where $\kappa = |\lambda/T|^{1/2}$. Again, the conditions $\psi(0) = \psi(L) = 0$ lead to $A = B = 0$ and so $\psi = 0$. Therefore, no eigenvector having non-positive eigenvalue exists.

The case $\lambda > 0$ allows a solution of the form $\psi = A\sin(\kappa x) + B\cos(\kappa x)$. The condition $\psi(0) = 0$ gives $B = 0$ while $\psi(L) = 0$ imposes $A\sin(\kappa L) = 0$. A non-trivial solution ψ exists only if $\kappa L = i\pi$ where i is integer.

Thus, the normal modes are

$$\psi_i(x) = \sin\left(i\frac{\pi}{L}x\right), \qquad i = 1, 2, \ldots \tag{6.5}$$

and their eigenvalues are $\lambda_i = T(i\pi/L)^2$. The sequence of natural frequencies is obtained by solving $m\omega_i^2 = \lambda_i$. We get

$$\omega_i = i\frac{\pi}{L}\sqrt{\frac{T}{m}}, \qquad i = 1, 2, \ldots \tag{6.6}$$

The frequency span between two successive modes is $\delta\omega = \pi/L \times (T/m)^{1/2}$. This is a constant and therefore the number of modes below ω is given by $N(\omega) = [\omega/\delta\omega]$ where

$[x]$ denotes the integer part of the real x. With an uncertainty of ± 1, we have

$$N(\omega) = \omega \frac{L}{\pi} \sqrt{\frac{m}{T}} \tag{6.7}$$

By deriving with respect to ω, we get the modal density,

$$n(\omega) = \frac{L}{\pi c} \tag{6.8}$$

where $c = (m/T)^{1/2}$ is the sound speed. The modal density of a string is constant.

Simply supported beam

Let us study a second simple example in one dimension which leads to a different result. The normal modes of a beam are the eigenvectors of the operator $EI\mathrm{d}^4./\mathrm{d}x^4$. In the case of a simply supported beam of length L they are functions ψ such that

$$EI \frac{\mathrm{d}^4 \psi}{\mathrm{d}x^4} = \lambda \psi \tag{6.9}$$

under the prescribed boundary conditions

$$\psi(0) = \psi''(0) = \psi(L) = \psi''(L) = 0 \tag{6.10}$$

They are (Soedel, 2004, p. 80)

$$\psi_i(x) = \sin\left(i \frac{\pi}{L} x\right), \qquad i = 1, 2, \ldots \tag{6.11}$$

where the eigenvalue is $\lambda_i = EI(i\pi/L)^4$. The sequence of natural frequencies $m\omega_i^2 = \lambda_i$ becomes

$$\omega_i = i^2 \frac{\pi^2}{L^2} \sqrt{\frac{EI}{m}}, \qquad i = 1, 2, \ldots \tag{6.12}$$

The number of modes below ω is the largest integer N such as $\omega_N \leq \omega$. Approximately,

$$N(\omega) = \frac{L}{\pi} \left(\frac{m}{EI}\right)^{1/4} \sqrt{\omega} \tag{6.13}$$

Then, by deriving with respect to ω,

$$n(\omega) = \frac{L}{2\pi} \left(\frac{m}{EI}\right)^{1/4} \frac{1}{\sqrt{\omega}} \tag{6.14}$$

We can observe that the modal density of a beam is a decreasing function of the frequency.

By introducing the group velocity $c_g = 2(EI/m)^{1/4}\sqrt{\omega}$,

$$n(\omega) = \frac{L}{\pi c_g} \tag{6.15}$$

This expression is quite similar to eqn (6.8) except that the speed appearing in the above relationship is the group speed.

Clamped beam

We now consider a slightly different problem which will result in the same modal density but with a further approximation. The beam is clamped at both ends. The spectral problem is now

$$EI\frac{d^4\psi}{dx^4} = \lambda\psi \tag{6.16}$$

with the boundary conditions

$$\psi(0) = \psi'(0) = \psi(L) = \psi'(L) = 0 \tag{6.17}$$

The mode shapes have the form

$$\psi(x) = A(\cos\kappa x - \cosh\kappa x) + B(\sin\kappa x - \sinh\kappa x) \tag{6.18}$$

where A and B are the solution to the system,

$$\begin{pmatrix} \cos\kappa L - \cosh\kappa L & \sin\kappa L - \sinh\kappa L \\ \sin\kappa L + \sinh\kappa L & -\cos\kappa L + \cosh\kappa L \end{pmatrix}\begin{pmatrix} A \\ B \end{pmatrix} = \begin{pmatrix} 0 \\ 0 \end{pmatrix} \tag{6.19}$$

A non-trivial solution exists only if the determinant is zero,

$$\cos\kappa L \cosh\kappa L = 1 \tag{6.20}$$

The associated eigenvalues are $\lambda_i = EI\kappa_i^4$ where κ_i are the successive solutions of eqn (6.20).

The solutions of eqn (6.20) may be determined numerically. The first solutions are $\kappa_1 L = 4.730$, $\kappa_2 L = 7.853$, $\kappa_3 L = 10.996$, $\kappa_4 L = 14.137$. For larger orders, the approximation $\cosh\kappa L \gg 1$ leads to simplifying eqn (6.20) to $\cos\kappa L = 0$ whose solutions are $\kappa_i L = (2i + 1)\pi/2$. This asymptotic solution gives three digits for $i = 4$ and is indeed better for larger i (Fig. 6.1).

We have therefore determined the sequence of natural frequencies $m\omega_i^2 = \lambda_i$ as

$$\omega_i = \left(i + \frac{1}{2}\right)^2 \frac{\pi^2}{L^2}\sqrt{\frac{EI}{m}}, \qquad \text{for } i \geq 4 \tag{6.21}$$

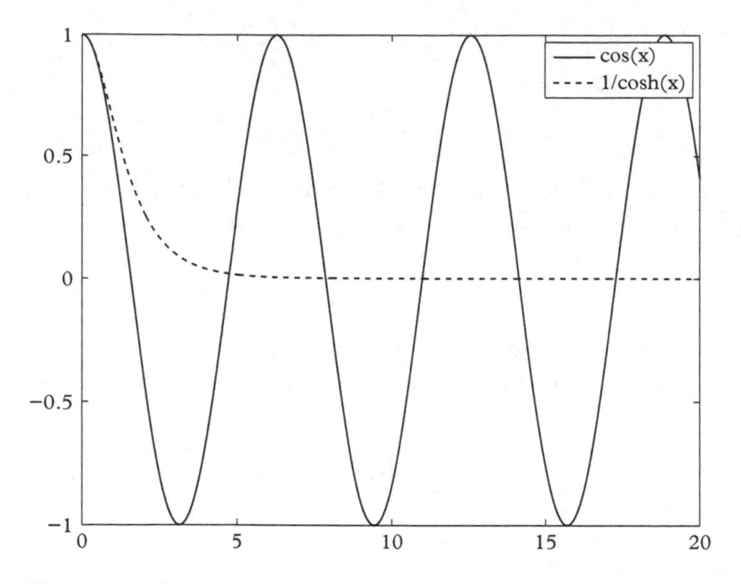

Figure 6.1 *Solutions of cos x cosh x = 1.*

The number of modes below ω is

$$N(\omega) = \frac{L}{\pi}\left(\frac{m}{EI}\right)^{1/4}\sqrt{\omega} - \frac{1}{2} \tag{6.22}$$

Again by deriving with respect to ω, one obtains

$$n(\omega) = \frac{L}{\pi c_g} \tag{6.23}$$

where $c_g = 2(EI/m)^{1/4}\sqrt{\omega}$ is the group velocity. It is now apparent that eqn (6.23), which is valid for simply supported and clamped boundary conditions, will also hold for other boundary conditions. But generally, eqn (6.23) is not exact but gives a reasonable value as soon as ω is sufficiently large. This is therefore an asymptotic approximation of the modal density.

Rectangular membrane

Let us now study a two-dimensional example. For a rectangular membrane of dimension $L_x \times L_y$ with fixed edges $x = 0, L_x$ and $y = 0, L_y$, the normal modes verify

$$-T\left(\frac{\partial^2 \psi}{\partial x^2} + \frac{\partial^2 \psi}{\partial y^2}\right) = \lambda\psi \tag{6.24}$$

with

$$\psi(0,y) = \psi(L_x,y) = \psi(x,0) = \psi(x,L_y) = 0 \tag{6.25}$$

for any x, y. In the above T is the tension per unit length of the membrane.

The normal modes are

$$\psi_{ij}(x,y) = \sin\left(i\frac{\pi}{L_x}x\right)\sin\left(j\frac{\pi}{L_y}y\right), \qquad i,j = 1,2,\ldots \tag{6.26}$$

and the eigenvalues are

$$\lambda_{ij} = T\left[\left(i\frac{\pi}{L_x}\right)^2 + \left(j\frac{\pi}{L_y}\right)^2\right] \tag{6.27}$$

The natural frequencies $m\omega_{ij}^2 = \lambda_{ij}$ therefore depend on two integers i, j,

$$\omega_{ij} = \pi c\sqrt{\frac{i^2}{L_x^2} + \frac{j^2}{L_y^2}} \qquad i,j = 1,2,\ldots \tag{6.28}$$

where $c = (T/m)^{1/2}$ is the sound speed. For any couple of integers i, j correspond to one and only one mode. But in certain circumstances, such as with a square membrane, it may transpire that two different couples of integers have the same natural frequency. We then have a degenerated mode.

Circular membranes

We now consider a circular membrane of radius R. The edge of the membrane is fixed. Then, the sequence of natural frequencies ω_{ij} depends on two integers i and j which vary from 0 to infinity. For any fixed value of i, the transcendental equation (Meirovitch, 1967),

$$\mathcal{J}_i(\lambda) = 0 \tag{6.29}$$

where $\mathcal{J}_i$ denotes the Bessel function of the first kind and order i, admits a sequence λ_{ij}, $j = 1, 2, \ldots$ of solutions. The natural frequencies ω_{ij} are then related to the frequency parameter λ_{ij} by

$$\omega_{ij} = \frac{\lambda_{ij}c}{R}, \qquad i = 0,1,\ldots \quad j = 1,2,\ldots \tag{6.30}$$

Only one mode corresponds to the solutions ω_{0j}. But when $i \neq 0$, there are two modes for each value of j. Modes are degenerated.

Rectangular plates

A rectangular plate of dimension $L_x \times L_y$ with simply supported edges has natural frequencies which follow the law,

$$\omega_{ij} = \pi^2 \sqrt{\frac{D}{m}} \left(\frac{i^2}{L_x^2} + \frac{j^2}{L_y^2} \right), \qquad i,j = 1,2,\ldots \tag{6.31}$$

For any couple of integers i,j correspond to one and only one mode. As for rectangular membranes, some modes can be degenerated.

Circular plates

Let us now turn to the case of clamped circular plates. The plate has radius R and as usual bending stiffness D and mass per unit area m. Then for any integer $i \geq 0$, the solutions to the transcendental equation (Graff, 1991, p. 252)

$$I_i(\lambda)\mathcal{J}_i'(\lambda) - \mathcal{J}_i(\lambda)I_i'(\lambda) = 0 \tag{6.32}$$

give a sequence of frequency parameters λ_{ij}. In the above, $\mathcal{J}_i$ and I_i are respectively the Bessel function and modified Bessel function of first kind and order i and $\mathcal{J}_i'$ and I_i' are their derivatives. The natural frequencies of a clamped circular plate are

$$\omega_{ij} = \frac{\lambda_{ij}^2}{R^2} \sqrt{\frac{D}{m}} \tag{6.33}$$

As for circular membranes, only one mode has the natural frequency ω_{0j} while two modes have the same natural frequency ω_{ij} when $i > 0$.

Acoustical volumes

The acoustical natural frequencies of a three-dimensional parallelepipedic volume of dimension $L_x \times L_y \times L_z$ are given by

$$\omega_{ijk}^2 = \pi^2 c^2 \left(\frac{i^2}{L_x^2} + \frac{j^2}{L_y^2} + \frac{k^2}{L_z^2} \right), \qquad i,j,k = 1,2,\ldots \tag{6.34}$$

when the Dirichlet boundary condition (null acoustical pressure) is assumed on the walls. In the case of Neumann boundary condition (null normal velocity), the sequence becomes

$$\omega_{ijk}^2 = \pi^2 c^2 \left(\frac{i^2}{L_x^2} + \frac{j^2}{L_y^2} + \frac{k^2}{L_z^2} \right), \qquad i,j,k = 0,1,\ldots \tag{6.35}$$

The number of modes below a given frequency ω thus depends on the applied boundary conditions.

We shall distinguish between the null mode ω_{000}, an edge mode for which two indices are zero like ω_{100} or ω_{020}, a side mode with a null index like ω_{102}, and, finally, an oblique mode with three non-zero indices like ω_{211}.

To count the number of modes below the frequency ω, we follow the method given in Blevins (2006). The condition $\omega \geq \omega_{ijk}$ reads

$$1 \geq \left(\frac{i}{i_{\max}}\right)^2 + \left(\frac{j}{j_{\max}}\right)^2 + \left(\frac{k}{k_{\max}}\right)^2 \tag{6.36}$$

where $i_{\max} = L_x\omega/\pi c$, $j_{\max} = L_y\omega/\pi c$, and $k_{\max} = L_z\omega/\pi c$. The number $N(\omega)$ of modes below ω is therefore the number of index triplets verifying the above condition. Since index i (resp. j and k) cannot be greater than $I = [i_{\max}]$ (resp. $\mathcal{J} = [j_{\max}]$ and $K = [k_{\max}]$), N may be written as

$$N(\omega) = \sum_{k=1}^{K}\sum_{j=1}^{\mathcal{J}}\sum_{i=1}^{I} \delta_{ijk} \quad \text{where } \delta_{ijk} = \begin{cases} 1 \text{ if } \omega \geq \omega_{ijk} \\ 0 \text{ otherwise} \end{cases} \tag{6.37}$$

In eqn (6.37), the summations run from 1 which corresponds to Dirichlet boundary conditions. The Neumann problem is solved below. The innermost summations may be reduced. For a given index pair (j, k), the maximum index i is

$$I(j,k) = \left[i_{\max}\sqrt{1 - \left(\frac{j}{j_{\max}}\right)^2 - \left(\frac{k}{k_{\max}}\right)^2} \right] \tag{6.38}$$

In a similar fashion, for a fixed value of k the index j ranges from 1 to $\mathcal{J}(k)$ where

$$\mathcal{J}(k) = \left[j_{\max}\sqrt{1 - \left(\frac{k}{k_{\max}}\right)^2} \right] \tag{6.39}$$

With these notations, the number of modes below ω with Dirichlet boundary conditions becomes

$$N(\omega) = \sum_{k=1}^{K}\sum_{j=1}^{\mathcal{J}(k)} I(j,k) \tag{6.40}$$

In the case of Neumann boundary conditions, we get

$$N(\omega) = \sum_{k=0}^{K}\sum_{j=0}^{\mathcal{J}(k)} (I(j,k) + 1) \tag{6.41}$$

Let us remark finally that the order of summations is well suited when $L_x > L_y > L_z$.

6.3 Asymptotic Modal Densities

We now turn to the derivation of general formulas for modal densities (Courant and Hilbert, 1953; Morse and Ingard, 1968). Although some of them may be guessed from the analytic natural frequency sequences studied in Section 6.2 (one-dimensional case), the others are more difficult to predict. Further readings for more complex systems are provided by Langley (1996) and Boutillon and Ege (2013).

One-dimensional subsystems

Let us consider a one-dimensional system of length L. A resonance occurs when the number of half wavelengths within the system is an integer. Since the wavelength is given by $2\pi/\kappa$, this condition reads

$$\frac{\kappa}{\pi}L = N \tag{6.42}$$

Another way to derive this condition is to remark that a resonance occurs when the phase shift for a round trip of length $2L$ is a multiple of 2π. This condition is sometimes referred to as the **phase-closure principle** (Mead, 1994). When writing eqn (6.42) we have assumed that the phase shift due to the reflection at ends is zero. This is true for some particular boundary conditions but not in general. However, the phase shift induced by reflection is constant. When the order of mode N is large, the relative error in κ is negligible and we can consider that to neglect the boundary effect on phase shift is a reasonable approximation in high frequency range.

Then, deriving eqn (6.42) with respect to ω gives the modal density $n = \mathrm{d}N/\mathrm{d}\omega$,

$$n = \frac{L}{\pi}\frac{\mathrm{d}\kappa}{\mathrm{d}\omega} \tag{6.43}$$

We recognize the reciprocal of the group velocity,

$$n(\omega) = \frac{L}{\pi c_{\mathrm{g}}} \tag{6.44}$$

This is the asymptotic modal density of any one-dimensional system.

Two-dimensional subsystems

Now, let us consider a two-dimensional rectangular system of dimension $a \times b$. The wavenumber is now a vector $\boldsymbol{\kappa}$ in the x, y-plane whose components are noted κ_x and κ_y. A resonance occurs when the wave phase is shifted by a multiple of 2π after a round trip in both x- and y-directions. Two conditions are therefore imposed for the apparition of a resonance,

$$2\kappa_x a = 2\pi p \tag{6.45}$$

$$2\kappa_y b = 2\pi q \tag{6.46}$$

where p and q are integers. The magnitude of the wavenumber vector is $\kappa = (\kappa_x^2 + \kappa_y^2)^{1/2}$. Since $\kappa = \omega / c_p$, it yields

$$\omega = c_p \sqrt{\left(\frac{\pi p}{a}\right)^2 + \left(\frac{\pi q}{b}\right)^2} \tag{6.47}$$

If we plot the curve $\omega(p, q)$ in the p, q-plane as in Fig. 6.2, we see that the number N of modes below ω is the number of points (p, q) inside the quarter ellipse delimited by $p \geq 0$, $q \geq 0$. This is approximately given by the surface under the quarter ellipse for which the equation is

$$1 = \left(\frac{\pi c_p}{\omega a}\right)^2 p^2 + \left(\frac{\pi c_p}{\omega b}\right)^2 q^2 \tag{6.48}$$

Then,

$$N = \frac{1}{4} \times \pi \frac{\omega a}{\pi c_p} \frac{\omega b}{\pi c_p} = \frac{ab}{4\pi} \kappa^2 \tag{6.49}$$

The modal density is simply

$$n = \frac{dN}{d\omega} = \frac{ab}{2\pi} \kappa \frac{d\kappa}{d\omega} \tag{6.50}$$

Finally, by introducing the phase speed and the group speed, we get

$$n(\omega) = \frac{S\omega}{2\pi c_p c_g} \tag{6.51}$$

where $S = ab$ is the subsystem surface. This relationship established for rectangular shapes remains valid for any geometry of area S.

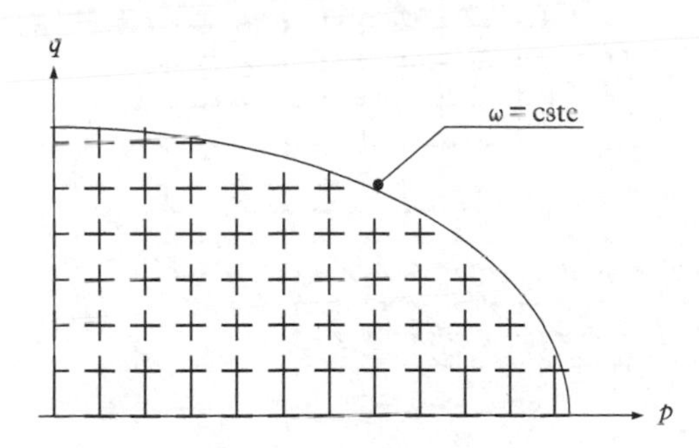

Figure 6.2 *Number of eigenfrequencies below ω for two-dimensional subsystems.*

Three-dimensional subsystems

The generalization for three-dimensional systems is straightforward. The wavenumber κ is a vector of components κ_x, κ_y, and κ_z. Three conditions are imposed for the existence of a resonance,

$$2\kappa_x a = 2\pi p \tag{6.52}$$

$$2\kappa_y b = 2\pi q \tag{6.53}$$

$$2\kappa_z c = 2\pi r \tag{6.54}$$

where p, q, r are integers. The magnitude of the wavenumber vector is $\kappa = (\kappa_x^2+\kappa_y^2+\kappa_z^2)^{1/2}$ and $\omega = \kappa c_p$. Hence,

$$\omega = c_p\sqrt{\left(\frac{\pi p}{a}\right)^2 + \left(\frac{\pi q}{b}\right)^2 + \left(\frac{\pi r}{c}\right)^2} \tag{6.55}$$

The number of modes $N(\omega)$ whose natural frequency is below ω is the number of points having integer coordinates inside the ellipsoid with the restrictions $p \geq 0$, $q \geq 0$, and $r \geq 0$ (Fig. 6.3). This is approximately one-eighth of the ellipsoid volume of equation

$$1 = \left(\frac{\pi c_p}{\omega a}\right)^2 p^2 + \left(\frac{\pi c_p}{\omega b}\right)^2 q^2 + \left(\frac{\pi c_p}{\omega c}\right)^2 r^2 \tag{6.56}$$

We get

$$N = \frac{1}{8} \times \frac{4}{3}\pi \frac{\omega a}{\pi c_p}\frac{\omega b}{\pi c_p}\frac{\omega c}{\pi c_p} = \frac{abc}{6\pi^2}\kappa^3 \tag{6.57}$$

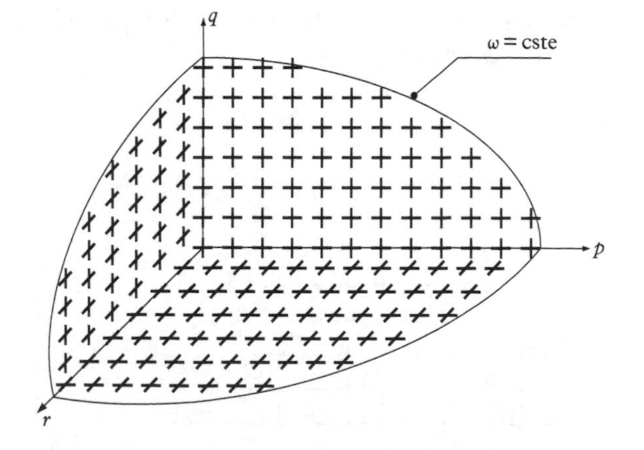

Figure 6.3 *Number of eigenfrequencies below ω for three-dimensional subsystems.*

By deriving,

$$n = \frac{\mathrm{d}N}{\mathrm{d}\omega} = \frac{abc}{6\pi^2}\kappa^2\frac{\mathrm{d}\kappa}{\mathrm{d}\omega} \tag{6.58}$$

And finally,

$$n(\omega) = \frac{V\omega^2}{2\pi^2 c_\mathrm{p}^2 c_\mathrm{g}} \tag{6.59}$$

where $V = abc$ is the volume of the cavity. Three-dimensional systems are usually non-dispersive. This is the case in acoustics, electromagnetism, and linear elastodynamics. In this case, eqn (6.59) reduces to

$$n = \frac{V\omega^2}{2\pi^2 c^3} \tag{6.60}$$

where c is the speed of sound, speed of light, etc.

6.4 Effect of Boundaries

The previous asymptotic relationships were derived without taking into account the effect of boundaries. The resulting values of modal densities are valid in the high frequency limit where the difference induced by various boundary conditions is negligible. For instance, when studying simply supported and clamped beams, we have seen in eqns (6.13) and (6.22) that the number of modes below ω only differs by the constant 1/2 and therefore the modal densities were not affected. But, when considering parallelepipedic volumes, although the sequence of normal modes was the same for both Neumann and Dirichlet boundary conditions, the starting values of the indices i, j, k in eqns (6.34) and (6.35) were not the same. The sequence induced by Neumann boundary conditions includes side and edge modes and therefore leads to a higher modal density. Since the number of side modes below ω increases with ω, it may be suspected that the difference in the number of modes is a term proportional to the surface of rooms.

Structural components

Let us first consider a rectangular plane structure of dimensions $L_x \times L_y$. Two extreme situations may be considered. For the first one, the sequence of normal modes is

$$\omega_{ij}^2 \propto \left(\frac{i^2}{L_x^2} + \frac{j^2}{L_y^2}\right) \qquad i, j = 1, 2, \ldots \tag{6.61}$$

This case applies for rectangular membranes, two-dimensional cavities with Dirichlet's boundary conditions, or simply supported rectangular plates for instance. We shall call them Dirichlet-type problems. The situation is sketched in Fig. 6.4(a). All modes have

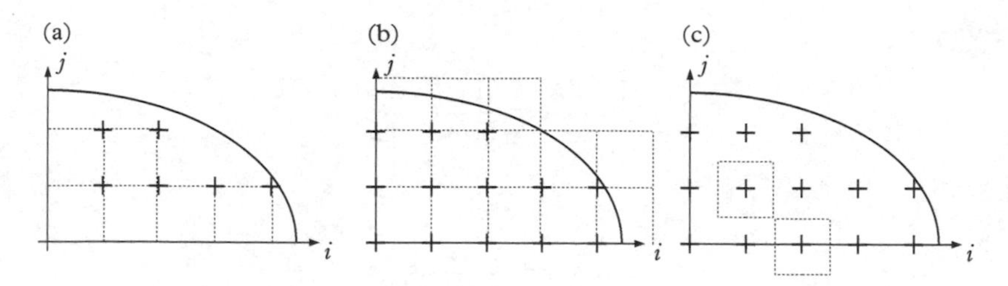

Figure 6.4 *Influence of boundary conditions on mode count and actual volume occupied by modes (broken-line boxes). (a) Dirichlet boundary conditions. (b) Neumann boundary conditions. (c) Boxes centred on modes.*

non-zero indices so that the vector (i, j) is always oblique. We shall call them oblique modes. In the second case, which we shall call Neumann-type problems, the sequence becomes

$$\omega_{ij}^2 \propto \left(\frac{i^2}{L_x^2} + \frac{j^2}{L_y^2} \right) \qquad i, j = 0, 1, \ldots \tag{6.62}$$

Some examples of such a situation are two-dimensional cavities with Neumann boundary conditions or membranes with free edges. Modes localized on the i, j-axis as shown in Fig. 6.4(b) will be referred to as edge modes. The two instances therefore differ by the presence of edge modes.

These two situations do not cover all cases. For instance clamped or free ends of plates are neither a Dirichlet-type nor Neumann-type problem. Furthermore, irregularities in the first eigenfrequencies induced by non-rectangular shapes (such as a circle) may induce discrepancies with above-normal mode sequences even in the case of simply supported plates.

In the reasoning of the previous subsection, the number of modes was estimated by the ellipse area, each mode occupying a unit area cell. But it is clear that in Dirichlet-type problems (Fig. 6.4(a)), the ellipse area slightly overestimates the number of modes (this is seen by positioning modes on the top-right corner of unit cells), while in Neumann-type problems there is an underestimation of the actual number of modes. In both cases the bias stems from the presence or absence of edge modes.

A better estimation is obtained by centring each mode in its unit cell (Fig. 6.4(c)). Then, when calculating the area of the quarter ellipse, all oblique modes have been included but only half of a unit cell is counted near the axis. Therefore, if edge modes are present (Neumann-type problem) the other half part of border unit cells must be added to get a better estimation. And indeed, they must be subtracted in the absence of edge modes (Dirichlet-type problem). Therefore the correcting term is half the modal density of a one-dimensional system of length L_x. The same reasoning applies for the j-axis, and the total corrected term is $1/2 \times (L_x + L_y)/\pi c_g$. In the absence of edge modes,

$$n(\omega) = \frac{L_x L_y \omega}{2\pi c_p c_g} - \frac{L_x + L_y}{2\pi c_g} \tag{6.63}$$

By introducing the area $S = L_x L_y$ and perimeter $P = 2(L_x + L_y)$, we get

$$n(\omega) = \frac{S\omega}{2\pi\, c_p c_g} - \frac{P}{4\pi\, c_g} \tag{6.64}$$

for Dirichlet-type problems. Similarly,

$$n(\omega) = \frac{S\omega}{2\pi\, c_p c_g} + \frac{P}{4\pi\, c_g} \tag{6.65}$$

for Neumann-type problems.

Rooms

For parallelepipedic acoustical rooms of dimensions $L_x \times L_y \times L_z$, the reasoning is quite similar. The correction must take into account all modes localized on the planes $i = 0$, $j = 0$, and $k = 0$. Oblique modes now denote modes for which all indices are non-zero, side modes when one index is zero, and edge modes when two indices are zero. By the centred cell technique, we see that one-eighth of the ellipsoid volume includes all oblique modes but only half the side modes and a quarter of the edge modes. All missing terms must be restored. In a Neumann-type problem, the modal density is the sum of the volume term given in eqn (6.59), half of modal densities of all side modes given in eqn (6.63), and three-quarters of modal densities of edge modes given in eqn (6.44). It yields

$$n(\omega) = \frac{V\omega^2}{2\pi^2 c^3} + \frac{1}{2}\left(\frac{L_x L_y \omega}{2\pi c^2} - \frac{L_x + L_y}{2\pi c} + \frac{L_y L_z \omega}{2\pi c^2} - \frac{L_y + L_z}{2\pi c} + \frac{L_x L_z \omega}{2\pi c^2} - \frac{L_x + L_z}{2\pi c}\right) + \frac{3}{4}\left(\frac{L_x + L_y + L_z}{\pi c}\right)$$

By introducing the cavity area $S = 2(L_x L_y + L_y L_z + L_x L_z)$ and the perimeter $P = 4(L_x + L_y + L_z)$, we get (Maa, 1939)

$$n(\omega) = \frac{V\omega^2}{2\pi^2 c^3} + \frac{S\omega}{8\pi c^2} + \frac{P}{16\pi c} \tag{6.66}$$

This is the modal density of an acoustical cavity with Neumann boundary conditions.

Similarly, in a Dirichlet-type problem, the modal density is the sum of the volume term given in eqn (6.59) minus half the modal densities of all side modes given in eqn (6.63) and minus a quarter of modal densities of edge modes given in eqn (6.44). It yields

$$n(\omega) = \frac{V\omega^2}{2\pi^2 c^3} - \frac{1}{2}\left(\frac{L_x L_y \omega}{2\pi c^2} - \frac{L_x + L_y}{2\pi c} + \frac{L_y L_z \omega}{2\pi c^2} - \frac{L_y + L_z}{2\pi c} + \frac{L_x L_z \omega}{2\pi c^2} - \frac{L_x + L_z}{2\pi c}\right) - \frac{1}{4}\left(\frac{L_x + L_y + L_z}{\pi c}\right)$$

And finally (Roe, 1941)

$$n(\omega) = \frac{V\omega^2}{2\pi^2 c^3} - \frac{S\omega}{8\pi c^2} + \frac{P}{16\pi c} \tag{6.67}$$

for Dirichlet boundary conditions.

Example

To illustrate the utility of the correcting terms in eqns (6.64), (6.65), (6.66), and (6.67), let us take an example. We consider a room of dimensions 10 × 15 × 30 feet. This is the traditional example first proposed by Bolt (1939) and then developed (Maa, 1939; Courant and Hilbert, 1953; Morse and Ingard, 1968; Blevins, 2006). The sequences of acoustical eigenfrequencies are given by eqns (6.34) and (6.35), and the exact number of modes is given by eqns (6.40) and (6.41). Figure 6.5 shows the exact count of eigenfrequencies within a frequency band of width 10 Hz and their values given by the asymptotic estimations. The importance of the correcting terms is especially clear at low frequencies.

As remarked by Blevins (2006) the correcting terms may give incorrect values. If we consider a flat room of dimension $L_x \times L_y \times \epsilon$, the volume $V = L_x L_y \epsilon$ is zero when $\epsilon \to 0$, while $S = 2(L_x L_y + \epsilon L_x + \epsilon L_y + \epsilon^2) \to 2L_x L_y$ and $P = 4(L_x + L_y + \epsilon) \to 4(L_x + L_y)$. So, for the limiting case of an infinitely flat room with Neumann boundary conditions, the modal density is

$$\lim_{\epsilon \to 0} n(\omega) = \frac{2L_x L_y \omega}{8\pi c^2} + \frac{4(L_x + L_y)}{16\pi c} \tag{6.68}$$

But the area of the equivalent two-dimensional system is $S' = L_x L_y$ and the perimeter is $P' = 2(L_x + L_y)$.

$$\lim_{\epsilon \to 0} n(\omega) = \frac{S'\omega}{4\pi c^2} + \frac{P'}{8\pi c} \tag{6.69}$$

This equation clearly contradicts eqn (6.65) by a factor of 2. A similar disagreement is also observed in the limiting case of long enclosures (by factor 4).

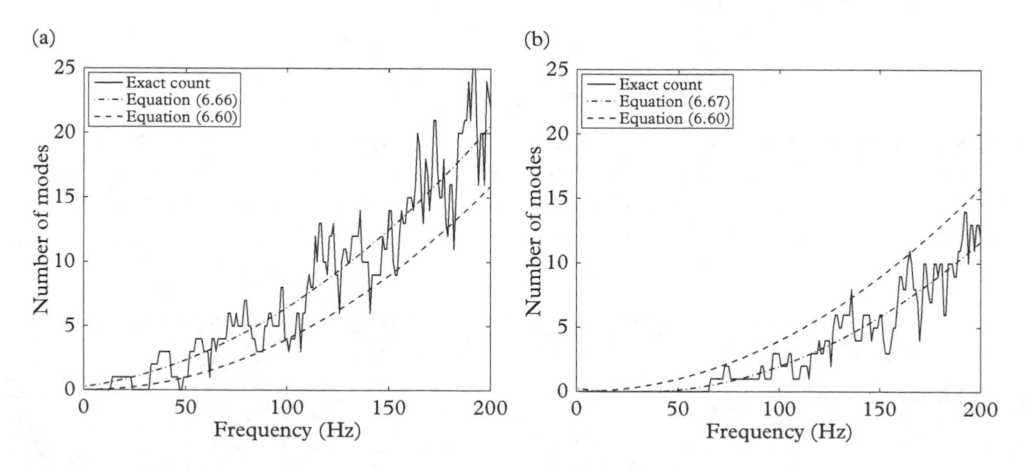

Figure 6.5 *Number of normal modes with frequencies between f − 5 and f + 5 in a room 10 × 15 × 20 feet. The solid line is an exact count by eqns (6.41), (6.40); dashed lines are asymptotic relationships. (a) Neumann boundary conditions; (b) Dirichlet boundary conditions.*

6.5 Comments

The concept of modal density is certainly one the most fruitful in statistical physics. It appears for instance in the derivation of Planck's law of black-body radiation. A cavity filled by an electromagnetic field in thermal equilibrium has a spectral energy density imposed by the modal density of the cavity. Another example is the density of states of a quantum particle in a box. The calculation of this density which reduces to a modal density is an important step in the determination of the structure and partition functions.

In acoustics, modal density is an important quantity in the context of reverberation theory. However, in contrast with statistical physics, the number of resonant modes is relatively low and the asymptotic value $V\omega^2/2\pi^2c^3$ may sometimes be a poor approximation. This is why Bolt (1939) and Maa (1939) proposed some correcting terms which account for the effect of boundaries. Similar formulas for structures are given by Bogomolny and Hugues (1998) and Xie et al. (2004).

The fact that modes have their own frequencies enabled us to introduce the modal density as the number of modes within a frequency band. At the same time, modes have a wavenumber which is a vector and therefore has a direction. It is therefore also possible to introduce a specific modal density as the number of modes within a certain frequency band and solid angle. This is the path followed by Bourgine (1973).

7

Diffuse Field

In this chapter, we introduce the most important concept in statistical theories of wave motion: the **diffuse field**.

7.1 Diffuse Reflection

When impinging on a boundary, rays are reflected inside the domain. This reflection process depends on the nature of the boundary. It is characterized by two angles, the incidence angle φ and reflection angle θ, both measured with the normal to the boundary. In the framework of geometrical acoustics, this phenomenon is local in the sense that the law which links θ and φ only depends on the local properties of the boundary.

Numerous laws of reflection may arise but two extreme cases are of special interest. A deterministic law such as specular reflection imposes an emission angle θ univocally determined by the incident angle φ, while a statistical law gives a probability density function for the ray to be reflected in direction θ having incidence φ. Among the latter, a memoryless law is a probability that does not depend on incidence. It may be shown (Joyce, 1975) by an argument on entropy that the only memoryless law is the so-called **Lambert's law** or **diffuse reflection law**, whose probability is

$$p_{2D}(\theta) = \frac{1}{2}\cos\theta \tag{7.1}$$

for two-dimensional domains and

$$p_{3D}(\theta) = \frac{1}{\pi}\cos\theta \tag{7.2}$$

for three-dimensional domains. It may readily be checked that these probabilities are normalized.

7.2 Mean Free Path

The notion of mean free path is encountered in various fields of physics. When a large number of particles or rays propagate in a finite domain, the **mean free path** is the

Foundation of Statistical Energy Analysis in Vibroacoustics. First Edition. A. Le Bot.

mean distance between two successive reflections. See Kosten (1960) for an elementary geometrical approach to mean free path.

Two-dimensional domain

Let us start by considering the case of a bounded two-dimensional domain Ω whose boundary is noted Γ. The distance travelled by a ray between two reflections is the distance between any two points of Γ. If $\mathbf{q}$ denotes the departure point and $\mathbf{p}$ the arrival point, then the mean free path is the average of the chords $R = |\mathbf{q} - \mathbf{p}|$ for all pairs $\mathbf{p}, \mathbf{q}$. However, performing the average, we must take into account the fact that emission angles are not equiprobable. Thus, the mean free path is the average of chords weighted by Lambert's law,

$$\bar{l} = \int_0^{2\pi} \mathrm{d}\alpha \int_\Gamma R \times \frac{\cos\theta}{2P} \, \mathrm{d}\Gamma \tag{7.3}$$

where P is the length of Γ and α the ray angle measured with a fixed direction (see Fig. 7.1). In these integrals, θ is the angle between the ray and the normal to the boundary and the integrand is set to zero for backward directions $|\theta| > \pi/2$. But $\mathrm{d}\Gamma \cos\theta$ is the infinitesimal width between two successive rays of same direction α. Therefore $\mathrm{d}\Omega = R\,\mathrm{d}\Gamma \cos\theta$ is an infinitesimal surface and

$$\bar{l} = \frac{1}{2P} \int_0^{2\pi} \mathrm{d}\alpha \int \mathrm{d}\Omega \tag{7.4}$$

The inner integral is S, the area of the domain Ω. It yields

$$\bar{l} = \frac{\pi S}{P} \tag{7.5}$$

This is the mean free path of a two-dimensional domain of area S and perimeter P.

Three-dimensional domain

The generalization to three-dimensional domains is straightforward. The average of chords must be performed over the boundary Γ of area S and the sphere of all solid angles (Kuttruff, 2000, p. 124),

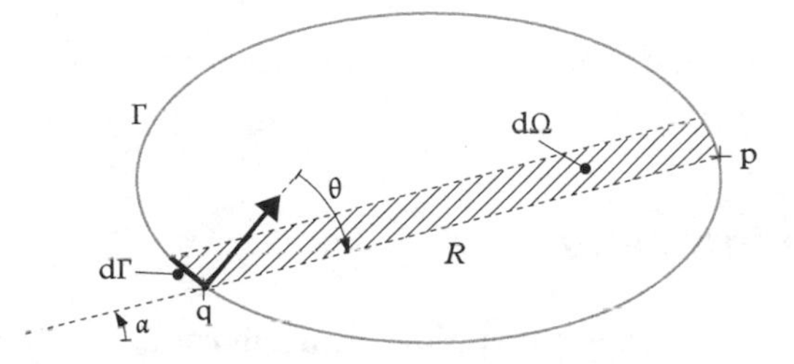

Figure 7.1 *Mean free path.*

$$\bar{l} = \int_{4\pi} \mathrm{d}\omega \int_{\Gamma} R \times \frac{\cos\theta}{\pi S} \mathrm{d}\Gamma \tag{7.6}$$

where $\mathrm{d}\omega$ is the infinitesimal solid angle. By developing $\mathrm{d}\omega = \sin\alpha\, \mathrm{d}\alpha \mathrm{d}\beta$ in spherical coordinates,

$$\bar{l} = \frac{1}{\pi S} \int_0^{\pi} \sin\alpha d\alpha \int_0^{2\pi} d\beta \int_{\Gamma} R\cos\theta\, \mathrm{d}\Gamma \tag{7.7}$$

Again the integrand is zero for $|\theta| > \pi/2$ and the inner integral is the volume domain V. We get

$$\bar{l} = \frac{4V}{S} \tag{7.8}$$

This is the mean free path of a three-dimensional domain of volume V and area S.

7.3 Energy Density

When a point source of unit power emits rays isotropically, the energy density at a distance R from the source is $\mathrm{e}^{-mR}/2\pi c_g R$ in a two-dimensional domain and $\mathrm{e}^{-mR}/4\pi c_g R^2$ in a three-dimensional domain. The attenuation factor m is related to the damping loss factor η, the frequency ω, and the group velocity c_g by $m = \eta\omega/c_g$. In acoustics, attenuation in air is expressed in dB per 100 m and is strongly dependent on the frequency and air humidity.

Define Ω_i a domain of propagation, bounded or not, and Γ_i its boundary. We denote by ρ_i the power density of sources located in Ω_i. Rays are emitted from sources ρ_i, propagate in Ω_i, and are reflected when reaching the boundary Γ_i. So to account for reflection of energy, we introduce a fictitious source density σ_i located on Γ_i. These sources do not radiate isotropically but proportionally to $\cos\theta$ following Lambert's law. By linear superposition, the energy density $W_i(\mathbf{r})$ at any point $\mathbf{r}$ in Ω_i is the sum of the direct field $\rho_i \mathrm{e}^{-mR}/2\pi c_{g_i} R$ and reflected field $\sigma_i \cos\theta \mathrm{e}^{-mR}/2\pi c_{g_i} R$,

$$W_i(\mathbf{r}) = \int_{\Omega_i} \rho_i(\mathbf{s}) \frac{\mathrm{e}^{-mR}}{2\pi c_{g_i} R} \mathrm{d}\Omega_s + \int_{\Gamma_i} \sigma_i(\mathbf{p}) \cos\theta \frac{\mathrm{e}^{-mR}}{2\pi c_{g_i} R} \mathrm{d}\Gamma_p \tag{7.9}$$

Of course, for a propagation in three dimensions the term $4\pi R^2$ must be used in place of $2\pi R$.

7.4 Radiative Intensity

We may also introduce the **specific radiative intensity** $I_i(\mathbf{r}, \alpha)$ defined as the energy flow per unit solid angle and unit area normal to the rays. Let $\mathbf{r}$ be a point in Ω_i, α a

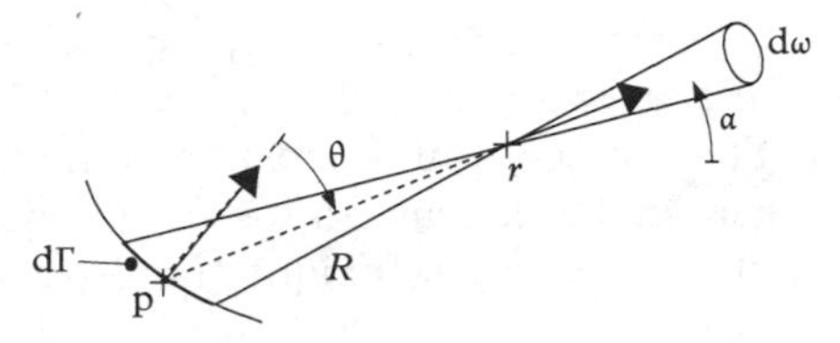

Figure 7.2 *Radiative intensity.*

direction, and $d\omega$ an infinitesimal solid angle about α. The incoming cone whose vertex is $\mathbf{r}$ and angle $d\omega$ intersects Γ_i by an infinitesimal surface $d\Gamma$ about a point $\mathbf{q}$ (Fig. 7.2). The infinitesimal power received at $\mathbf{r}$ per unit area normal to the ray is

$$d\mathcal{P}(\mathbf{r},\alpha) = \int_{d\omega} \rho_i(\mathbf{s}) \frac{e^{-mR}}{2\pi R} d\Omega_s + \sigma_i(\mathbf{q}) \cos\theta \frac{e^{-mR}}{2\pi R} d\Gamma \tag{7.10}$$

but in polar coordinates $d\Omega = R\,dR\,d\omega$. The infinitesimal surface embodied in the cone verifies $d\Gamma \cos\theta = R\,d\omega$ where $R = |\mathbf{q}-\mathbf{r}|$. The radiative intensity $I_i = d\mathcal{P}/d\omega$ is therefore

$$I_i(\mathbf{r},\alpha) = \int_0^R \rho_i(\mathbf{s}) \frac{e^{-ms}}{2\pi} ds + \sigma_i(\mathbf{q}) \frac{e^{-mR}}{2\pi} \tag{7.11}$$

where the first integral is performed over the line from $\mathbf{r}$ to $\mathbf{q}$ and s is a curvilinear abscissa along this line while $R = |\mathbf{q}-\mathbf{r}|$. The expression of the radiative intensity is the same in three dimensions except for a factor 4π in place of 2π.

7.5 Radiosity

We have not yet specified how to find the surface sources σ_i. The **radiosity** B is the energy leaving the surface per unit time and area. Consider the half-sphere of outgoing directions centred on a point $\mathbf{p}$ of the boundary. The radiosity is the flux of $\sigma_i \cos\theta e^{-m\epsilon}/2\pi\epsilon$ where ϵ is the radius of the sphere. Since the infinitesimal surface of the sphere is $\epsilon\, d\theta$, we get

$$B = \frac{\sigma_i}{2\pi} \int_{-\frac{\pi}{2}}^{\frac{\pi}{2}} \cos\theta \, d\theta = \frac{\sigma_i}{\pi} \tag{7.12}$$

in the limit $\epsilon \to 0$. In dimension three, the factor is 4 instead of π.

The energy balance at the boundary imposes that B is equal to the incoming energy that is the integral $\int d\mathcal{P}(\mathbf{p},\varphi) \cos\varphi \, d\omega$ where φ is the incidence angle at $\mathbf{p}$. It yields (Le Bot, 1998, 2002)

$$\frac{\sigma_i(\mathbf{p})}{\pi} = \int_{\Omega_i} \rho_i(\mathbf{s}) \frac{e^{-mR}}{2\pi R} \cos\varphi \, d\Omega_s + \int_{\Gamma_i} \sigma_i(\mathbf{q}) \cos\theta \frac{e^{-mR}}{2\pi R} \cos\varphi \, d\Gamma_q \tag{7.13}$$

where $R = |\mathbf{s}-\mathbf{p}|$ in the first integral and $R = |\mathbf{q}-\mathbf{p}|$ in the second integral. This is a Fredholm equation of the second kind with unknown σ_i. In three dimensions, the factor π is replaced by 4. The solution to this equation combined with eqn (7.9) gives the energy density everywhere in the domain. In room acoustics, it may be shown that its solution is equivalent to the ray-tracing technique (Le Bot and Bocquillet, 2000).

7.6 Emergence of Diffuse Field

We now examine the case of a highly reverberant domain. Here, rays are reflected many times before disappearing by absorption. During their travel, they lose the memory of their initial point and a state of perfect mixing of energy is reached. All points of reflection on the boundary are equivalent and the boundary source σ_i is independent of the position $\mathbf{p}$.

Let us integrate over Γ_i both sides of eqn (7.13). We get

$$\int_{\Gamma_i} \frac{\sigma_i(\mathbf{p})}{\pi}\, d\Gamma_{\mathbf{p}} = \int_{\Gamma_i}\int_{\Omega_i} \rho_i(\mathbf{s}) \frac{e^{-mR}}{2\pi R} \cos\varphi \, d\Omega_{\mathbf{s}} d\Gamma_{\mathbf{p}} + \int_{\Gamma_i\times\Gamma_i} \sigma_i(\mathbf{q}) \cos\theta \frac{e^{-mR}}{2\pi R} \cos\varphi \, d\Gamma_{\mathbf{q}} d\Gamma_{\mathbf{p}}$$

where $R = |\mathbf{s}-\mathbf{p}|$ in the first integral and $R = |\mathbf{q}-\mathbf{p}|$ in the second integral. The incidence angle at $\mathbf{p}$ is denoted φ, and θ is the emission angle at $\mathbf{q}$. By virtue of the light damping assumption, $mR \ll 1$ and the term e^{-mR} is of order of unity. Therefore it varies slowly compared with the rest of the integrand. It may be set to its mean value $e^{-m\bar{l}}$ and leave the integrals. Furthermore, imposing a constant value on ρ_i and σ_i gives

$$\frac{\sigma_i}{\pi} P_i = e^{-m\bar{l}} \left(\rho_i \int_{\Gamma_i}\int_{\Omega_i} \frac{\cos\varphi}{2\pi R}\, d\Omega d\Gamma + \sigma_i \int_{\Gamma_i\times\Gamma_i} \frac{\cos\theta \cos\varphi}{2\pi R}\, d\Gamma d\Gamma \right) \tag{7.14}$$

where P_i is the perimeter of Γ_i. We will now calculate both integrals in succession.

For the first term in parentheses, the integrals may be permuted. The inner integral is

$$\int_{\Gamma_i} \frac{\cos\varphi}{2\pi R}\, d\Gamma = \int_0^{2\pi} \frac{1}{2\pi R} R\, d\alpha = 1 \tag{7.15}$$

where the change of variable $R\,d\alpha = d\Gamma \cos\varphi$ (see Fig. 7.3(a)) has been performed. The inner integral of the second term is

$$\int_{\Gamma_i} \frac{\cos\theta \cos\varphi}{2\pi R}\, d\Gamma = \int_{-\frac{\pi}{2}}^{\frac{\pi}{2}} \frac{\cos\theta\cos\varphi}{2\pi R} \times \frac{R\, d\theta}{\cos\varphi} = \frac{1}{\pi} \tag{7.16}$$

by the change of variable $R\,d\theta = d\Gamma \cos\varphi$ (see Fig. 7.3(b)).

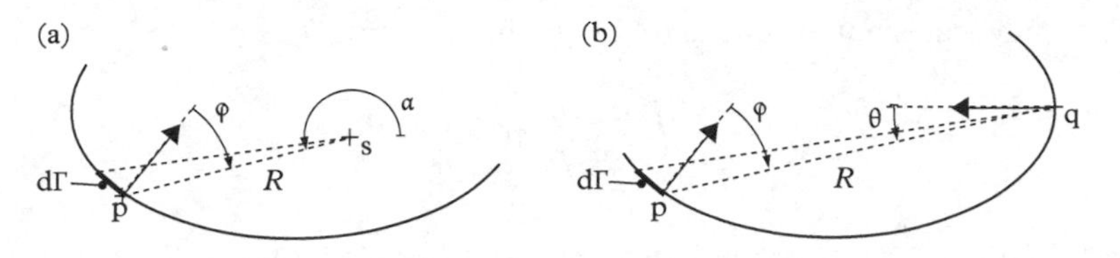

Figure 7.3 *Change of variables in (a) eqn (7.15); (b) eqn (7.16).*

Substituting eqns (7.15) and (7.16) into (7.14) yields

$$\frac{\sigma_i}{\pi}P_i = \mathrm{e}^{-m\bar{l}}\left(\rho_i S_i + \frac{\sigma_i}{\pi}P_i\right) \tag{7.17}$$

where S_i is the area of Ω_i. Introducing the mean free path $\bar{l} = \pi S_i/P_i$ and solving in σ_i gives

$$\sigma_i = \frac{\mathrm{e}^{-m\bar{l}}}{1-\mathrm{e}^{-m\bar{l}}}\rho_i\bar{l} \tag{7.18}$$

We have obtained σ_i in terms of $\rho_i\bar{l}$. Since $m\bar{l}$ is small compared with unity, we have

$$\rho_i\bar{l} \ll \sigma_i \tag{7.19}$$

This inequality implies that the direct field is always negligible, as we shall see now.

The energy density for constant ρ_i and σ_i is found from eqn (7.9),

$$W_i(\mathbf{r}) = \rho_i \int_{\Omega_i} \frac{\mathrm{d}\Omega}{2\pi c_{g_i} R} + \sigma_i \int_{\Gamma_i} \frac{\cos\theta}{2\pi c_{g_i} R}\,\mathrm{d}\Gamma \tag{7.20}$$

Note that we have considered $\mathrm{e}^{-m\bar{l}} \sim 1$ in the above expression. Both integrals which depend on the position $\mathbf{r}$ may be calculated. By expanding $\mathrm{d}\Omega = R\,\mathrm{d}R\mathrm{d}\alpha$ in polar coordinates, the first integral becomes

$$\int_{\Omega_i} \frac{\mathrm{d}\Omega}{2\pi R} = \int_0^{2\pi}\int_0^{l(\mathbf{r},\alpha)} \frac{R\,\mathrm{d}R\mathrm{d}\alpha}{2\pi R} = \frac{1}{2\pi}\int_0^{2\pi} l(\mathbf{r},\alpha)\,\mathrm{d}\alpha$$

where $l(\mathbf{r},\alpha)$ denotes the distance from $\mathbf{r}$ to Γ_i in direction α. But $l(\mathbf{r},\alpha) + l(\mathbf{r},-\alpha)$ is the boundary-to-boundary distance passing through $\mathbf{r}$ in direction α. The integral may therefore be interpreted as one-half of the mean boundary-to-boundary distance through $\mathbf{r}$. Of course, this is of the order of the mean free path,

$$\int_{\Omega_i} \frac{\mathrm{d}\Omega}{2\pi R} \sim \frac{\bar{l}}{2} \tag{7.21}$$

The second integral is equivalent to (7.15) by interpreting θ as φ,

$$\int_{\Gamma_i} \frac{\cos\theta}{2\pi R}\,\mathrm{d}\Gamma = 1 \tag{7.22}$$

From (7.21) and (7.22), we observe that the first term of (7.20) is $\rho_i\bar{l}/2c_{g_i}$ and the second term is σ_i/c_{g_i}. But we have shown that the first term is negligible. It yields

$$W_i = \frac{\sigma_i}{c_{g_i}} \tag{7.23}$$

Thus, the energy density is constant within the domain. The vibrational field is homogeneous.

The radiative intensity for constant ρ_i and σ_i is, from eqn (7.11),

$$I_i(\mathbf{r},\alpha) = \rho_i \int_0^{l(\mathbf{r},-\alpha)} \frac{\mathrm{d}R}{2\pi} + \frac{\sigma_i}{2\pi} \tag{7.24}$$

where again the exponentials have been taken as unity. The first term is of order of $\rho_i\bar{l}/2\pi$ and is therefore negligible. It comes

$$I_i = \frac{\sigma_i}{2\pi} \tag{7.25}$$

The radiative intensity depends on neither the position nor the direction. We may now give a definition of a **diffuse field**. A vibrational field is said to be diffuse if it is homogeneous and isotropic.

7.7 Exchange of Energy Between Subsystems

When two adjacent subsystems i and j have a common boundary, they exchange energy through this boundary. We consider the situation where both subsystems are in diffuse field state (diffuse field assumption) and for which the common edge is small compared with other reflective boundaries (light coupling assumption).

The energy-conducting connection may characterized by its transmission coefficient. We introduce the transmission efficiency T_{ij} from i to j as the ratio of incident power per unit length of boundary to transmitted power. The transmission efficiency is usually calculated for an incident plane wave in which case it only depends on the incident angle φ (many examples of canonical problems will be solved in Chapter 9).

By linear superposition, the energy passing from i to j is the sum of contributions of all rays hitting the common boundary,

$$P_{i\to j} = \int_L \int_{-\frac{\pi}{2}}^{\frac{\pi}{2}} T_{ij}(\varphi) I(\mathbf{p}, \varphi) \cos\varphi \, d\varphi d\Gamma \tag{7.26}$$

where L is the common boundary. Also by linear superposition, the net exchanged energy P_{ij} is the difference of $P_{i\to j}$ and $P_{j\to i}$,

$$P_{ij} = P_{i\to j} - P_{j\to i} \tag{7.27}$$

When the vibrational field is diffuse, we may also introduce the mean transmission efficiency,

$$\bar{T}_{ij} = \frac{1}{2} \int_{-\frac{\pi}{2}}^{\frac{\pi}{2}} T_{ij}(\varphi) \cos\varphi \, d\varphi \tag{7.28}$$

which, of course, no longer depends on incidence.

By substituting eqns (7.25) and (7.28) into (7.26),

$$P_{i\to j} = L \frac{\bar{T}_{ij}}{\pi} \sigma_i \tag{7.29}$$

By (7.27) and (7.23),

$$P_{ij} = L \frac{c_{g_i} \bar{T}_{ij}}{\pi S_i} E_i - L \frac{c_{g_j} \bar{T}_{ji}}{\pi S_j} E_j$$

where we have substituted the total vibrational energy $E_i = W_i S_i$ where S_i denotes the area of Ω_i. We may now introduce the coupling loss factor,

$$\eta_{ij} = \frac{L c_{g_i}}{\pi \omega S_i} \bar{T}_{ij} \tag{7.30}$$

and obtain

$$P_{ij} = \omega \eta_{ij} E_i - \omega \eta_{ji} E_j \tag{7.31}$$

This is the coupling power proportionality. Let us summarize the result obtained in this section. *Two domains filled by a diffuse field of rays and connected by a small aperture (weak coupling) exchange energy proportionally to their vibrational energies.*

The coupling power proportionality has been derived already, in Chapter 3, eqn (3.19) by the modal approach of statistical energy analysis. The present demonstration (from Le Bot, 2007) is rather based on the wave approach of statistical energy analysis with the key assumption of diffuse field. Other approaches are possible (Langley, 1989; Maidanik and Dickey, 1990; Tanner, 2009).

7.8 Diffuse Field Conditions

We now investigate the conditions which lead to a diffuse field.

Mode count

The general framework of geometrical acoustics stipulates that energy propagates as rays. The geometrical approximation is valid in the high frequency limit (Assumption 7). If we introduce the mode count N as the number of modes below ω, the mode count and the modal density are related by

$$N = \int_0^{\omega} n(\omega)\, \mathrm{d}\omega \tag{7.32}$$

The condition of high frequencies then reads

$$N \gg 1 \tag{7.33}$$

The number of modes below ω must be large.

Normalized attenuation factor

In the derivation of the coupling power proportionality (7.31), we have shown that the direct field is negligible. This approximation stems from the light damping assumption (Assumption 5) which enabled us to set that the term $m\bar{l}$ is small compared with unity. We may therefore introduce the **normalized attenuation factor**,

$$\bar{m} = \frac{\eta\omega}{c_g}\bar{l} \tag{7.34}$$

which must be small,

$$\bar{m} \ll 1 \tag{7.35}$$

A light attenuation ensures that rays will travel for a long time and will be reflected many times before they vanish. This is a favourable condition for mixing energy.

Modal overlap

The frequency response function of a multimode linear system exhibits peaks at the resonance frequencies, separated by troughs (see Fig. 7.4). This shape of well-separated peaks is representative of a dynamic dominated by individual modes. This behaviour is typical of low frequencies. But when the frequency increases, the peaks begin to overlap and the frequency response function becomes more regular, with hills and valleys.

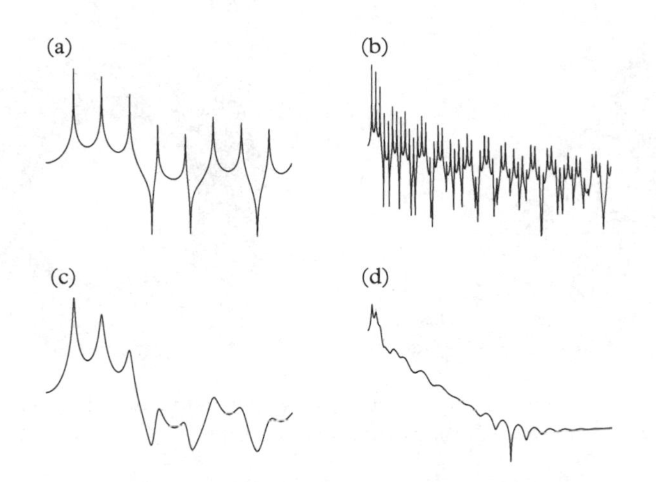

Figure 7.4 *Frequency response functions for: (a) low modal density n; (b) high modal density; (c) high modal overlap M; and (d) high modal overlap and modal density.*

Ultimately, the response is smooth indicating that modes have lost their individuality in the sense that their exact frequency has no influence on the shape of the response. The factor that controls this transition is the so-called **modal overlap**. The modal overlap M is defined by comparing the width of a resonance peak, the half-power bandwidth $\eta\omega$, and the mean interval between natural frequencies $1/n(\omega)$,

$$M = \eta\omega n(\omega) \tag{7.36}$$

A large modal overlap $\eta\omega > 1/n$ is a favourable condition for the reliability of a probabilistic-based estimation of the response. We therefore write

$$M \gg 1 \tag{7.37}$$

A high modal overlap ensures that no mode dominates the dynamics.

Ergodic billiards

The former conditions are generally not sufficient to ensure the emergence of diffuse field. Mixing of rays is not performed by all geometries, and some counter-examples are known.

In the mathematical theory of billiards, **ergodicity** is the property that almost all rays spend equal time in the vicinity of each position and direction. This is of course a condition favourable to diffuseness. For instance, Figure 7.5 shows the trajectory of a ray after 200 reflections in first a triangle and then a circle. In the triangle the ray passes through the vicinity of all points and directions. A unique ray leads to a uniform

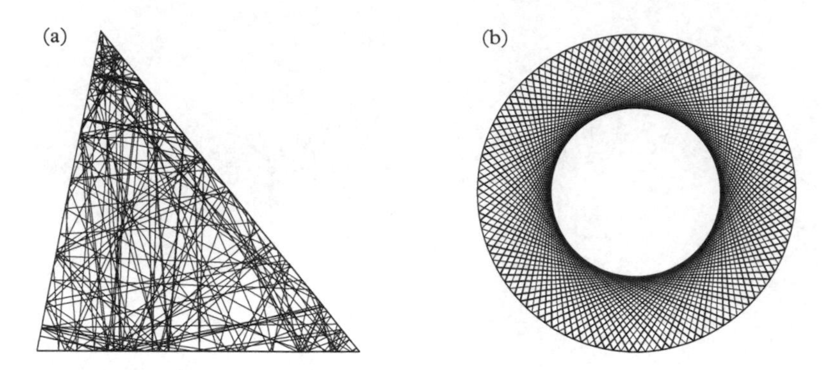

Figure 7.5 *Ray tracing in billiards. (a) homogeneous and isotropic field in a triangle; (b) formation of a caustic in a circle.*

and isotropic distribution of vibrational energy. But in circles, a ray is confined in the outer zone of a caustic. The interior of the caustic is never explored by the ray and the vibrational field is obviously not diffuse.

Various geometries have been studied. Let us mention some results. In a rectangle with an irrational length to width ratio, a ray explores the entire rectangle but takes only four directions. The resulting field is homogeneous but not isotropic with a single ray. But if the source emits several rays isotropically, the resulting field is also isotropic. As has been discussed, the case of a circle leads to a non-homogeneous and non-isotropic field. And in triangles only a few results are known.

In general, complexity in shape is a favourable condition to disorder. We can say roughly that the more complex the geometry the more ergodic the billiard. See Polack (1993) for a rapid overview of the billiards theory applied to acoustics.

The establishment of diffuse field in billiards may also be frustrated by effects of coherent rays. Generally, rays arriving at a receiver point have taken different paths during their propagation from the source and their relative phases are uncorrelated. But when the source and receiver coincide, for each ray a time-reversed version exists which has the same phase. Both rays interfere constructively giving rise to a local increase of energy. This is the so-called **enhanced backscatter** (Wright and Weaver, 2010, Chapter 8). See Weaver and Lobkis (2000) for experimental evidence of the effect.

7.9 Diffuse Field in Plates

Let us examine the emergence of diffuse field in a vibrating plate. Consider a rectangular plate in flexural vibration. The plate has dimensions $a \times b$ and is excited by a harmonic point force $F\mathrm{e}^{\imath\omega t}$ at x_0, y_0. The vibrational response $U(x,y)\mathrm{e}^{\imath\omega t}$ is governed by the steady-state Love equation,

$$\Delta^2 U - \kappa_0^4(1-\imath\eta)U = \frac{F}{D}\delta(x-x_0)\delta(y-y_0) \tag{7.38}$$

where D is the bending stiffness, $\kappa_0 = (m\omega^2/D)^{1/4}$ the wavenumber, m the mass per unit area, and η the damping loss factor. The boundary conditions impose any two values among the deflection, rotation of cross-section, moment, and transverse force at the edges. For instance,

$$U(0,y) = U(a,y) = U(x,0) = U(x,b) = 0 \tag{7.39}$$

$$\partial_x U(0,y) = \partial_x U(a,y) = \partial_y U(x,0) = \partial_y U(x,b) = 0 \tag{7.40}$$

for clamped edges.

The only physical parameters of this set of equations are the wavenumber κ_0, the damping loss factor η, the length a, and the width b; that is four physical parameters. Their only physical unit is the length (the time and mass do not appear in these parameters). The theorem of Vaschy–Buckingham gives the number of dimensionless parameters of this problem; $4 - 1 = 3$. These dimensionless parameters can be chosen arbitrarily provided that they are independent. Several choices are possible. The most natural way to construct dimensionless numbers is to choose the mean free path as characteristic length. The mean free path in a rectangle is, from eqn (7.5),

$$\bar{l} = \frac{\pi ab}{2(a+b)} \tag{7.41}$$

We may now introduce a dimensionless wavenumber,

$$\kappa = \frac{\kappa_0 \bar{l}}{2\pi} = \frac{ab\sqrt{\omega}}{4(a+b)}\left(\frac{m}{D}\right)^{\frac{1}{4}} \tag{7.42}$$

which is interpreted as the number of wavelengths per mean free path. The following shape ratio,

$$\epsilon = \frac{a+b}{\sqrt{\pi ab}} \tag{7.43}$$

is defined as the ratio between the perimeter of the plate to that of a circle of the same area. Finally the third dimensionless number may be simply the damping loss factor η.

This set of dimensionless parameters enable us to rewrite the governing equation and the related boundary conditions in dimensionless form. Let us introduce the dimensionless abscissa $x' = x/\bar{l}$, ordinate $y' = y/\bar{l}$, deflection $U'(x',y') = U(x'\bar{l}, y'\bar{l})/\bar{l}$, and force $F' = F\bar{l}/D$. Equation (7.38) becomes

$$\Delta^2 U' - (2\pi\kappa)^4(1 - \imath\eta)U' = F'\delta(x' - x'_0)\delta(y' - y'_0) \tag{7.44}$$

and eqns (7.39) and (7.40)

$$U'(0,y') = U'(a',y') = U'(x',0) = U'(x',b') = 0 \tag{7.45}$$

$$\partial_{x'} U'(0,y') = \partial_{x'} U'(a',y') = \partial_{y'} U'(x',0) = \partial_{y'} U'(x',b') = 0 \tag{7.46}$$

The problem of a vibrating plate is mathematically fully determined by the only three dimensionless parameters κ, ϵ, and η. Since the diffuse field state is a phenomenon that emerges in the sense that it can be observed from the solution to Love's equation, the region of diffuse field is necessarily confined into the three-dimensional space κ, ϵ, and η. However, the choice of κ, ϵ, and η as coordinates is arbitrary. And any other choice of independent dimensionless parameters is possible. In particular, the set of dimensionless parameters N, M, and $\bar{m}$ introduced in Section 7.8 is acceptable provided that a one-to-one map can be found,

$$\kappa, \eta, \epsilon \longrightarrow N, M, \bar{m}. \tag{7.47}$$

According to eqn (6.51), the modal density is

$$n(\omega) = \frac{ab}{4\pi}\sqrt{\frac{m}{D}} \tag{7.48}$$

and the number of modes below ω is

$$N = \int_0^{\omega} n(\omega)\, \mathrm{d}\omega = \frac{ab\omega}{4\pi}\sqrt{\frac{m}{D}} \tag{7.49}$$

The modal overlap is

$$M = \eta\omega n = \eta\omega\frac{ab}{4\pi}\sqrt{\frac{m}{D}} \tag{7.50}$$

The normalized attenuation factor is

$$\bar{m} = \frac{\eta\omega}{c_g}\bar{l} = \frac{\eta\sqrt{\omega}}{(D/m)^{1/4}}\frac{\pi\, ab}{4(a+b)} \tag{7.51}$$

These transformation relations are easily found by combining eqns (7.42), (7.43), and (7.49)–(7.51),

$$N = 4\kappa^2\epsilon^2 \tag{7.52}$$

$$M = 4\eta\kappa^2\epsilon^2 \tag{7.53}$$

$$\bar{m} = \pi\eta\kappa \tag{7.54}$$

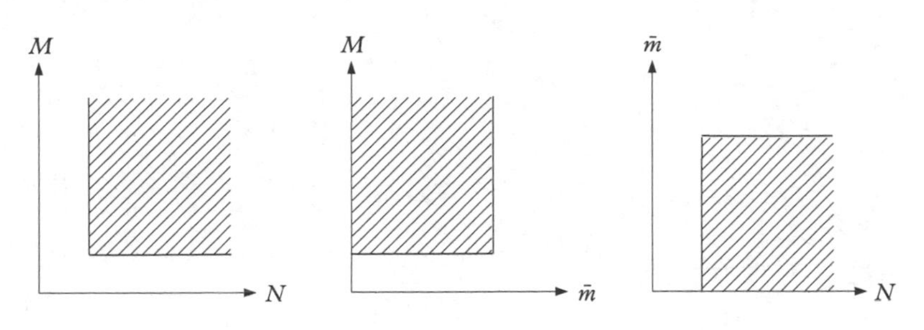

Figure 7.6 *Diffuse field region (hatched zone) in the N, M, $\bar{m}$-space.*

In the space N, M, $\bar{m}$, the region of diffuse state is delimited by the three conditions (7.33), (7.37), and (7.35). The surfaces which delimit this region have equations

$$N = 1 \tag{7.55}$$

$$M = 1 \tag{7.56}$$

$$\bar{m} = 1 \tag{7.57}$$

The region of diffuse field is shown in Fig. 7.6.

By the transformation (7.52), these equations become, in the κ, ϵ, η-space,

$$\kappa^2\epsilon^2 = \frac{1}{4} \tag{7.58}$$

$$\eta\kappa^2\epsilon^2 = \frac{1}{4} \tag{7.59}$$

$$\eta\kappa = \frac{1}{\pi} \tag{7.60}$$

respectively for $N = 1$, $M = 1$, and $\bar{m} = 1$. These equations define the varieties which delimit the domain of diffuse field in the κ, ϵ, η-space.

In the κ, η-plane, the condition (7.58) ($N = 1$) reads $\kappa \propto 1$, that is a vertical line whose position depends on ϵ. The condition (7.59) ($M = 1$) reads $\eta \propto \kappa^{-2}$, and the condition (7.60) ($\bar{m} = 1$) becomes $\eta \propto \kappa^{-1}$ a hyperbolic line.

In the κ, ϵ-plane, the condition (7.58) ($N = 1$) leads to $\epsilon \propto \kappa^{-1}$; the condition (7.59) ($M = 1$) also gives $\epsilon \propto \kappa^{-1}$ but usually at a higher level since the proportionality coefficient is like η^{-1}; and (7.60) ($\bar{m} = 1$) gives $\kappa \propto 1$, a vertical line.

In the η, ϵ-plane, the condition (7.58) ($N = 1$) gives $\epsilon \propto 1$ that is a horizontal line; the condition (7.59) ($M = 1$) has equation $\epsilon \propto \eta^{-1/2}$; and finally (7.60) ($\bar{m} = 1$) gives $\eta \propto 1$, a vertical line. This domain is represented in Fig. 7.7.

The reader may refer to Le Bot and Cotoni (2010) for more details on the above analysis. See also Weaver (1984) for a discussion of diffuse field in thick plates with SH and Lamb waves.

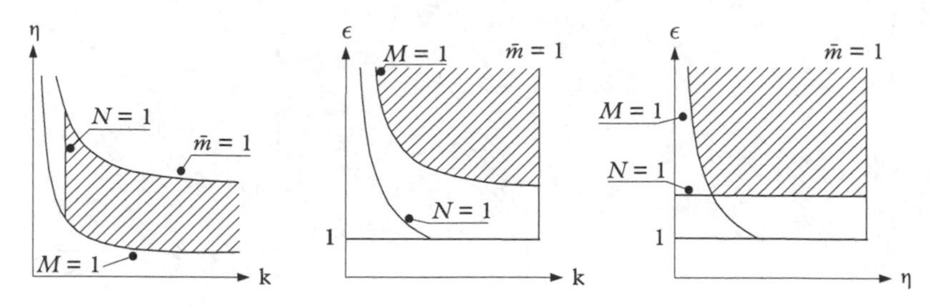

Figure 7.7 *Diffuse field region (hatched zone) in the κ, ϵ, η-space.*

Example

Figure 7.8 shows the diffuse field domain of a rectangular plate with a shape ratio $\epsilon = 1.13$. The grey levels and contour lines represent σ_{E}, the standard deviation of the vibrational energy normalized by the mean energy. The black colour indicates a low σ_{E} representative of a homogeneous energy field. The white colour represents a highly fluctuating field for which the energy varies rapidly inside the domain.

The energy field is obtained as follows. Equation (7.44) has been solved with simply supported boundary conditions for each pair of source x'_0, y'_0 and receiver x', y'. For each η, κ, the frequency response function $H'(\kappa') = U'/F'$ is finely computed in the octave band $\kappa' \in \left[\kappa/2^{1/4}, \kappa.2^{1/4}\right]$. The dimensionless kinetic energy at x', y' in the frequency band is computed by

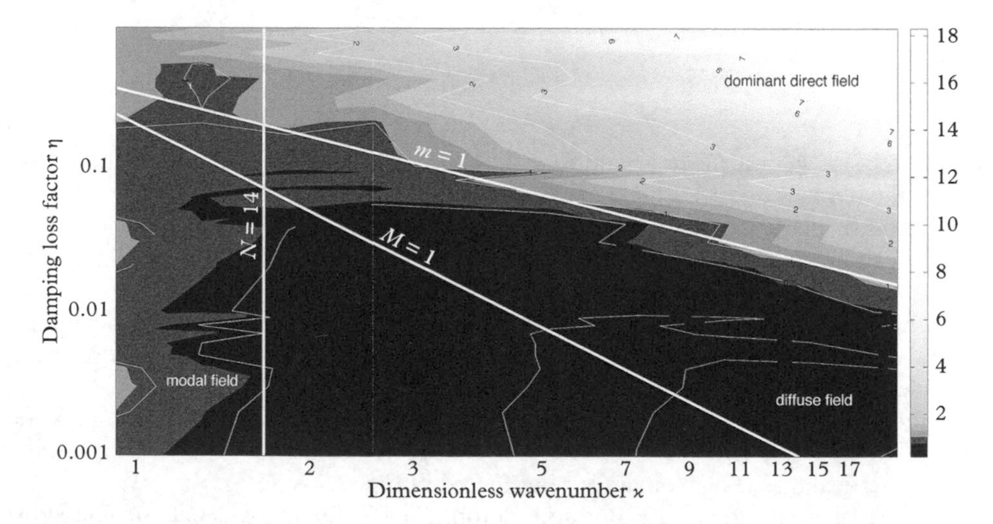

Figure 7.8 *Domain of diffuse field in the κ, η-plane for a rectangular plate ($\epsilon = 1.13$) with isovalues of σ_{E} (from Lafont et al. 2014 by permission of the Royal Society).*

$$\int \omega^2 |H'|^2 \, d\omega = \int 2\kappa'^5 |H'|^2 \, d\kappa' \tag{7.61}$$

where the change of variable $\omega = \kappa^2$ has been performed. The standard deviation is then estimated for 30,000 receivers chosen at random on the plate.

Different zones in the κ, η-plane may be observed from Fig. 7.8. On the left side (for low κ), the field is dominated by one or few modes and is consequently highly irregular. This is the modal domain. In the top right-hand corner (bright zone), the frequency is high so that the geometrical approximation applies. But absorption of rays is also important ($\bar{m} \gg 1$). Therefore the energy received at any point is dominated by the direct field emanating from the source. The reverberant field is negligible. The dark zone in the bottom right-hand corner is defined by $\sigma_E < 0.7$. This is the domain of diffuse field. We can observe that the left limit of the domain is a vertical line which matches with the value $N/\sqrt{2} = 10$, i.e. ten modes per octave band (white vertical line). The upper limit is more or less an inclined regular line corresponding to the white line $\bar{m} = 1$ (a straight line in a log scale). The line $M = 1$ (bottom white line) is less important to delimit the domain of diffuse field because the number of modes contained in the octave band is large. The frequency average is therefore not very sensitive to the position of the receiver. This would not be the case in pure tone where the magnitude of a resonant peak is highly dependent on the receiver position for a low modal overlap. We may conclude by these observations that the combination of Assumptions 5 (light damping) and 7 (high frequency) leads to diffuse field (Lafont et al., 2014).

8

Injected Power

In this chapter, we shall derive useful relationships to evaluate the power supplied to a subsystem by external sources.

8.1 Mobility

Definition

Consider a vibrating structure excited by a harmonic point force $F(\omega)$. At the point of excitation, the vibrational velocity in the direction of force is $V(\omega)$ and when the governing equation is linear the structure may be considered as a linear filter with input $F(\omega)$ and output $V(\omega)$. The corresponding frequency response function is called **driving-point mobility** (Appendix D) and is noted

$$Y(\omega) = \frac{V(\omega)}{F(\omega)} \tag{8.1}$$

When the receiver point at which V is evaluated differs from the source point where F is applied, we shall speak of **transfer mobility**.

The mobility is the Fourier transform of impulse response (time-evolution of the velocity after a unit shock excitation). Since the Fourier transform of a real function is Hermitian,

$$\overline{Y}(\omega) = Y(-\omega) \tag{8.2}$$

The **conductance** is defined as the real part of the mobility,

$$G(\omega) = \text{Re}[Y(\omega)] \tag{8.3}$$

Mobility and conductance are useful to evaluate the power supplied to a structure by a point force.

Foundation of Statistical Energy Analysis in Vibroacoustics. First Edition. A. Le Bot.

Mobility of finite structures

For a continuous finite structure, the mobility may be expressed in terms of natural modes. Let a system whose governing is

$$m\frac{\partial^2 u}{\partial t^2} + c\frac{\partial u}{\partial t} + \mathcal{K}u = Fe^{\imath\omega t}\delta(x - x_0) \tag{8.4}$$

where the right-hand side is a harmonic point force at frequency ω and position x_0, with $\mathcal{K}$ being the stiffness operator whose natural modes are ψ_α. But

$$u(x,t) = \sum_{\alpha \geq 0} U_\alpha(t)\psi_\alpha(x) \tag{8.5}$$

and by orthonormality,

$$m\ddot{U}_\alpha + c\dot{U}_\alpha + m\omega_\alpha^2 U_\alpha = Fe^{\imath\omega t}\psi_\alpha(x_0) \tag{8.6}$$

The forced solution is

$$U_\alpha(t) = H_\alpha(\omega)Fe^{\imath\omega t}\psi_\alpha(x_0) \tag{8.7}$$

where

$$H_\alpha(\omega) = \frac{1}{m\omega_\alpha^2\left[1 + 2\imath\zeta_\alpha\frac{\omega}{\omega_\alpha} - \left(\frac{\omega}{\omega_\alpha}\right)^2\right]} \tag{8.8}$$

and $2\zeta_\alpha\omega_\alpha = c/m$. Hence,

$$u(x,t) = \sum_{\alpha \geq 0} H_\alpha(\omega)Fe^{\imath\omega t}\psi_\alpha(x_0)\psi_\alpha(x) \tag{8.9}$$

By deriving and applying at x_0,

$$\dot{u}(x_0,t) = \sum_{\alpha \geq 0} \imath\omega H_\alpha(\omega)Fe^{\imath\omega t}\psi_\alpha^2(x_0) \tag{8.10}$$

The mobility at x_0 is finally found by taking the ratio $\dot{u}(x_0,t)/Fe^{\imath\omega t}$,

$$Y(x_0,\omega) = \sum_{\alpha \geq 0} \imath\omega H_\alpha(\omega)\psi_\alpha^2(x_0) \tag{8.11}$$

The above equation shows that the mobility may be calculated at any point as soon as the natural frequencies and modes are completely determined.

8.2 Power Supplied to a Structure

Two cases are considered: a point force and a pressure force field.

Point excitation

The power supplied by a point source to a continuous structure is $\langle P\rangle = \langle fv\rangle$ where f is the random force and v the velocity at the contact point in the direction of force. Let S_{fv} be the cross-power spectral density between force and velocity,

$$\langle P\rangle = \frac{1}{2\pi}\int_{-\infty}^{\infty} S_{fv}(\omega)\,\mathrm{d}\omega \tag{8.12}$$

Since $S_{fv} = YS_{ff}$ where $Y(x_0, \omega)$ is the mobility at the contact point x_0,

$$\langle P\rangle = \frac{1}{2\pi}\int_{-\infty}^{\infty} S_{ff}(\omega)\,Y(x_0, \omega)\,\mathrm{d}\omega \tag{8.13}$$

If f has a constant power spectral density in the frequency band $\Delta\omega$,

$$S_{ff}(\omega) = \begin{cases} S_0 & \text{if } |\omega| \in \Delta\omega \\ 0 & \text{otherwise} \end{cases} \tag{8.14}$$

Hence,

$$\langle P\rangle = \frac{S_0}{2\pi}\int_{\Delta\omega} [Y(x_0, -\omega) + Y(x_0, \omega)]\,\mathrm{d}\omega \tag{8.15}$$

But $Y(x_0, -\omega) = \overline{Y}(x_0, \omega)$,

$$\langle P\rangle = \frac{S_0}{\pi}\int_{\Delta\omega} \mathrm{Re}[Y(x_0, \omega)]\,\mathrm{d}\omega \tag{8.16}$$

The expectation of the square of f is obtained by integration of S_{ff} divided by 2π giving $\langle f^2\rangle = S_0\Delta\omega/\pi$. So,

$$\langle P\rangle = \langle f^2\rangle \times \frac{1}{\Delta\omega}\int_{\Delta\omega} \mathrm{Re}[Y(x_0, \omega)]\,\mathrm{d}\omega \tag{8.17}$$

The power supplied to a structure by a point force is therefore equal to the product of the mean square force $\langle f^2\rangle$ by the frequency-average of the conductance. The question of the power supplied by a point force then reduces to the determination of the frequency-average conductance at the contact point.

Distributed excitation

Suppose now that $p(x,t)$ is a random pressure field. The total power supplied to the structure is $\langle P\rangle = \int\langle pv\rangle\, dA$ where the integral is performed over the structure surface. Introducing the cross-power spectral density $S_{pv}(\omega)$ between pressure at x and velocity at x,

$$\langle P\rangle = \int_A \frac{1}{2\pi}\int_{-\infty}^{\infty} S_{pv}(\omega)\,\mathrm{d}\omega\mathrm{d}A \tag{8.18}$$

But the continuous version of eqn (1.84) reads

$$S_{pv}(\omega) = \int_A Y(x,x',\omega)S_{pp'}(\omega)\,\mathrm{d}A' \tag{8.19}$$

where $Y(x,x',\omega)$ is the transfer mobility between a force at x' and velocity at x, and $S_{pp'}$ the power spectral density of pressure respectively at x and x'. The integration is performed with respect to x'. It yields

$$\langle P\rangle = \frac{1}{2\pi}\int_A\int_{-\infty}^{\infty}\int_A Y(x,x',\omega)S_{pp'}(\omega)\,\mathrm{d}A'\mathrm{d}\omega\mathrm{d}A \tag{8.20}$$

Consider now the case of a rain-on-the-roof pressure field with power spectral density S_0 in $\Delta\omega$. The cross-power is

$$S_{pp'}(\omega) = \begin{cases} S_0\delta(x-x') & \text{if } |\omega|\in\Delta\omega \\ 0 & \text{otherwise}\end{cases} \tag{8.21}$$

and the innermost integral of eqn (8.20) reduces. Hence,

$$\langle P\rangle = \frac{S_0}{\pi}\int_A\int_{\Delta\omega}\mathrm{Re}[Y(x,\omega)]\,\mathrm{d}\omega\mathrm{d}A \tag{8.22}$$

where we have noted $Y(x,\omega) = Y(x,x,\omega)$ the point mobility. Further, integrating S_{pp} over both space and frequency and dividing by 2π gives $\langle p^2\rangle = S_0\Delta\omega/\pi A$.

$$\langle P\rangle = \langle p^2\rangle A^2 \times \frac{1}{A\Delta\omega}\int_A\int_{\Delta\omega}\mathrm{Re}[Y(x,\omega)]\,\mathrm{d}\omega\mathrm{d}A \tag{8.23}$$

Again, the power supplied by a pressure field to a structure is the product of the mean square force $\langle p^2\rangle A^2$ by the mean conductance. Prediction of injected power reduces to the knowledge of the mean mobility.

8.3 Statistical Estimation of Mobility

For a finite structure whose natural modes are ψ_α and frequencies ω_α, the point mobility at x_0 is given by eqn (8.11). This allows a further calculation of the frequency-average mobility introduced in eqns (8.17) and (8.23). By eqn (8.11),

$$\frac{1}{\Delta\omega}\int_{\Delta\omega} \mathrm{Re}\,[Y]\;\mathrm{d}\omega = \frac{1}{\Delta\omega}\int_{\Delta\omega}\sum_{\alpha\geq 0}\psi_\alpha^2(x_0)\mathrm{Re}\,[\imath\omega H_\alpha(\omega)]\;\mathrm{d}\omega \tag{8.24}$$

After re-arranging,

$$\frac{1}{\Delta\omega}\int_{\Delta\omega} \mathrm{Re}\,[Y]\;\mathrm{d}\omega = \frac{1}{\Delta\omega}\sum_{\alpha\geq 0}\psi_\alpha^2(x_0)\times\int_{\Delta\omega}\mathrm{Re}\,[\imath\omega H_\alpha(\omega)]\;\mathrm{d}\omega \tag{8.25}$$

Two cases may arise concerning the frequency integral. For a non-resonant mode $\omega_\alpha \notin \Delta\omega$ the contribution of the integral to the sum is simply neglected (Assumption 6). For a resonant mode $\omega_\alpha \in \Delta\omega$ the integration bandwidth $\Delta\omega$ is sufficiently large compared with the resonance peak $2\zeta_\alpha\omega_\alpha$ (Assumption 5) so that the frequency limits may be extended to infinity.

$$\frac{1}{\Delta\omega}\int_{\Delta\omega} \mathrm{Re}\,[Y]\;\mathrm{d}\omega = \frac{1}{\Delta\omega}\sum_{\alpha\in\Delta\omega}\psi_\alpha^2(x_0)\times\int_0^\infty\mathrm{Re}\,[\imath\omega H_\alpha(\omega)]\;\mathrm{d}\omega \tag{8.26}$$

By eqn (C.34 (Appendix C)) the value of the integral is $\pi/2m$ where m is the mass per unit surface. Therefore,

$$\frac{1}{\Delta\omega}\int_{\Delta\omega} \mathrm{Re}\,[Y]\;\mathrm{d}\omega = \frac{\pi}{2m\Delta\omega}\sum_{\alpha\in\Delta\omega}\psi_\alpha^2(x_0) \tag{8.27}$$

The frequency-average mobility therefore depends on the statistical term $\sum\psi_\alpha^2$, encountered in Chapter 4.

Let us now calculate the frequency and space average of the mobility,

$$\frac{1}{A\Delta\omega}\int_A\int_{\Delta\omega} \mathrm{Re}\,[Y]\;\mathrm{d}\omega\mathrm{d}A = \frac{\pi}{2mA\Delta\omega}\sum_{\alpha\in\Delta\omega}\int_A\psi_\alpha^2(x_0)\;\mathrm{d}A \tag{8.28}$$

But $\int\psi_\alpha^2\,\mathrm{d}A = 1$, hence,

$$\frac{1}{A\Delta\omega}\int_A\int_{\Delta\omega} \mathrm{Re}\,[Y]\;\mathrm{d}\omega\mathrm{d}A = \frac{\pi}{2mA\Delta\omega}\sum_{\alpha\in\Delta\omega}1 \tag{8.29}$$

The sum is therefore the number ΔN of resonant modes. Introducing the modal density $n = \Delta N / \Delta\omega$ in the bandwidth and the total mass $M = mA$,

$$\frac{1}{A\Delta\omega}\int_A \int_{\Delta\omega} \mathrm{Re}\,[Y]\;\mathrm{d}\omega\mathrm{d}A = \frac{\pi\, n(\omega)}{2M} \tag{8.30}$$

Surprisingly, the final expression of frequency and space-averaged mobility is very simple. All the system complexity, such as wave type, system shape, boundary conditions, and so on, no longer appears in the final expression. The average mobility just depends on the modal density and total mass. This highlights the role of modal density as a measure of the capacity to absorb energy.

8.4 Mobility of Infinite Systems

Next, we calculate the mobilities of some infinite systems, summarized in Table 8.1. Other formulas are available (see Cremer and Heckl 1988, Chapter IV; Fahy and Gardonio 2007, Chapter 2; and Goyder and White 1980a, 1980b, 1980c).

Table 8.1 *Driving-point mobility of various infinite systems with E Young's modulus, G shear modulus, D bending stiffness, T tension, S cross-section area, I second moment of area, J moment of inertia per unit length, and m mass per unit length or area.*

System	Motion	Excitation	Diagram	Mobility
String	transversal	force	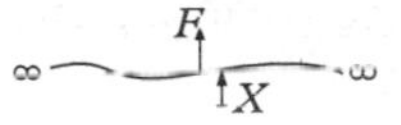	$\frac{\dot{X}(\omega)}{F(\omega)} = \frac{1}{2\sqrt{mT}}$
Rod	longitudinal	force	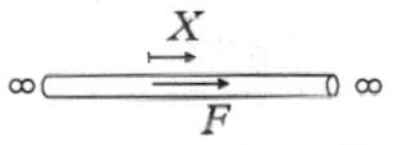	$\frac{\dot{X}(\omega)}{F(\omega)} = \frac{1}{2\sqrt{mES}}$
Rod	longitudinal	force		$\frac{\dot{X}(\omega)}{F(\omega)} = \frac{1}{\sqrt{mES}}$
Bar	torsional	moment	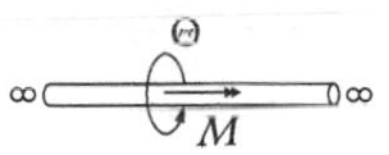	$\frac{\dot{\Theta}(\omega)}{M(\omega)} = \frac{1}{2\sqrt{GJ}}$
Beam	flexural	force	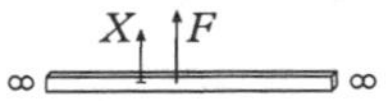	$\frac{\dot{X}(\omega)}{F(\omega)} = \frac{1-\imath}{4m^{3/4}(EI)^{1/4}\sqrt{\omega}}$
Beam	flexural	moment	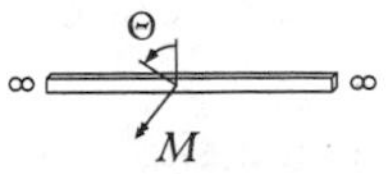	$\frac{\dot{\Theta}(\omega)}{M(\omega)} = \frac{(1+\imath)\sqrt{\omega}}{4m^{1/4}(EI)^{3/4}}$
Plate	flexural	force	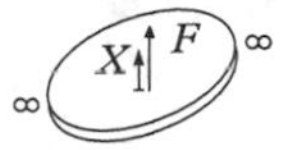	$\frac{\dot{X}(\omega)}{F(\omega)} = \frac{1}{8\sqrt{mD}}$

String

As a first example, we consider a vibrating string of infinite extent. The governing equation of a vibrating string was introduced in Section 4.2. A harmonic point force of magnitude F and frequency ω is applied at x_0. The force field is therefore $f(x,t) = Fe^{\imath\omega t}\delta(x-x_0)$. The transverse deflection is $u(x,t) = U(x,\omega)e^{\imath\omega t}$ where $U(x,\omega)$ is complex. So, substituting $f(x,t)$ and $u(x,t)$ into eqn (4.27) gives

$$\frac{d^2U}{dx^2} + k^2U = -\frac{F}{T}\delta(x-x_0) \tag{8.31}$$

where the wavenumber k is

$$k^2 = \frac{m\omega^2}{T} \tag{8.32}$$

Since the vibrational velocity is related to the deflection by $V(x,\omega) = \imath\omega U(x,\omega)$, the mobility is $Y_\infty(\omega) = \imath\omega U(x_0,\omega)/F(\omega)$. To explain the mobility, we need to introduce the harmonic Green function of an infinite string. The Green function $g(x,x_0,\omega)$ is the fundamental solution to the Helmholtz equation,

$$\frac{\partial^2 g}{\partial x^2} + k^2 g = -\delta(x-x_0) \tag{8.33}$$

where x_0 denotes the source point and x the receiver point. We have not yet specified the boundary conditions. Since the string is assumed to be infinite, the only waves that can emanate from a source point are those which verify the Sommerfeld condition,

$$\lim_{|x|\to\infty}\left(\frac{\partial}{\partial|x|} + \imath k\right) g(x,x_0,\omega) = 0 \tag{8.34}$$

(The sign before $\imath k$ depends on the time convention, here $e^{\imath\omega t}$.) The Green function is

$$g(x,x_0,\omega) = \frac{1}{2\imath k}e^{-\imath k|x-x_0|} \tag{8.35}$$

To check that the above expression does give the Green function of an infinite string, we first derive with respect to x,

$$\frac{\partial g}{\partial x} = -\frac{1}{2}e^{-\imath k|x-x_0|}\operatorname{sgn}(x-x_0) \tag{8.36}$$

where we have introduced the sign function as the derivative of the absolute value function $|x|' = \operatorname{sgn}(x)$. By deriving once again,

$$\frac{\partial^2 g}{\partial x^2} = \frac{\imath k}{2}e^{-\imath k|x-x_0|} - \delta(x-x_0) \tag{8.37}$$

since the derivative of the sign function is twice the Dirac function. In the right-hand side, we may recognize $-k^2 g$ which proves eqn (8.33).

The mobility of an infinite string is therefore

$$Y_\infty(\omega) = \frac{\iota\omega}{T} g(x_0, x_0) \tag{8.38}$$

By substituting eqns (8.32) and (8.35) into eqn (8.38),

$$Y_\infty(\omega) = \frac{1}{2\sqrt{mT}} \tag{8.39}$$

The point mobility of an infinite string is purely real and independent of frequency.

This result merits a comment. The fact that the imaginary part is zero means that no reactive power is injected in the system when exciting by a harmonic point force. But the most important remark is that this result matches well with the statistical estimation of the mean mobility of eqn (8.30). The modal density of a string is $n = L/\pi c$ where $c = (T/m)^{1/2}$ and L is the string length. Furthermore the total mass is $M = mL$, so that eqn (8.30) gives the mean conductance $\pi n/2M = \pi/2mL \times L/\pi\,(T/m)^{1/2}$ in a remarkable agreement with eqn (8.39). The point mobility of an infinite string is therefore equal to the mean conductance of a finite string.

Rod

The case of longitudinal vibrations in rods is formally equivalent to that of transverse vibrations in strings. The analogy is $T \longleftrightarrow ES$ and $m \longleftrightarrow \rho S$ where ρ is the material density, E the Young modulus, and S the cross-section area. So, if the excitation force is applied to a single point and if the rod has an infinite extent, then the point mobility is

$$Y_\infty(\omega) = \frac{1}{2S\sqrt{\rho E}} \tag{8.40}$$

Another interesting case is when the rod is semi-infinite. The force $Fe^{\iota\omega t}$ is applied at the free end of a rod.

$$\frac{d^2U}{dx^2} + k^2 U = 0 \tag{8.41}$$

where the wavenumber k is defined by $k^2 = \rho\omega^2/E$. The force is applied at $x = 0$,

$$F = -ES\frac{dU}{dx}(0, \omega) \tag{8.42}$$

The solution to the two above equations with the Sommerfeld condition is

$$U(x, \omega) = \frac{F}{ESk} e^{-\iota kx} \tag{8.43}$$

The point mobility is $Y_{\infty/2} = \imath\omega U(0,\omega)/F$,

$$Y_{\infty/2}(\omega) = \frac{1}{S\sqrt{\rho E}} \tag{8.44}$$

The point mobility of a semi-infinite rod is twice the point mobility of the infinite rod.

Beam

For transverse vibrations in beams subjected to a harmonic point force $Fe^{\imath\omega t}$ at x_0, the governing equation is

$$\frac{d^4U}{dx^4} - k^4U = \frac{F}{EI}\delta(x - x_0) \tag{8.45}$$

where EI is the bending stiffness and the wavenumber k is

$$k^4 = \frac{m\omega^2}{EI} \tag{8.46}$$

m being the mass per unit length.

The harmonic Green function $g(x, x_0, \omega)$ satisfies

$$\frac{\partial^4 g}{\partial x^4} - k^4 g = \delta(x - x_0) \tag{8.47}$$

jointly with the Sommerfeld condition. We find

$$g(x, x_0, \omega) = \frac{1}{4\imath k^3}e^{-\imath k|x-x_0|} - \frac{1}{4k^3}e^{-k|x-x_0|} \tag{8.48}$$

This may be verified by a direct substitution. For instance, the second derivative of g is

$$\frac{\partial^2 g}{\partial x^2} = \frac{\imath}{4k}e^{-\imath k|x-x_0|} - \frac{1}{4k}e^{-k|x-x_0|} \tag{8.49}$$

Two additional derivatives give eqn (8.47). The mobility of an infinite beam is therefore

$$Y_\infty(\omega) = \frac{\imath\omega}{EI}g(x_0, x_0) = \frac{\omega}{4EIk^3}(1 - \imath) \tag{8.50}$$

Substituting eqn (8.46),

$$Y_\infty(\omega) = \frac{1 - \imath}{4m^{3/4}(EI)^{1/4}\sqrt{\omega}} \tag{8.51}$$

Again, the asymptotic modal density $n = L/\pi c_g$ where $c_g = 2(EI/m)^{1/4}\sqrt{\omega}$ and the total mass $M = mL$ give the ratio $\pi n/2M = 1/4m^{3/4}(EI)^{1/4}\sqrt{\omega}$ in perfect agreement with the real part of Y_∞.

Another case of interest is a point excitation by a moment. For a moment density is $Me^{\imath\omega t}\delta(x - x_0)$, and the governing equation becomes

$$\frac{d^4U}{dx^4} - k^4U = -\frac{M}{EI}\delta'(x - x_0) \tag{8.52}$$

where δ' denotes the derivative of the Dirac function. The mechanical power is the product of the moment M and the rotation speed of cross-section dV/dx. The relevant mobility Y is the response of the filter whose input is M and output dV/dx and is therefore defined as the ratio $dV/dx \times 1/M$.

By deriving eqn (8.47) with respect to x, it is clear that the Green function associated with this problem is $-g'$. Since dV/dx is obtained by deriving again with respect to x and t, the mobility is finally,

$$Y_\infty(\omega) = -\frac{\imath\omega}{EI}\frac{\partial^2 g}{\partial x^2}(x_0, x_0) = \frac{-\imath\omega}{4EIk}(\imath - 1) \tag{8.53}$$

or,

$$Y_\infty(\omega) = \frac{(1+\imath)\sqrt{\omega}}{4m^{1/4}(EI)^{3/4}} \tag{8.54}$$

This is the mobility of an infinite beam excited by a moment.

Plate

The governing equation of an infinite plate excited by a unit harmonic force $e^{\imath\omega t}$ at the origin is

$$\Delta^2 U - \kappa^4 U = \frac{1}{D}\delta(x)\delta(y) \tag{8.55}$$

where D is the bending stiffness, $\kappa = (m\omega^2/D)^{1/4}$ the wavenumber, and m the mass per unit area. The mobility is $Y(\omega) = \imath\omega U(0, 0, \omega)$.

By taking the spatial Fourier transform of U,

$$\widehat{U}(k_1, k_2, \omega) = \int_{-\infty}^{\infty}\int_{-\infty}^{\infty} U(x, y, \omega)e^{-\imath(k_1 x + k_2 y)}\,dxdy \tag{8.56}$$

the governing equation becomes

$$\left[(k_1^2 + k_2^2)^2 - \kappa^4\right]\widehat{U}(k_1, k_2, \omega) = \frac{1}{D} \tag{8.57}$$

And, by inverse Fourier transform,

$$U(0,0,\omega) = \frac{1}{4\pi^2 D}\int_{-\infty}^{\infty}\int_{-\infty}^{\infty}\frac{\mathrm{d}k_1\,\mathrm{d}k_2}{\left[(k_1^2+k_2^2)^2-\kappa^4\right]} \tag{8.58}$$

The integrand only depends on the norm of the vector (k_1, k_2). In polar coordinates,

$$U(0,0,\omega) = \frac{1}{2\pi D}\int_{0}^{\infty}\frac{k\,\mathrm{d}k}{k^4-\kappa^4} \tag{8.59}$$

By the change of variable $z = k^2$,

$$U(0,0,\omega) = \frac{1}{4\pi D}\int_{0}^{\infty}\frac{\mathrm{d}z}{z^2-\kappa^4} \tag{8.60}$$

and by symmetry,

$$U(0,0,\omega) = \frac{1}{8\pi D}\int_{-\infty}^{\infty}\frac{\mathrm{d}z}{z^2-\kappa^4} \tag{8.61}$$

To evaluate the integral by the residue theorem, we may introduce a damping by substituting $\kappa^4 \mapsto \kappa^4(1-\imath\eta)^2$,

$$U(0,0,\omega,\eta) = \frac{1}{8\pi D}\int_{-\infty}^{\infty}\frac{\mathrm{d}z}{z^2-\kappa^4(1-\imath\eta)^2} \tag{8.62}$$

The rational function has two poles, $\kappa^2(1-\imath\eta)$ and $-\kappa^2(1-\imath\eta)$. Choosing a contour as in Fig. 8.1 and remarking that the residue at $\kappa^2(1-\imath\eta)$ is $1/2\kappa^2(1-\imath\eta)$ gives

$$U(0,0,\omega,\eta) = \frac{-2\imath\pi}{16\pi D\kappa^2(1-\imath\eta)} \tag{8.63}$$

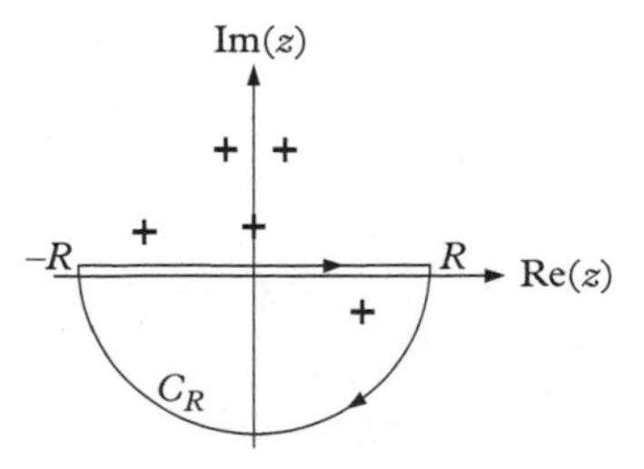

Figure 8.1 *Integration path of eqn (8.62).*

The minus sign in the numerator comes from the choice of integration path, which goes around the pole in the clockwise direction. Taking the limit $\eta \to 0$, multiplying by $\imath\omega$, and simplifying leads to

$$Y_\infty(\omega) = \frac{1}{8\sqrt{mD}} \tag{8.64}$$

The mobility of an infinite plate is purely real and is independent of frequency. Again, this result matches with the mean conductance $\pi n/2M$ where the modal density is $n = S\omega/2\pi c_\mathrm{p} c_\mathrm{g}$ with $c_\mathrm{g} = 2c_\mathrm{p} = 2(D/m)^{1/4}\sqrt{\omega}$ and the total mass $M = mS$.

General case

In all previous examples, we have remarked that the real part of the point mobility of an infinite system is $\pi n/2M$ where n is the asymptotic expression of the modal density and M the total mass of the same but finite system (of any extent since the ratio n/M does not depend on the surface). This result is in fact quite general (Bourgine, 1973) as we shall see now.

Let us consider an infinite two-dimensional structure whose governing equation is eqn (8.4),

$$m\frac{\partial^2 u}{\partial t^2} + \mathcal{K}u = F\mathrm{e}^{\imath\omega_0 t}\delta(x)\delta(y) \tag{8.65}$$

where $\mathcal{K}$ is the stiffness operator. The dispersion equation is

$$D(\kappa_1, \kappa_2) - m\omega^2 = 0 \tag{8.66}$$

where $D(\kappa_1, \kappa_2)$ is the polynomial function obtained from $\mathcal{K}$ by the substitution $\partial/\partial x \leftrightarrow \imath\kappa_1$, $\partial/\partial y \leftrightarrow \imath\kappa_2$. For instance, we have $D(\kappa_1, \kappa_2) = D(\kappa_1^2 + \kappa_2^2)^2$ for a plate where D denotes the bending stiffness in the right-hand side, while $D(\kappa_1, \kappa_2) = -T(\kappa_1^2 + \kappa_2^2)$ for a membrane. The point mobility is $Y_\infty(\omega_0) = \dot{u}(0, 0, t)/F\mathrm{e}^{\imath\omega_0 t}$. By a Fourier transform,

$$Y_\infty(\omega_0) = \frac{1}{4\pi^2}\int_{-\infty}^{\infty}\int_{-\infty}^{\infty} \frac{\imath\omega_0}{D(\kappa_1, \kappa_2) - m\omega_0^2}\,\mathrm{d}\kappa_1\mathrm{d}\kappa_2 \tag{8.67}$$

In polar coordinates $\mathrm{d}\kappa_1\mathrm{d}\kappa_2 = \kappa\,\mathrm{d}\kappa\mathrm{d}\varphi$,

$$Y_\infty(\omega_0) = \frac{1}{4\pi^2}\int_0^{\infty}\int_0^{2\pi} \frac{\imath\omega_0}{D(\kappa, \varphi) - m\omega_0^2}\kappa\,\mathrm{d}\kappa\mathrm{d}\varphi \tag{8.68}$$

Let ω be the positive root of eqn (8.66) for any (κ_1, κ_2). By the change of variables $(\kappa, \varphi) \mapsto (\omega, \varphi)$,

$$Y_\infty(\omega_0) = \frac{1}{4\pi^2} \int_0^{2\pi} \int_0^\infty \frac{\iota\omega_0}{m(\omega^2 - \omega_0^2)} \kappa(\omega) \frac{\mathrm{d}\kappa}{\mathrm{d}\omega}(\omega)\, \mathrm{d}\omega \mathrm{d}\varphi \tag{8.69}$$

By the residue theorem with the integration path shown in Fig. 8.1,

$$\int_{-\infty}^{\infty} \frac{f(\omega)}{\omega^2 - \omega_0^2(1 - \iota\eta)^2}\, \mathrm{d}\omega = -2\iota\pi \frac{f(\omega_0(1 - \iota\eta))}{2\omega_0(1 - \iota\eta)} \tag{8.70}$$

for any $\eta > 0$. Taking the limit $\eta \to 0$ and substituting the above integral into eqn (8.69),

$$Y_\infty(\omega_0) = \frac{1}{4m} \kappa(\omega_0) \frac{\mathrm{d}\kappa}{\mathrm{d}\omega}(\omega_0) = \frac{\omega_0}{4mc_\varphi c_g} \tag{8.71}$$

where we have introduced the phase speed c_φ and group speed c_g. Let us multiply both numerator and denominator by S (whose actual value is of no importance). By eqn (6.51), the modal density is $n = S\omega_0/2\pi c_\varphi c_g$ while the total mass is $M = mS$. Substituting gives

$$Y_\infty(\omega_0) = \frac{\pi n}{2M} \tag{8.72}$$

Although the present proof has been detailed in two dimensions, it may be easily adapted to one and three dimensions. In all cases, we would find that the conductance (real part of Y_∞) is $\pi n/2M$. This result is consistent with all special cases previously studied.

Equivalence mean conductance–infinite system conductance

We have proved that the infinite system conductance is $\pi n/2M$ where n is the asymptotic expression of the modal density. Of course the ratio n/M does not depend on the extent of the system since we found in Chapter 6 that the asymptotic modal density is proportional to length, surface, or volume as well as M. But eqn (8.30) also gives the mean conductance of finite systems as $\pi n/2M$. We may therefore write

$$\frac{1}{A\Delta\omega} \int_A \int_{\Delta\omega} \mathrm{Re}\,[Y]\, \mathrm{d}\omega \mathrm{d}A = \mathrm{Re}\,[Y_\infty(\omega)] \tag{8.73}$$

That is, the space and frequency average of the conductance of a finite system is equal to the conductance of the equivalent infinite system.

8.5 Comments

Skudrzyk (1958) has shown that the point mobility of a finite system may be approximated by the mobility of the same system of infinite extent. This result was originally formulated without invoking any average and therefore constitutes a stronger result than the one established in eqn (8.73). Skudrzyk's calculation, well developed in his book (Skudrzyk, 1968), starts from eqn (8.11) in which the discrete sum is replaced by an integral over ω_α after multiplying by the modal density. The obtained integral is tractable in some special cases and gives the expected result. Although the result is stronger, it is based on more restrictive assumptions.

The method developed in this text rather follows the approach adopted by Manning (1994), except that the ensemble average is here replaced by a spatial average. The final result given in eqn (8.30) gives no local information since we have chosen to express it on a frequency and space average of mobility, as in Cremer and Heckl (1988, Chapter IV), and Wijker (2009, Appendix I). Nevertheless, this is sufficient for a prediction of power supplied by a rain-on-the-roof force field which is the most interesting situation in the strict framework of statistical energy analysis as presented in the earlier chapters of this book.

According to eqn (8.30), the mean conductance is $\pi n/2M$, and this must be connected with the power supplied to a single resonator $S_i/2m_i$, given in eqn (2.112). If the continuous structure is viewed as a set of N resonators of same mass m_i excited by uncorrelated random forces of same power spectral density S_i then the total power supplied to the set is $\sum_i S_i/2m_i = NS_i/2m_i$. But $S_i = \pi \langle f^2 \rangle / N\Delta\omega$ where $\langle f^2 \rangle$ is the total square force applied to the structure. Furthermore, by considering that the total mass is $M = m_i N$ and the modal density $n = N/\Delta\omega$, the supplied power becomes $\pi \langle f^2 \rangle n/2M$, which is the expected result.

9

Coupling Loss Factor

The effective determination of coupling loss factors in all cases of interest is usually difficult using a direct modal approach. In this chapter, we rather adopt the wave approach to derive asymptotic relationships for coupling loss factors.

9.1 General Method

The general problem is the following. We consider two interacting subsystems with vibrational energies E_i for $i = 1, 2$. The coupling power proportionality (7.31) allows us to split the exchanged power P_{12} as a difference of two terms $P_{1\to 2}$ and $P_{2\to 1}$ as in eqn (7.27). The power flowing from subsystem 1 to subsystem 2 is

$$P_{1\to 2} = \omega\eta_{12}E_1 \tag{9.1}$$

while the power flowing from 2 to 1 is $P_{2\to 1} = \omega\eta_{21}E_2$. We have shown in Chapter 7 that the coupling power proportionality is valid when the field is diffuse. In the context of the wave approach of statistical energy analysis, we may raise this condition to the rank of assumption,

Assumption 8 *Vibrational fields are diffuse.*

The method consists in considering the special situation where subsystem 1 provides energy to subsystem 2 but not the converse. In that case $P_{2\to 1} = 0$. The coupling loss factor η_{12} is then determined by calculating separately $P_{1\to 2}$ and E_1 and taking the ratio $P_{1\to 2}/\omega E_1$. The reciprocal coupling loss factor η_{21} is next determined by reciprocity,

$$\eta_{12}n_1 = \eta_{21}n_2 \tag{9.2}$$

where n_i for $i = 1, 2$ are the modal densities.

To find such a situation, we apply the principle of locality. This principle, valid at the limit of high frequencies, states that the transmission process between subsystems only depends on the immediate vicinity of the boundary. In particular, this process remains

Foundation of Statistical Energy Analysis in Vibroacoustics. First Edition. A. Le Bot.

unchanged if we replace the two subsystems with two equivalent subsystems of infinite extent, provided that the local characteristics of the boundary are not modified.

Since the subsystems have infinite extent, a single wave propagating towards the boundary is split into reflected and transmitted waves. These waves then move away from the boundary and never come back. This canonical problem may generally be solved analytically giving access to reflection and transmission coefficients. The energy carried out by the waves and the power passing through the boundary are totally determined by the knowledge of these coefficients.

9.2 Point-Coupled Subsystems

We first recall a coupling loss factor, obtained in Chapter 3 in eqn (3.50) by the modal approach.

Two subsystems of mass M_1 and M_2 coupled by a stiffness K, a gyroscopic constant G, and a coupling inertia M have the coupling loss factor

$$\eta_{12} = \frac{\pi\,[(K^2 + M\omega^2) + \omega^2 G]n_2}{2\omega^3 M_1 M_2} \tag{9.3}$$

where n_2 is the modal density of subsystem 2. It is clear by interchanging the subscripts that reciprocity applies. This coupling loss factor appears in Newland (1968, eqn (25)) but is also mentioned in numerous other texts (including Lyon, 1975; and Keane and Price, 1987).

To derive eqn (9.3) by the wave approach, we start by assuming the mobility expression of subsystem 2 to be of infinite extent. Since we have established in Chapter 8 that the mobility of infinite systems is equal to the mean mobility, eqn (8.30) gives

$$Y_2 = \frac{\pi n_2}{2M_2} \tag{9.4}$$

If subsystem 1 exerts a random force f on subsystem 2, then the transmitted power is

$$P_{1\to 2} = \langle f^2 \rangle Y_2 \tag{9.5}$$

where $\langle f^2 \rangle$ is the expectation of the square force. When the coupling is realized through elastic, gyroscopic, and inertial constants respectively equal to K, G, and M, the force applied by subsystem 1 is

$$f = Ku_1 + G\dot{u}_1 - M\ddot{u}_1 \tag{9.6}$$

where u_1 is the deflection of subsystem 1 at the coupling point. If a positive displacement of 1 induces a repulsive force on 2 then a negative acceleration of 1 results from a negative force acting on 1 and therefore a positive force on 2 by the third Newton law.

Squaring the above expression, taking the expectation, and remarking that $\langle u_1 \dot{u}_1 \rangle = \langle \dot{u}_1 \ddot{u}_1 \rangle = 0$ by eqn (1.48),

$$\langle f^2 \rangle = \left(K^2 \langle u_1^2 \rangle + G^2 \langle \dot{u}_1^2 \rangle - 2MK \langle u_1 \ddot{u}_1 \rangle + M^2 \langle \ddot{u}_1^2 \rangle \right) \tag{9.7}$$

But if the energy is confined in the frequency bandwidth centred on ω then $\langle \dot{u}_1^2 \rangle \simeq \omega^2 \langle u_1^2 \rangle$, $\langle u_1 \ddot{u}_1 \rangle \simeq -\omega^2 \langle u_1^2 \rangle$, and $\langle \ddot{u}_1^2 \rangle \simeq \omega^4 \langle u_1^2 \rangle$. Hence,

$$\langle f^2 \rangle = \left[(K + M\omega^2)^2 + G^2 \omega^2 \right] \langle u_1^2 \rangle \tag{9.8}$$

Furthermore, by Assumption 8, the vibrational energy is equally spread over subsystem 1. Then $\langle u_1^2 \rangle$ has the same value at all positions and the total energy E_1 written as twice the kinetic energy becomes

$$E_1 = M_1 \omega^2 \langle u_1^2 \rangle \tag{9.9}$$

Thus, by dividing $P_{1 \to 2}$ and ωE_1, we obtain the coupling loss factor of point-connected subsystems,

$$\eta_{12} = \frac{\pi \left[(K + M\omega^2)^2 + G^2 \omega^2 \right] n_2}{2\omega^3 M_1 M_2} \tag{9.10}$$

This completes the proof of eqn (9.3).

9.3 One-Dimensional Coupled Subsystems

To illustrate the power of the wave approach of statistical energy analysis, we continue by the quite general situation of two one-dimensional subsystems coupled together at their extremities (Fig. 9.1). In the first subsystem a single wave propagates toward the coupling. The intensity I_0 and energy density W_0 carried out by this wave are related by

$$I_0 = c_{g_1} W_0 \tag{9.11}$$

where c_{g_1} is the group velocity in subsystem 1. When the incident wave impinges on the connection point, a wave is reflected in subsystem 1 and a wave is transmitted to subsystem 2. This transmission process is characterized by the transmission efficiency

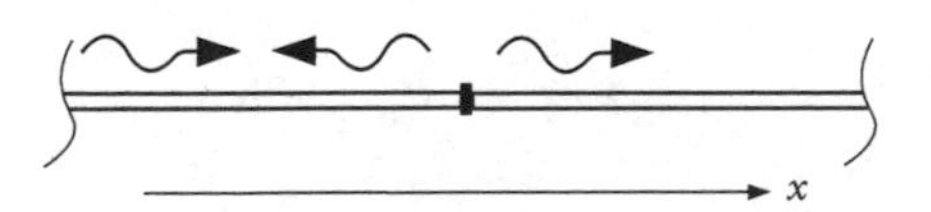

Figure 9.1 *Wave transmission in coupled one-dimensional subsystems.*

T defined as the ratio of transmitted intensity to incident intensity. Similarly, we may introduce the reflection efficiency R as the ratio of reflected intensity to incident intensity. Of course, the conservation of energy at the interface (Assumption 1) imposes $R+T = 1$. The transmitted power is therefore

$$P_{1\to 2} = TI_0 \tag{9.12}$$

To evaluate the energy of subsystem 1, we use the hypothesis that the field is diffuse, that is to say the energy density does not depend on the position. If subsystem 1 has length L_1 and denoting W_1 the energy density of the reflected wave, the total vibrational energy is the sum of that of incident and reflected waves,

$$E_1 = (W_0 + W_1)L_1 \tag{9.13}$$

But $W_1 = RW_0 = (1-T)W_0$. Combining eqns (9.11)–(9.13), the coupling loss factor is finally obtained by dividing $P_{1\to 2}$ and ωE_1

$$\eta_{12} = \frac{c_{g_1} T}{\omega(2-T)L_1} \tag{9.14}$$

This simple reasoning is valid for any one-dimensional subsystem independently of the kind (string, beam, rod, acoustical duct, etc.). It highlights that the determination of the coupling loss factor reduces to that of the transmission efficiency T. The problem is therefore to compute T in all situations of interest.

9.4 Beams Coupled at Their Edges

As an example of application, we start by examining again the problem of Section 4.3 from the wave point of view.

Beams connected by a torsional spring

Consider a beam of mass per unit length m and bending stiffness EI where E is Young's modulus and I the moment of inertia. A flexural wave having the form $e^{\iota(\omega t-\kappa x)}$ can exist only if $m\omega^2 = EI\kappa^4$ is verified. So if we denote $\kappa_i = (m_i\omega^2/E_iI_i)^{1/4}$ the wavenumber of beam i at frequency ω, only two propagating waves $e^{\iota(\omega t\pm\kappa_i x)}$ and two evanescent waves $e^{(\iota\omega t\pm\kappa_i x)}$ exist.

To derive the coupling loss factors through eqn (9.14), we must evaluate the power being transferred when a propagating wave impinges the junction. Suppose that a unit incident wave $e^{\iota(\omega t+\kappa_1 x)}$ travels in beam 1 towards the junction. At the junction, two waves are reflected (propagating and evanescent) and two waves are transmitted to beam 2 (Fig. 9.2). In beam 1 the transverse deflection is therefore

$$u_1(x,t) = e^{\iota\omega t}\left(e^{\iota\kappa_1 x} + re^{-\iota\kappa_1 x} + r'e^{-\kappa_1 x}\right) \tag{9.15}$$

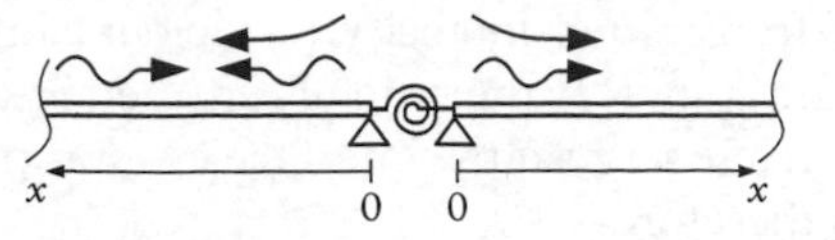

Figure 9.2 *Wave transmission in beams through a torsional spring.*

where we have noted r and r' the complex reflection coefficients respectively associated with the propagating and evanescent waves. In beam 2, the deflection is

$$u_2(x,t) = e^{\imath\omega t}\left(te^{-\imath\kappa_2 x} + t'e^{-\kappa_2 x}\right) \tag{9.16}$$

where t and t' are called the transmission coefficients. These coefficients are determined by applying the coupling conditions (4.72) and (4.73). First, the condition $u_1 = u_2 = 0$ at $x = 0$ imposes

$$r + r' = -1 \tag{9.17}$$

$$t + t' = 0 \tag{9.18}$$

and the coupling condition $E_1I_1u_1'' = -E_2I_2u_2'' = K(u_1' - u_2')$ at $x = 0$ gives

$$E_1I_1\kappa_1^2(-1 - r + r') = K\left(\imath\kappa_1 - \imath\kappa_1 r - \kappa_1 r' + \imath\kappa_2 t + \kappa_2 t'\right) \tag{9.19}$$

$$E_2I_2\kappa_2^2(-t + t') = -K\left(\imath\kappa_1 - \imath\kappa_1 r - \kappa_1 r' + \imath\kappa_2 t + \kappa_2 t'\right) \tag{9.20}$$

Solving these four equations gives

$$r = -\frac{2E_1I_1\kappa_1^2 + K\kappa_1(1+\imath) + K\kappa_2\theta(1-\imath)}{2E_1I_1\kappa_1^2 + K(\kappa_1 + \theta\kappa_2)(1-\imath)} \tag{9.21}$$

$$t = \frac{2\imath K\kappa_1\theta}{2E_1I_1\kappa_1^2 + K(\kappa_1 + \theta\kappa_2)(1-\imath)} \tag{9.22}$$

where

$$\theta = \frac{E_1I_1\kappa_1^2}{E_2I_2\kappa_2^2} \tag{9.23}$$

The transmitted power is the power travelling in beam 2. The active power in beam 2 is $1/2 \times \mathrm{Re}[E_2I_2(u_2'''\overline{\dot{u}_2} - u_2''\overline{\dot{u}_2'})]$ by eqn (5.121) at any point of the beam. When the damping is light, we may choose a point far from the junction where the evanescent wave is negligible. We obtain for the time-average transmitted power,

$$P_{1\to 2} = E_2I_2\kappa_2^3\omega\,|\,t\,|^2 \tag{9.24}$$

The time-average energy density in beam 1 is $1/4 \times \left(m_1 |\dot{u}_1|^2 + E_1 I_1 |u_1''|^2\right)$ by eqns (5.119) and (5.120). If the beam has length L_1 and always by neglecting the contribution of evanescent waves, the time-averaged vibrational energy of beam 1 is

$$E_1 = \frac{1}{2} E_1 I_1 \kappa_1^4 \left(1 + |r|^2\right) L_1 \tag{9.25}$$

The factor $\omega\eta_{12}$ is defined by eqn (9.1) as the ratio of $P_{1\to 2}$ to E_1,

$$\omega\eta_{12} = 2\frac{E_2 I_2 \kappa_2^3 \omega}{E_1 I_1 \kappa_1^4 L_1} \times \frac{|t|^2}{1 + |r|^2} \tag{9.26}$$

In order to compare with the result obtained in Section 4.3, let us consider the special case of light coupling $K \ll E_i I_i \kappa_i$ for $i = 1, 2$. A development in powers of $K/E_i I_i \kappa_i$ gives

$$\eta_{12} = \frac{K^2}{E_1 I_1 L_1 E_2 I_2 \kappa_1^2 \kappa_2} \tag{9.27}$$

This result is valid up to order two in $K/E_1 I_1 \kappa_1$. After substituting the expression of κ_1, it yields (Crandall and Lotz, 1971)

$$\eta_{12} = \frac{K^2}{L_1 \left(m_1 E_1 I_1\right)^{1/2} m_2^{1/4} \left(E_2 I_2\right)^{3/4} \omega^{3/2}} \tag{9.28}$$

This is exactly the same expression that we obtained in Section 4.3 by the modal approach of statistical energy analysis.

This example highlights that the modal and wave approaches of statistical energy analysis lead to the same results. This duality between modal and wave approaches is certainly one of the most fascinating aspects of the theory.

L-junction of beams

In this second example, we shall investigate the effect of the presence of several waves. The system is composed of two beams rigidly connected at a right angle as shown in Fig. 9.3.

A flexural wave travels in beam 1 toward the junction. Let $\kappa_{Bi} = (m_i \omega^2 / E_i I_i)^{1/4}$ be the flexural wavenumber at frequency ω in beam i. The incident field is

$$v_0(x, t) = e^{\imath(\omega t - \kappa_{B1} x)} \tag{9.29}$$

where v_0 denotes the transverse deflection of beam 1 caused by this incident wave. Generally, we shall denote u and v the two components of displacement of any point of the beam respectively in the x- and y-directions.

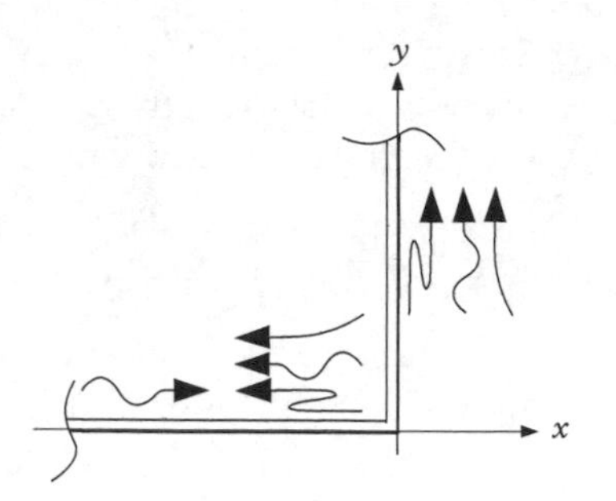

Figure 9.3 *Wave transmission in L-coupled beams.*

Since the junction is a right-angle, a transverse displacement in beam 1 induces a longitudinal displacement in beam 2, both being in the y-direction. Conversely, a longitudinal displacement of beam 1 at the junction corresponds to a transverse displacement of beam 2. So, all kinds of movement are generated during the reflection and transmission process. Longitudinal and flexural waves cannot be separated. We therefore introduce three reflection coefficients $r_{B\beta}$ and three transmission coefficients $t_{B\beta}$ with the subscript $\beta = B, E$, or L respectively for propagating bending (or flexural) wave, evanescent bending wave, and longitudinal wave. The reflection and transmitted fields are

$$u_1(x,t) = r_{BL} e^{\iota(\omega t + \kappa_{L1} x)} \tag{9.30}$$

$$v_1(x,t) = r_{BB} e^{\iota(\omega t + \kappa_{B1} x)} + r_{BE} e^{\iota \omega t + \kappa_{B1} x} \tag{9.31}$$

$$u_2(y,t) = t_{BB} e^{\iota(\omega t - \kappa_{B2} y)} + t_{BE} e^{\iota \omega t - \kappa_{B2} y} \tag{9.32}$$

$$v_2(y,t) = t_{BL} e^{\iota(\omega t - \kappa_{L2} y)} \tag{9.33}$$

where $\kappa_{Li} = (m_i \omega^2 / E_i S_i)^{1/2}$ is the longitudinal wavenumber in beam i and S_i the cross-section area.

At the junction, three compatibility conditions on displacement and three equilibrium equations must be imposed. The first two of the following equations express the equality of displacements at $x = y = 0$ projected on respectively x- and y-axis.

$$u_1(0,t) = u_2(0,t) \tag{9.34}$$

$$v_0(0,t) + v_1(0,t) = v_2(0,t) \tag{9.35}$$

$$\frac{\partial v_0}{\partial x}(0,t) + \frac{\partial v_1}{\partial x}(0,t) = -\frac{\partial u_2}{\partial y}(0,t) \tag{9.36}$$

The third equation states the equality of cross-section rotations at the junction. But due to the relative orientation of axis, a minus sign appears in the right-hand side.

Equilibrium of forces and moments gives three further equations. First, the projection of internal forces on the x-axis reads as follows. The normal force in beam 1 applied by the right part to the left part is $N_1 = E_1 S_1 u_1'$. The shear force in beam 2 applied by the upper part to the lower part is $V_2 = -E_2 I_2 u_2'''$. Equality $N_1 = V_2$ gives the first equation,

$$E_1 S_1 \frac{\partial u_1}{\partial x}(0,t) = -E_2 I_2 \frac{\partial^3 u_2}{\partial y^3}(0,t) \tag{9.37}$$

The second equation is determined in a similar fashion. We write $V_1 = N_2$,

$$-E_1 I_1 \left(\frac{\partial^3 v_0}{\partial x^3}(0,t) + \frac{\partial^3 v_1}{\partial x^3}(0,t) \right) = E_2 S_2 \frac{\partial v_2}{\partial y}(0,t) \tag{9.38}$$

The last equation is equilibrium of moments. The bending moment in beam 1 is $M_1 = E_1 I_1 v_1''$ but $M_2 = -E_2 I_2 u_2''$ in beam 2 due to the orientation of the frame. Equality $M_1 = M_2$ reads

$$E_1 I_1 \left(\frac{\partial^2 v_0}{\partial x^2}(0,t) + \frac{\partial^2 v_1}{\partial x^2}(0,t) \right) = -E_2 I_2 \frac{\partial^2 u_2}{\partial y^2}(0,t) \tag{9.39}$$

We therefore obtain a set of six linear equations on the reflection and transmission coefficients,

$$\begin{aligned} 0 &= r_{BL} - t_{BB} - t_{BE} \\ -1 &= r_{BB} + r_{BE} - t_{BL} \\ \imath\kappa_{B1} &= \kappa_{B1}(\imath r_{BB} + r_{BE}) - \kappa_{B2}(\imath t_{BB} + t_{BE}) \\ 0 &= E_1 S_1 \imath\kappa_{L1} r_{BL} + E_2 I_2 \kappa_{B2}^3 (\imath t_{BB} - t_{BE}) \\ -E_1 I_1 \imath\kappa_{B1}^3 &= E_1 I_1 \kappa_{B1}^3 (-\imath r_{BB} + r_{BE}) - E_2 S_2 \imath\kappa_{L2} t_{BL} \\ E_1 I_1 \kappa_{B1}^2 &= E_1 I_1 \kappa_{B1}^2 (-r_{BB} + r_{BE}) + E_2 I_2 \kappa_{B2}^2 (-t_{BB} + t_{BE}) \end{aligned} \tag{9.40}$$

which uniquely determines r_{BB}, r_{BL}, t_{BB}, and t_{BL}.

The transmission efficiencies, defined in terms of ratios of powers, can now be derived from these coefficients. But the mechanism of transmission involves several waves, which leads us to introduce several transmission and reflection efficiencies. Let us denote by R_{BB}, R_{BL}, T_{BB}, and T_{BL} the reflection and transmission efficiencies from a bending wave to a bending or longitudinal wave. They are defined as the ratios of related outgoing to ingoing powers. The longitudinal active intensity is $1/2 \times ES\mathrm{Re}(u_i'\overline{\dot{u}_i})$ while the bending active intensity is $1/2 \times EI\mathrm{Re}(v'''\overline{\dot{v}_i} - v_i''\overline{\dot{v}_i'})$. For travelling waves of magnitudes u_i and v_i they are respectively $ES\kappa_{Li}\omega|u_i|^2$ and $2EI\kappa_{Bi}^3\omega|v_i|^2$. By taking their ratios, we get the efficiencies,

$$R_{BB} = |r_{BB}|^2 \tag{9.41}$$

$$R_{BL} = \frac{E_1 S_1 \kappa_{L1}}{2E_1 I_1 \kappa_{B1}^3}|r_{BL}|^2 \tag{9.42}$$

$$T_{BB} = \frac{E_2 I_2 \kappa_{B2}^3}{E_1 I_1 \kappa_{B1}^3} \, | \, t_{BB} \, |^2 \tag{9.43}$$

$$T_{BL} = \frac{E_2 S_2 \kappa_{L2}}{2 E_1 I_1 \kappa_{B1}^3} \, | \, t_{BL} \, |^2 \tag{9.44}$$

Conservation of energy at the junction imposes the condition,

$$R_{BB} + R_{BL} + T_{BB} + T_{BL} = 1 \tag{9.45}$$

The system is composed of two beams. But from the point of view of statistical energy analysis, one must introduce two subsystems per beam, one for each type of wave. A subsystem is therefore identified by two subscripts: $i = 1, 2$ is the beam index and $\alpha = B, L$ the wave type. By applying eqn (9.14) with the above reflection and transmission efficiencies, we get

$$\eta_{B1,L1} = \frac{c_{g_1} R_{BL}}{\omega(2 - R_{BL}) L_1} \tag{9.46}$$

$$\eta_{B1,B2} = \frac{c_{g_1} T_{BB}}{\omega(2 - T_{BB}) L_1} \tag{9.47}$$

$$\eta_{B1,L2} = \frac{c_{g_1} T_{BL}}{\omega(2 - T_{BL}) L_1} \tag{9.48}$$

The further three coupling loss factors $\eta_{L1,B1}$, $\eta_{L1,B2}$, and $\eta_{L1,L2}$ may be determined by repeating the analysis with an incident longitudinal wave. Other coupling loss factors are obtained by permuting the indices 1 and 2 in the former coefficients or, equivalently, by applying reciprocity. The total number of coupling loss factors is twelve.

9.5 Two-Dimensional Coupled Subsystems

Let us consider a pair of two-dimensional subsystems coupled along a line as shown in Fig. 9.4. The plates have vibrational energies E_i. To evaluate the energy flow incident on the boundary, we use again the assumption of diffuse field. The repartition of energy in subsystem 1 is homogeneous and isotropic. Therefore, the energy per unit area and unit angle is $E_1/2\pi S_1$ where S_1 is the area of plate 1. If c_{g_1} denotes the group speed, the radiative intensity defined as the power per unit angle and unit length normal to the ray is

$$I = \frac{c_{g_1} E_1}{2\pi S_1} \tag{9.49}$$

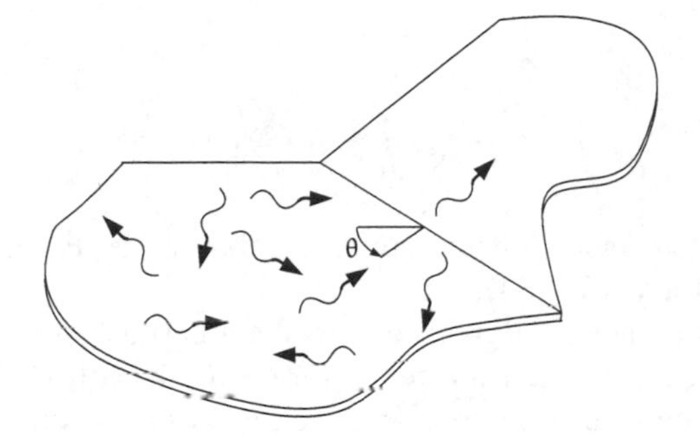

Figure 9.4 *Wave transmission in coupled two-dimensional subsystems.*

Now, let us consider a point of the common boundary. The power incident on the boundary at this point is

$$I\cos\theta = \frac{c_{g_1}E_1}{2\pi S_1}\cos\theta \tag{9.50}$$

where θ is taken as the angle normal to the boundary. During the interaction process, the incident power is partly reflected and partly transmitted. We introduce the reflection and transmission efficiencies $R(\theta)$ and $T(\theta)$ respectively defined as the ratios of reflected and incident powers and transmitted and incident powers. Energy conservation imposes the condition,

$$R(\theta) + T(\theta) = 1 \tag{9.51}$$

Furthermore, by symmetry $R(-\theta) = R(\theta)$ and $T(-\theta) = T(\theta)$. The total power being transmitted from 1 to 2 is obtained by integrating $T(\theta)I\cos\theta$ over all incidences and over the boundary,

$$P_{1\to 2} = \int_{-\pi/2}^{\pi/2}\int_L T(\theta)I\cos\theta\,\mathrm{d}L\mathrm{d}\theta \tag{9.52}$$

But $I\cos\theta$ is constant along the boundary and therefore the inner integral reduces to multiplying by L. Substituting eqn (9.50) into eqn (9.52) leads to

$$P_{1\to 2} = L\frac{c_{g_1}E_1}{2\pi S_1}\int_{-\pi/2}^{\pi/2} T(\theta)\cos\theta\,\mathrm{d}\theta \tag{9.53}$$

Applying the symmetry of the integrand factor,

$$P_{1\to 2} = L\frac{c_{g_1}E_1}{\pi S_1}\int_{0}^{\pi/2} T(\theta)\cos\theta\,\mathrm{d}\theta \tag{9.54}$$

Hence,

$$\eta_{12} = \frac{Lc_{g_1}}{\pi \omega S_1} \int_0^{\pi/2} T(\theta) \cos\theta \, d\theta \tag{9.55}$$

This is the coupling loss factor of line-coupled subsystems. Equation (9.55) has been proved by Lyon and Eichler (1964).

Compared with the case of dimension one, similarities are apparent except that the transmission appears through the term $T/(2-T)$ in eqn (9.14) but simply T in eqn (9.55). This is due to a difference in application of the diffuse field assumption. The reflected field has been taken as $1-T$ in dimension one but 1 in the present reasoning. The advantage is that reciprocity applies to (9.55) but not to (9.14). That would be the case by substituting $T/(2-T)$ with $T/2$ into eqn (9.14). This approximation is valid when $T \ll 1$ which is consistent with the light coupling condition. See Craik (1999) for a short discussion on this topic.

Equation (9.55) shows that the calculation of coupling loss factors reduces to that of transmission efficiencies.

9.6 Coupled Plates

The effective determination of transmission efficiency requires a complete description of systems and coupling conditions. We now specialize the discussion to coupled plates.

Waves in thin plates

The z-axis is chosen normal to the plate while the x- and y-axes belong to the middle plane of the plate. The x-, y-, and z-components of the displacement of a point are respectively noted u, v, and w. The in-plane motion is therefore $u\mathbf{x} + v\mathbf{y}$ while $w\mathbf{z}$ is the out-of-plane motion. The governing equations are (Guyader, 2002)

$$\frac{Eh}{(1-\nu^2)} \left[\frac{\partial^2 u}{\partial x^2} + \frac{1-\nu}{2} \frac{\partial^2 u}{\partial y^2} + \frac{1+\nu}{2} \frac{\partial^2 v}{\partial x \partial y} \right] = m \frac{\partial^2 u}{\partial t^2} \tag{9.56}$$

$$\frac{Eh}{(1-\nu^2)} \left[\frac{\partial^2 v}{\partial y^2} + \frac{1-\nu}{2} \frac{\partial^2 v}{\partial x^2} + \frac{1+\nu}{2} \frac{\partial^2 u}{\partial y \partial x} \right] = m \frac{\partial^2 v}{\partial t^2} \tag{9.57}$$

$$-\frac{Eh^3}{12(1-\nu^2)} \left[\frac{\partial^4 w}{\partial x^4} + 2 \frac{\partial^4 w}{\partial x^2 \partial y^2} + \frac{\partial^4 w}{\partial y^4} \right] = m \frac{\partial^2 w}{\partial t^2} \tag{9.58}$$

where E is Young's modulus, ν the Poisson coefficient, h the thickness, ρ the material density, and $m = \rho h$ the mass per unit area.

The third equation governing the movement w is naturally decoupled. But movements u and v are coupled in the first two equations. In order to make apparent the waves attached to in-plane motion, we may decouple the equations on u and v by introducing the Helmholtz decomposition,

$$u = \frac{\partial \varphi}{\partial x} + \frac{\partial \psi}{\partial y} \tag{9.59}$$

$$v = \frac{\partial \varphi}{\partial y} - \frac{\partial \psi}{\partial x} \tag{9.60}$$

where φ and ψ are displacement potentials. Substituting this decomposition into the first two governing equations, summing and subtracting their derivatives with respect to x and y, and integrating the result yields

$$\frac{\partial^2 \varphi}{\partial x^2} + \frac{\partial^2 \varphi}{\partial y^2} - \frac{1}{c_L^2}\frac{\partial^2 \varphi}{\partial t^2} = 0 \tag{9.61}$$

$$\frac{\partial^2 \psi}{\partial x^2} + \frac{\partial^2 \psi}{\partial y^2} - \frac{1}{c_T^2}\frac{\partial^2 \psi}{\partial t^2} = 0 \tag{9.62}$$

where $c_L = [E/\rho(1-\nu^2)]^{1/2}$ and $c_T = [E/2\rho(1+\nu)]^{1/2}$. The potential φ is attached to a longitudinal wave travelling at speed c_L and ψ to a transverse shear wave at speed c_T. The corresponding dispersion equations are

$$\kappa^2 = \frac{\omega^2}{c_\alpha^2}, \quad \alpha = L, T \tag{9.63}$$

whose positive solutions are noted $\kappa_L = \omega/c_L$ and $\kappa_T = \omega/c_T$. These waves are non-dispersive. The dispersion equation of out-of-plane waves is obtained from eqn (9.58),

$$m\omega^2 = D\kappa^4 \tag{9.64}$$

where $D = Eh^3/12(1-\nu^2)$. This equation is similar to that of a beam in flexural motion, and, specifically, this is a dispersive wave. The four solutions are $\pm\kappa_B$ and $\pm i\kappa_B$ where $\kappa_B = (m\omega^2/D)^{1/4}$. This shows that evanescent waves exist. Thus, four wave types may arise in thin plates; two propagating in-plane waves and one propagating and one evanescent out-of-plane wave.

We also need the force–displacement relationships in thin plates. Let a face be normal to the x-axis as shown in Fig. 9.5. The internal forces and moments per unit length applied to the middle line of the section by the positive half-plate are respectively N the normal force, T the tangential shear force, Q the transverse shear force, M the bending moment, and L the twisting moment. They are obtained by integrating the relevant stress tensor component over the thickness. For instance $N = \int \sigma_{xx}\,\mathrm{d}z$, $L = \int -z\sigma_{xy}\,\mathrm{d}z$, $M = \int z\sigma_{xx}\,\mathrm{d}z$. Leaving all the details of their calculations, we arrive at the following result:

$$N = \frac{Eh}{(1-\nu^2)}\left(\frac{\partial u}{\partial x} + \nu\frac{\partial v}{\partial y}\right) \tag{9.65}$$

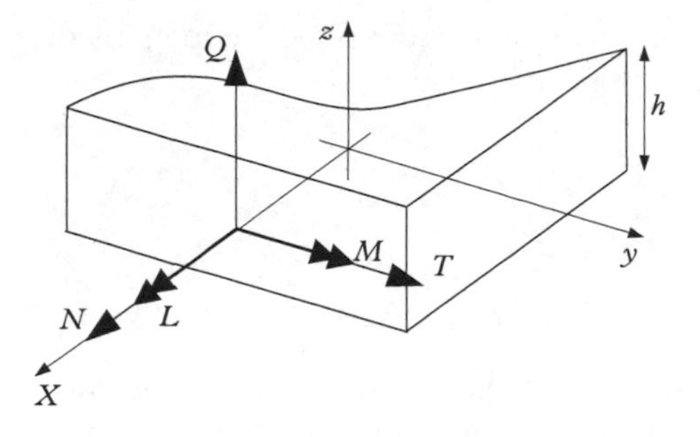

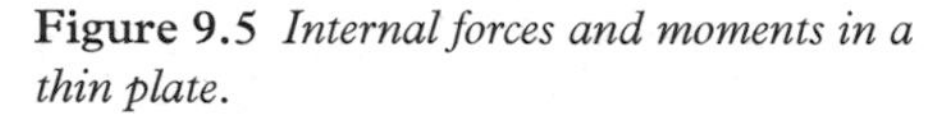

Figure 9.5 *Internal forces and moments in a thin plate.*

$$Q = -\frac{Eh^3}{12(1-\nu^2)}\left(\frac{\partial^3 w}{\partial x^3} + \frac{\partial^3 w}{\partial x \partial y^2}\right) \tag{9.66}$$

$$T = \frac{Eh}{2(1+\nu)}\left(\frac{\partial u}{\partial y} + \frac{\partial v}{\partial x}\right) \tag{9.67}$$

$$L = \frac{Eh^3}{12(1-\nu^2)}(1-\nu)\frac{\partial^2 w}{\partial x \partial y} \tag{9.68}$$

$$M = \frac{Eh^3}{12(1-\nu^2)}\left(\frac{\partial^2 w}{\partial x^2} + \nu\frac{\partial^2 w}{\partial y^2}\right) \tag{9.69}$$

Forces and moments applied to a face normal to the y-axis may easily be derived by permuting x and y, and simultaneously u and v in the above equations. However, they are not necessary in what follows.

Introducing eqns (9.59) and (9.60) into eqns (9.65) and (9.67), the expressions of normal and tangential shear forces in terms of potentials are

$$N = \frac{Eh}{(1-\nu^2)}\left(\frac{\partial^2 \varphi}{\partial x^2} + \nu\frac{\partial^2 \varphi}{\partial y^2} + (1-\nu)\frac{\partial^2 \psi}{\partial x \partial y}\right) \tag{9.70}$$

$$T = \frac{Eh}{2(1+\nu)}\left(2\frac{\partial^2 \varphi}{\partial x \partial y} + \frac{\partial^2 \psi}{\partial y^2} - \frac{\partial^2 \psi}{\partial x^2}\right) \tag{9.71}$$

Also of interest is the so-called effective shear force defined by $V = Q - \partial L/\partial y$. After substituting,

$$V = -\frac{Eh^3}{12(1-\nu^2)}\left(\frac{\partial^3 w}{\partial x^3} + (2-\nu)\frac{\partial^3 w}{\partial x \partial y^2}\right) \tag{9.72}$$

This is the net transverse force applied to the face when the moment per unit length L is replaced by equivalent forces L and $L + \partial_y L$ at distance dy. The fact that V is involved in the expressions of boundary conditions of free edges and balance equations at interfaces but not Q is a classical difficulty of thin plate theory (see for instance Ventsel and Krauthammer (2001) p. 27).

The structural intensity is a vector lying in the plate plane. Since we are only interested in the power crossing or reflected by an interface, say $y = 0$, we may give only the x-component of intensity,

$$I_x = -\left[N\frac{\partial u}{\partial t} + T\frac{\partial v}{\partial t} + M\frac{\partial^2 w}{\partial x \partial t} + L\frac{\partial^2 w}{\partial y \partial t} + Q\frac{\partial w}{\partial t}\right] \tag{9.73}$$

The expression of energy density in thin plates is

$$\begin{aligned} W = \frac{1}{2}m\left[\left(\frac{\partial u}{\partial t}\right)^2 + \left(\frac{\partial v}{\partial t}\right)^2 + \left(\frac{\partial w}{\partial t}\right)^2\right] + \frac{Eh}{4(1+\nu)}\left[\frac{\partial u}{\partial y} + \frac{\partial v}{\partial x}\right]^2 + \\ \frac{Eh}{2(1-\nu^2)}\left[\left(\frac{\partial u}{\partial x} + \nu\frac{\partial v}{\partial y}\right)\frac{\partial u}{\partial x} + \left(\frac{\partial v}{\partial y} + \nu\frac{\partial u}{\partial x}\right)\frac{\partial v}{\partial y}\right] + \\ \frac{Eh^3}{24(1-\nu^2)}\left[\left(\frac{\partial^2 w}{\partial x^2}\right)^2 + \left(\frac{\partial^2 w}{\partial y^2}\right)^2 + 2\nu\frac{\partial^2 w}{\partial x^2}\frac{\partial^2 w}{\partial y^2} + 2(1-\nu)\left(\frac{\partial^2 w}{\partial x \partial y}\right)^2\right] \end{aligned} \tag{9.74}$$

The first line is the kinetic energy while the other two lines constitute the elastic energy of respectively in-plane and out-of-plane motions. See Bouthier and Bernhard (1995) for a review of energetics of out-of-plane motion and Park et al. (2001) for in-plane motion.

Free edge

The simplest case of calculation of $T(\theta)$ is the so-called **mode conversion** phenomenon which occurs at boundaries of plates. When a bending wave hits a conservative boundary of a plate (a free edge for instance), a propagating and an evanescent wave are reflected. But evanescent waves do not transport energy away from the boundary which means that the reflection efficiency associated with bending waves is unit. However, when an in-plane wave is incident, something different occurs. Generally both longitudinal and transversal waves are reflected, which leads to a conversion of incident energy to longitudinal and transverse energies.

Consider a free edge along the y-axis. The boundary conditions are $N = T = 0$. According to eqns (9.70) and (9.71), they take the form

$$\frac{\partial^2 \varphi}{\partial x^2} + \nu\frac{\partial^2 \varphi}{\partial y^2} + (1-\nu)\frac{\partial^2 \psi}{\partial x \partial y} = 0 \tag{9.75}$$

$$2\frac{\partial^2 \varphi}{\partial x \partial y} + \frac{\partial^2 \psi}{\partial y^2} - \frac{\partial^2 \psi}{\partial x^2} = 0 \tag{9.76}$$

Let us consider first a single longitudinal plane wave impinging on the boundary with incidence θ. The incident field is

$$\varphi_0(x, y, t) = \mathrm{e}^{\imath(\omega t + \kappa_L x \cos\theta + \kappa_L y \sin\theta)} \tag{9.77}$$

Two plane waves are reflected, namely a longitudinal and a transverse wave. Let r_{LL} be the reflection coefficient of longitudinal waves and r_{LT} that of transverse waves. The reflected fields are

$$\varphi(x, y, t) = r_{LL}\mathrm{e}^{\imath(\omega t - \kappa_L x \cos\theta + \kappa_L y \sin\theta)} \tag{9.78}$$

$$\psi(x, y, t) = r_{LT}\mathrm{e}^{\imath(\omega t - \lambda_T x + \kappa_L y \sin\theta)} \tag{9.79}$$

In the reflected L-wave, we have prejudged that the angle of reflection equals the angle of incidence in agreement with specular reflection. This can be checked directly by substituting the fields (incident plus reflected) into eqns (9.75) and (9.76); any other reflected angle is not possible. In the same way, compatibility imposes that the y-dependence of these fields is the same for all fields. However, concerning the reflected T-wave, we have introduced the x-component λ_T of the wavenumber vector which, of course, must be determined. The dispersion equation of shear waves imposes

$$(\kappa_L \sin\theta)^2 + \lambda_T^2 = \kappa_T^2 \tag{9.80}$$

Let us remark that since the longitudinal speed is always greater than the shear wave, λ_T is real-valued. Among the two square roots which give λ_T, the only physical solution is that for which the wave moves away from the boundary (Sommerfeld's condition). With this constraint, we obtain

$$\lambda_T = \sqrt{\kappa_T^2 - \kappa_L^2 \sin^2\theta} \tag{9.81}$$

We are now in a position to apply the boundary conditions. Substituting the fields $\varphi_0 + \varphi$ and ψ into eqns (9.75) and (9.76) leads to a set of linear equations on r_{LL} and r_{LT},

$$\begin{pmatrix} -\kappa_L^2(\cos^2\theta + \nu\sin^2\theta) & (1-\nu)\kappa_L\lambda_T\sin\theta \\ 2\kappa_L^2\sin\theta\cos\theta & -\kappa_L^2\sin^2\theta + \lambda_T^2 \end{pmatrix}\begin{pmatrix} r_{LL} \\ r_{LT} \end{pmatrix} = \begin{pmatrix} \kappa_L^2(\cos^2\theta + \nu\sin^2\theta) \\ 2\kappa_L^2\sin\theta\cos\theta \end{pmatrix} \tag{9.82}$$

Let us introduce the reflection efficiencies: T_{LL} denotes the ratio of powers carried out by φ and φ_0, and T_{LT} the ratio of the powers of ψ and φ_0. Conservation of energy imposes

$$T_{LL}(\theta) + T_{LT}(\theta) = 1 \tag{9.83}$$

To get an explicit value of T_{LL}, one substitutes successively the fields φ_0 and φ into eqns (9.59), (9.60) and (9.70), (9.71) and then takes the ratio of active powers $1/2 \times \mathrm{Re}(N\bar{u} + T\bar{v})$. This yields

$$T_{LL}(\theta) = |r_{LL}|^2 \tag{9.84}$$

We may now solve eqn (9.82) and substitute the solution r_{LL}:

$$T_{LL}(\theta) = \left| \frac{(\kappa_T^2 - 2\kappa_L^2 \sin^2\theta)(\cos^2\theta + \nu\sin^2\theta) - (1-\nu)\kappa_L \sin\theta\sin 2\theta\sqrt{\kappa_T^2 - \kappa_L^2\sin^2\theta}}{(\kappa_T^2 - 2\kappa_1^2 \sin^2\theta)(\cos^2\theta + \nu\sin^2\theta) + (1-\nu)\kappa_L \sin\theta\sin 2\theta\sqrt{\kappa_T^2 - \kappa_L^2\sin^2\theta}} \right|^2$$

But $\kappa_L^2/\kappa_T^2 = (1-\nu)/2$. So, after substituting and simplifying,

$$T_{LL}(\theta) = \left| \frac{(\cos^2\theta + \nu\sin^2\theta)^2 - \frac{(1-\nu)^{3/2}}{\sqrt{2}}\sin\theta\sin 2\theta\sqrt{1 - \frac{1-\nu}{2}\sin^2\theta}}{(\cos^2\theta + \nu\sin^2\theta)^2 + \frac{(1-\nu)^{3/2}}{\sqrt{2}}\sin\theta\sin 2\theta\sqrt{1 - \frac{1-\nu}{2}\sin^2\theta}} \right|^2 \quad (9.85)$$

In turn, T_{LT} is given by $T_{LT} = 1 - T_{LL}$. Note that these reflection efficiencies depend on the Poisson coefficient but not other characteristics of the material.

The case of an incident shear wave is quite similar. The incident field is

$$\psi_0(x, y, t) = e^{\iota(\omega t + \kappa_T x \cos\theta' + \kappa_T y \sin\theta')} \quad (9.86)$$

The reflected fields may be found as plane waves whose magnitudes are noted r_{TL} and r_{TT}:

$$\varphi(x, y, t) = r_{TL} e^{\iota(\omega t - \lambda_L x + \kappa_T y \sin\theta')} \quad (9.87)$$

$$\psi(x, y, t) = r_{TT} e^{\iota(\omega t - \kappa_T x \cos\theta' + \kappa_T y \sin\theta')} \quad (9.88)$$

However, a qualitative difference exists with the previous situation where a longitudinal and a shear wave are created during reflection. When a shear wave is incident, the Snell law reads

$$\left(\kappa_T \sin\theta'\right)^2 + \lambda_L^2 = \kappa_L^2 \quad (9.89)$$

Since $c_L > c_T$, that is $\kappa_L < \kappa_T$, the solution of λ_L may not exist as a real number. Beyond a critical angle whose value $\theta_c = \arcsin(c_T/c_L)$, λ_L becomes complex indicating that the reflected longitudinal wave is evanescent. In this case, the reflection efficiency T_{TT} is unit and T_{TL} is zero.

The values of T_{TT} and T_{TL} when $\theta < \theta_c$ can be obtained by the same method as the previous case. However, instead of repeating a tedious calculation, we may employ the argument of reciprocity. Basically, the governing equations of motion in plates are invariant under the permutation $t \leftrightarrow -t$. Therefore, the solution (9.87), (9.88) remains a solution if the time is reversed. The situation is sketched in Fig. 9.6. When the time flows from past to future, a unit shear wave is incident and two waves of magnitudes r_{TL} and r_{TT} are reflected. But when the time flows from future to past the direction of propagation is reversed and both waves r_{TL} and r_{TT} become incident and combine to

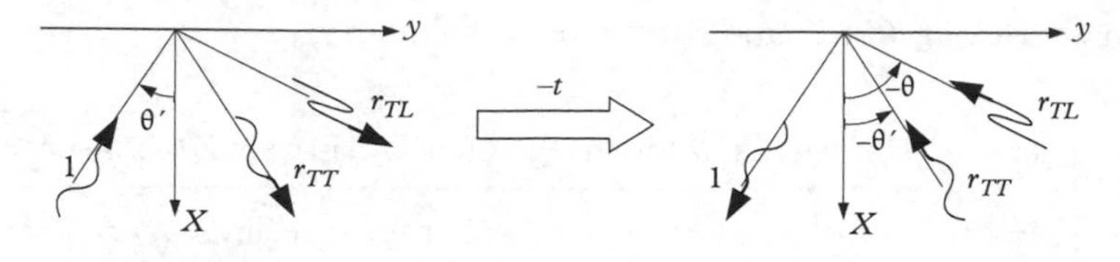

Figure 9.6 *When the time is reversed, the direction of waves is reversed but their magnitudes remain unchanged.*

give a unique shear wave of magnitude 1. Since no longitudinal wave is created in this process, we may assert that they also combine to give a longitudinal wave of magnitude 0. By linearity,

$$0 = r_{TL}(\theta')r_{LL}(-\theta) + r_{TT}(\theta')r_{TL}(-\theta') \tag{9.90}$$

where the incidence angle of the longitudinal (respectively shear) waves is $-\theta'$ (respectively $-\theta$). Both angles are linked by $\kappa_T \sin\theta' = \kappa_L \sin\theta$. But by symmetry $r_{LL}(-\theta) = r_{LL}(\theta)$ and $r_{TL}(-\theta') = r_{TL}(\theta')$. After simplifying,

$$r_{LL}(\theta) = -r_{TT}(\theta') \tag{9.91}$$

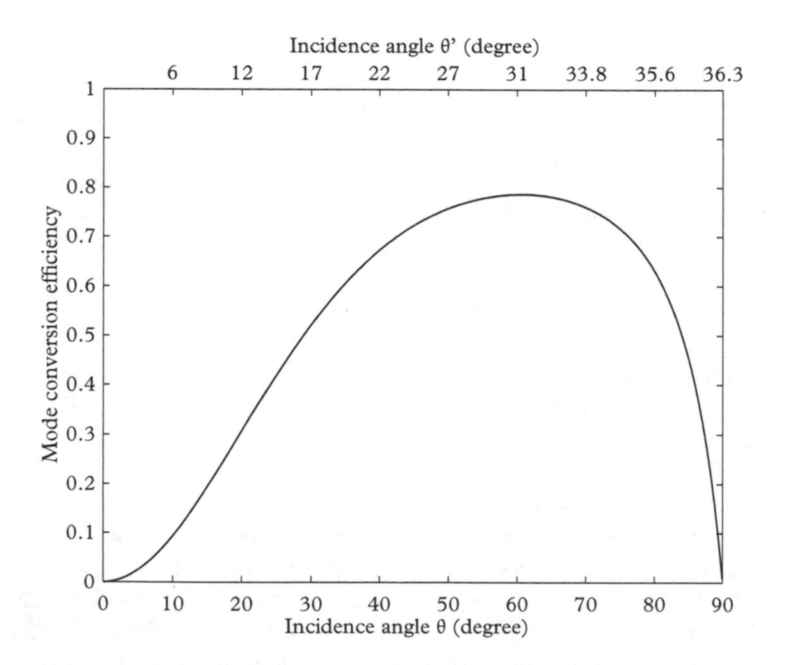

Figure 9.7 *Mode conversion efficiency at a free edge of plates for* $\nu = 0.3$. *Bottom axis: incident longitudinal wave. Top axis: incident shear wave.*

The reflection efficiency being $T_{TT} = |r_{TT}|^2$, we get

$$T_{TT}(\theta') = T_{LL}(\theta) \tag{9.92}$$

and by energy conservation $T_{TL} = 1 - T_{TT}$.

Figure 9.7 plots the reflection efficiency $T_{LT}(\theta) = T_{TL}(\theta')$ of the mode conversion for $\nu = 0.3$. The bottom axis is the angle θ (incident longitudinal wave) while the top axis is θ' (incident shear wave). The maximum efficiency for conversion of a longitudinal to shear wave occurs at $\theta = 60°$ and at $\theta' = 31°$ for the reciprocal conversion.

Plates coupled along a line

We now focus on the case of several plates joined at a common edge. Due to its practical importance, many authors have presented solutions for particular configurations. For instance Cremer and Heckl (1988) determined explicitly the transmission efficiencies for L-junctions and cross-junctions with the same plates on both sides. Budrin and Nikiforov (1964) and later Whöle et al. (1981) generalized the cross-junction solution. Finally, the case of an arbitrary number of plates joined through a beam but without assuming coincidence of middle planes of plates and the centroid/shear axis of the beam, is solved by Langley and Heron (1990). See also Hopkins (2003) for an application to masonry walls.

For the sake of simplicity, we restrict the present analysis to the case where all middle planes of plates intersect at a common line corresponding to the y-axis. The x-axis is chosen as arbitrarily normal to the y-axis while a local x_i-axis lies in the middle plane of plate i toward the inside. The angle between the x- and x_i-axis is noted α_i (see Fig. 9.8).

The equilibrium of forces and moment at the common line $x_i = 0$ imposes four conditions. If the normal, tangential, effective shear forces, and bending moment in plate i at $x_i = 0$ are respectively noted N_i, T_i, V_i, M_i, the four equilibrium conditions are

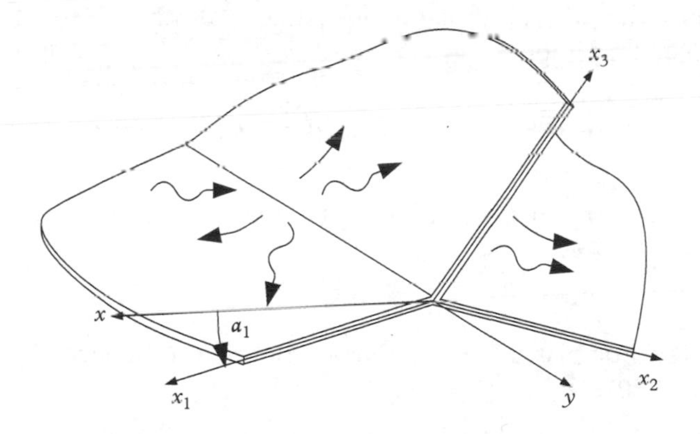

Figure 9.8 *Plates coupled along a common edge. The middle planes intersect at the y-axis.*

$$0 = \sum_{i=1}^{n} N_i \cos\alpha_i - V_i \sin\alpha_i \tag{9.93}$$

$$0 = \sum_{i=1}^{n} T_i \tag{9.94}$$

$$0 = \sum_{i=1}^{n} N_i \sin\alpha_i + V_i \cos\alpha_i \tag{9.95}$$

$$0 = \sum_{i=1}^{n} M_i \tag{9.96}$$

By substituting the force–field relationships (9.69), (9.70), (9.71), and (9.72), this yields

$$0 = \sum_{i=1}^{n} D_i'\left(\frac{\partial^2\varphi_i}{\partial x^2} + \nu_i\frac{\partial^2\varphi_i}{\partial y^2} + (1-\nu_i)\frac{\partial^2\psi_i}{\partial x\partial y}\right)\cos\alpha_i + D_i\left(\frac{\partial^3 w_i}{\partial x^3} + (2-\nu_i)\frac{\partial^3 w_i}{\partial x\partial y^2}\right)\sin\alpha_i \tag{9.97}$$

$$0 = \sum_{i=1}^{n} D_i''\left(2\frac{\partial^2\varphi_i}{\partial x\partial y} + \frac{\partial^2\psi_i}{\partial y^2} - \frac{\partial^2\psi_i}{\partial x^2}\right) \tag{9.98}$$

$$0 = \sum_{i=1}^{n} D_i'\left(\frac{\partial^2\varphi_i}{\partial x^2} + \nu_i\frac{\partial^2\varphi_i}{\partial y^2} + (1-\nu_i)\frac{\partial^2\psi_i}{\partial x\partial y}\right)\sin\alpha_i - D_i\left(\frac{\partial^3 w_i}{\partial x^3} + (2-\nu_i)\frac{\partial^3 w_i}{\partial x\partial y^2}\right)\cos\alpha_i \tag{9.99}$$

$$0 = \sum_{i=1}^{n} D_i\left(\frac{\partial^2 w_i}{\partial x^2} + \nu_i\frac{\partial^2 w_i}{\partial y^2}\right) \tag{9.100}$$

where we have noted $D_i = E_i h_i^3/12(1-\nu_i^2)$, $D_i' = E_i h_i/(1-\nu_i^2)$, and $D_i'' = E_i h_i/2(1+\nu_i)$ the different stiffnesses involved in the force–field relationships.

Furthermore, some compatibility conditions must be respected at the common line. The displacements u_i, v_i, w_i, and rotation $\partial_x w_i$ at $x_i = 0$ must be equal when projected in the global frame. This imposes

$$0 = (u_i\cos\alpha_i - w_i\sin\alpha_i) - (u_1\cos\alpha_1 - w_1\sin\alpha_1) \tag{9.101}$$

$$0 = v_i - v_1 \tag{9.102}$$

$$0 = (u_i\sin\alpha_i + w_i\cos\alpha_i) - (u_1\sin\alpha_1 + w_1\cos\alpha_1) \tag{9.103}$$

$$0 = \frac{\partial w_i}{\partial x} - \frac{\partial w_1}{\partial x} \tag{9.104}$$

for $i = 2, \ldots, n$. The first plate is conventionally chosen as reference. By substituting the fields (9.59) and (9.60),

$$0 = \left(\frac{\partial\varphi_i}{\partial x} + \frac{\partial\psi_i}{\partial y}\right)\cos\alpha_i - w_i\sin\alpha_i - \left(\frac{\partial\varphi_1}{\partial x} + \frac{\partial\psi_1}{\partial y}\right)\cos\alpha_1 + w_1\sin\alpha_1 \tag{9.105}$$

$$0 = \left(\frac{\partial \varphi_i}{\partial y} - \frac{\partial \psi_i}{\partial x}\right) - \left(\frac{\partial \varphi_1}{\partial y} - \frac{\partial \psi_1}{\partial x}\right) \tag{9.106}$$

$$0 = \left(\frac{\partial \varphi_i}{\partial x} + \frac{\partial \psi_i}{\partial y}\right) \sin\alpha_i + w_i \cos\alpha_i - \left(\frac{\partial \varphi_1}{\partial x} + \frac{\partial \psi_1}{\partial y}\right) \sin\alpha_1 - w_1 \cos\alpha_1 \tag{9.107}$$

$$0 = \frac{\partial w_i}{\partial x} - \frac{\partial w_1}{\partial x} \tag{9.108}$$

for $i = 2, \ldots, n$.

Let us assume an incident α-wave in plate 1 where $\alpha = B, L, T$ respectively denotes bending, longitudinal, and transverse waves. If the incidence angle on the common line is θ, then the incident field is

$$w_0(x, y, t) = \delta_{B\alpha} e^{\iota(\omega t + \kappa_{B1} x \cos\theta + \kappa_{B1} y \sin\theta)} \tag{9.109}$$

$$\varphi_0(x, y, t) = \delta_{L\alpha} e^{\iota(\omega t + \kappa_{L1} x \cos\theta + \kappa_{L1} y \sin\theta)} \tag{9.110}$$

$$\psi_0(x, y, t) = \delta_{T\alpha} e^{\iota(\omega t + \kappa_{T1} x \cos\theta + \kappa_{T1} y \sin\theta)} \tag{9.111}$$

where $\kappa_{Bi} = \left(m_i\omega^2/D_i\right)^{1/4}$ is the bending wavenumber, $\kappa_{Li} = \omega/c_{Li}$ the longitudinal wavenumber, and $\kappa_{Ti} = \omega/c_{Ti}$ the transverse wavenumber. The Kronecker symbol $\delta_{\alpha\beta}$ is unit when $\beta = \alpha$ and zero otherwise. Since the plates are not necessarily coplanar, the bending motion is coupled to longitudinal and transverse motion. So, generally, all types of wave can be transmitted and reflected.

The transmitted fields in plate i are

$$w_i(x, y, t) = t_{\alpha B,i} e^{\iota\omega t + \lambda_{Bi} x + \lambda_0 y} + t_{\alpha E,i} e^{\iota\omega t + \lambda_{Ei} x + \lambda_0 y} \tag{9.112}$$

$$\varphi_i(x, y, t) = t_{\alpha L,i} e^{\iota\omega t + \lambda_{Li} x + \lambda_0 y} \tag{9.113}$$

$$\psi_i(x, y, t) = t_{\alpha T,i} e^{\iota\omega t + \lambda_{Ti} x + \lambda_0 y} \tag{9.114}$$

We have introduced the transmission coefficients $t_{\alpha\beta,i}$, $\beta = B, E, L, T$ of an α-wave in plate 1 to a β-wave in plate i. The subscript E means evanescent. By compatibility, the exponentials have the same y-dependence noted $\lambda_0 = \iota\kappa_{\alpha 1} \sin\theta$. The x-dependence involve the parameters $\lambda_{\beta i}$ which are determined by the dispersion relationships,

$$\left[-(\kappa_{\alpha 1} \sin\theta)^2 + \lambda_{\beta i}^2\right]^2 = \kappa_{Bi}^4, \qquad \beta = B, E \tag{9.115}$$

$$-(\kappa_{\alpha 1} \sin\theta)^2 + \lambda_{Li}^2 = -\kappa_{Li}^2 \tag{9.116}$$

$$-(\kappa_{\alpha 1} \sin\theta)^2 + \lambda_{Ti}^2 = -\kappa_{Ti}^2 \tag{9.117}$$

Their solutions are

$$\lambda_{Bi} = -\sqrt{(\kappa_{\alpha 1} \sin\theta)^2 - \kappa_{Bi}^2} \tag{9.118}$$

$$\lambda_{Ei} = -\sqrt{(\kappa_{\alpha 1} \sin\theta)^2 + \kappa_{Bi}^2} \tag{9.119}$$

$$\lambda_{Li} = -\sqrt{(\kappa_{\alpha 1}\sin\theta)^2 - \kappa_{Li}^2} \tag{9.120}$$

$$\lambda_{Ti} = -\sqrt{(\kappa_{\alpha 1}\sin\theta)^2 - \kappa_{Ti}^2} \tag{9.121}$$

Note that the square root is determined such as the implicit Sommerfeld condition (outgoing wave) is respected. In particular $\sqrt{-1} = \imath$ when $\kappa_{\beta i} > \kappa_{\alpha 1}\sin\theta$.

The coefficients $t_{\alpha B,i}$, $t_{\alpha E,i}$, $t_{\alpha L,i}$, and $t_{\alpha T,i}$ are determined by the equilibrium conditions (9.97)–(9.100) and the compatibility conditions (9.105)–(9.108). The fields w_j, φ_j, ψ_j for $j = 2, \ldots, n$ of eqns (9.112)–(9.114) are substituted into these equations. However, the fields in plate 1 are $w_0 + w_1$, $\varphi_0 + \varphi_1$, $\psi_0 + \psi_1$. During this substitution, a x-derivative corresponds to a multiplication by $\lambda_{\beta i}$ and a y-derivative a multiplication by λ_0. For the sake of convenience, we introduce the following parameters:

$$\begin{aligned}
&a_i = D_i\left[\lambda_{Bi}^3 + (2-\nu_i)\lambda_{Bi}\lambda_0^2\right], && b_i = D_i\left[\lambda_{Ei}^3 + (2-\nu_i)\lambda_{Ei}\lambda_0^2\right] \\
&c_i = D_i'\left(\lambda_{Li}^2 + \nu_i\lambda_0^2\right), && d_i = D_i'(1-\nu_i)\lambda_{Ti}\lambda_0 \\
&e_i = 2D_i''\lambda_{Li}\lambda_0, && f_i = D_i''(\lambda_0^2 - \lambda_{Ti}^2) \\
&g_i = D_i\left(\lambda_{Bi}^2 + \nu_i\lambda_0^2\right), && h_i = D_i\left(\lambda_{Ei}^2 + \nu_i\lambda_0^2\right)
\end{aligned} \tag{9.122}$$

Let

$$\mathbf{A}_i = \begin{pmatrix} a_i\sin\alpha_i & b_i\sin\alpha_i & c_i\cos\alpha_i & d_i\cos\alpha_i \\ 0 & 0 & e_i & f_i \\ -a_i\cos\alpha_i & -b_i\cos\alpha_i & c_i\sin\alpha_i & d_i\sin\alpha_i \\ g_i & h_i & 0 & 0 \end{pmatrix}, \mathbf{B}_i = \begin{pmatrix} \sin\alpha_i & \sin\alpha_i & -\lambda_{L1}\cos\alpha_i & -\lambda_0\cos\alpha_i \\ 0 & 0 & -\lambda_0 & \lambda_{Ti} \\ -\cos\alpha_i & -\cos\alpha_i & -\lambda_{L1}\sin\alpha_i & -\lambda_0\sin\alpha_i \\ -\lambda_{Bi} & -\lambda_{Ei} & 0 & 0 \end{pmatrix}$$

be two 4×4 matrices, $\mathbf{X}_i = (t_{\alpha B,i},\ t_{\alpha E,i},\ t_{\alpha L,i},\ t_{\alpha T,i})^T$ the column vector of unknowns, and $\boldsymbol{\Delta}_\alpha = (\delta_{\alpha B},\ 0,\ \delta_{\alpha L},\ \delta_{\alpha T})^T$. Then eqns (9.97)–(9.100) and (9.105)–(9.108) give the following matrix equation:

$$\begin{pmatrix} \mathbf{A}_1 & \mathbf{A}_2 & \ldots & \mathbf{A}_n \\ \mathbf{B}_1 & -\mathbf{B}_2 & & \bigcirc \\ \vdots & & \ddots & \\ \mathbf{B}_1 & \bigcirc & & -\mathbf{B}_n \end{pmatrix} \begin{pmatrix} \mathbf{X}_1 \\ \mathbf{X}_2 \\ \vdots \\ \mathbf{X}_n \end{pmatrix} = \begin{pmatrix} -\mathbf{A}_1\boldsymbol{\Delta}_\alpha \\ -\mathbf{B}_1\boldsymbol{\Delta}_\alpha \\ \vdots \\ -\mathbf{B}_1\boldsymbol{\Delta}_\alpha \end{pmatrix} \tag{9.123}$$

The above block matrix has total dimension $4n \times 4n$. The first block row corresponds to the equilibrium conditions (9.97)–(9.100). The next $n - 1$ block rows correspond to the compatibility conditions (9.105)–(9.108). They contain only two non-zero entries.

The efficiencies are obtained by taking the ratios of active intensities normal to the boundary. Active intensities are extracted from eqn (9.73) by application of the rule (5.112). One gets $1/2 \times \mathrm{Re}(N\bar{\dot{u}} + T\bar{\dot{v}})$ for in-plane motion and $1/2 \times \mathrm{Re}(M\partial_x\bar{\dot{w}} + L\partial_y\bar{\dot{w}} + Q\bar{\dot{w}})$

for out-of-plane motion. On introducing the plane waves (9.112)–(9.114), the active intensities for respectively bending, longitudinal, and transverse waves are

$$I_{Bi} = -\tfrac{\omega}{2}\mathrm{Im}\left[D_i(\lambda_{Bi}^2 + \nu_i\lambda_0^2)\bar{\lambda}_{Bi} + D_i(1-\nu_i)\lambda_{Bi}\lambda_0\bar{\lambda}_0 - D_i(\lambda_{Bi}^3 + \lambda_{Bi}\lambda_0^2)\right] |t_{Bi}|^2$$

$$I_{Li} = -\tfrac{\omega}{2}\mathrm{Im}\left[D_i'(\lambda_{Li}^2 + \nu_i\lambda_0^2)\bar{\lambda}_{Li} + 2D_i''\lambda_{Li}\lambda_0\bar{\lambda}_0\right] |t_{Li}|^2$$

$$I_{Ti} = -\tfrac{\omega}{2}\mathrm{Im}\left[D_i'(1-\nu_i)\lambda_{Ti}\lambda_0\bar{\lambda}_0 - D_i''(\lambda_0^2 - \lambda_{Ti}^2)\bar{\lambda}_{Ti}\right] |t_{Ti}|^2$$

The above expressions simplify tremendously with the following considerations. First, λ_0 is purely imaginary, hence $\bar{\lambda}_0 = -\lambda_0$. Second, in eqns (9.118)–(9.121), $\lambda_{\beta i}$ is either real or purely imaginary. In the first case the expression in brackets is real and therefore $I_{\beta i} = 0$. In the second case $\bar{\lambda}_{\beta i} = -\lambda_{\beta i}$. Remarking that $2D_i'' = D_i'(1-\nu_i)$, one obtains

$$I_{Bi} = \omega D_i \kappa_{Bi}^2 \mathrm{Im}\,[\lambda_{Bi}]\,|t_{Bi}|^2,\; I_{Li} = \frac{\omega D_i'}{2}\kappa_{Li}^2 \mathrm{Im}\,[\lambda_{Li}]\,|t_{Li}|^2,\; I_{Ti} = \frac{\omega D_i''}{2}\kappa_{Ti}^2 \mathrm{Im}\,[\lambda_{Ti}]\,|t_{Ti}|^2 \quad (9.124)$$

Similarly, the active intensity of incident field is found by considering successively the three cases $\alpha = B, L, T$,

$$I_{B0} = -\omega D_1 \kappa_{B1}^3 \cos\theta, \quad I_{L0} = -\frac{\omega D_1'}{2}\kappa_{L1}^3 \cos\theta, \quad I_{T0} = -\frac{\omega D_1''}{2}\kappa_{T1}^3 \cos\theta \quad (9.125)$$

The reflection and transmission efficiencies between plates 1 and i are obtained by taking the ratios

$$R_{\alpha\beta} = \frac{I_{\beta 1}}{I_{\alpha 0}}, \text{ and } T_{\alpha\beta} = \frac{I_{\beta i}}{I_{\alpha 0}} \quad (9.126)$$

Figure 9.9 shows the reflection and transmission efficiencies for an L-junction computed by the above method. This example appears in Cremer and Heckl (1988, p. 442) and Langley and Heron (1990). Two plates are coupled at right angles. They have the same properties with $\nu = 0.3$ and the frequency is chosen so that $c_L/c_B = 0.3$. The three reflection and three transmission efficiencies of a bending wave are plotted in the same graph against the incidence angle. The efficiency value corresponds to the thickness of the zone in which it appears. Since $R_{BB} + R_{BL} + R_{BT} + T_{BB} + T_{BL} + T_{BT} = 1$ by energy conservation, the value of R_{BB} is the complement of the upper curve. We may observe in Fig. 9.9(a) that mode conversion occurs up to $\sin\theta = 0.3$. Beyond this critical angle for longitudinal waves, transverse waves are still generated up to $\sin\theta = 0.507$ which corresponds to the critical angle for transverse waves. Above this angle, only bending waves can be transmitted and reflected. Similar comments apply for the case of an incident longitudinal wave shown in Fig. 9.9(b), except that there is no critical angle and therefore mode conversion always exists. See Hopkins (2009) for the importance of mode conversion in building acoustics.

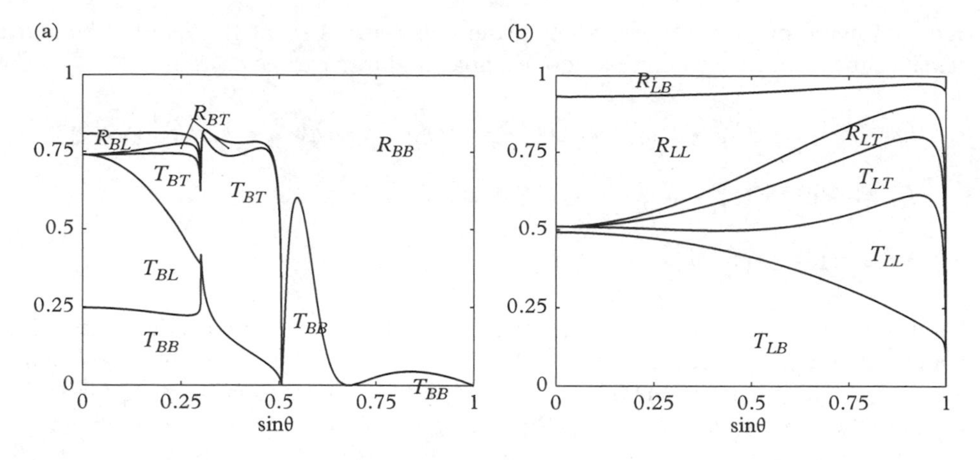

Figure 9.9 *Transmission and reflection efficiencies for an L-junction between two identical plates with $\nu = 0.3$ and $c_L/c_B = 0.3$. (a) the case of an incident bending wave; (b) the case of an incident longitudinal wave.*

An interesting property of these transmission curves is that they are generally peaked about certain values of incidence angle. This leads to selection of certain directions during transmission, the other ones being rejected. When several plates are placed in cascade, this selection process is repeated across the successive boundaries. This effect is called spatial filtering. It may be responsible for a strongly non-diffuse field in the last plate.

9.7 Beam Line Coupled to Plate

The coupling between plates and beams may arrive in various ways. The case of a beam normal to the plate plane and attached to the plate by an extremity is outlined in Lyon and Eichler (1964). But the case of a beam whose extremity is attached to a plate edge, with both lying in the same plane, is discussed in Lyon and Scharton (1965).

In this section, we rather discuss the case of a beam coupled to a plate along one of its sides. The situation is shown in Fig. 9.10. For the sake of conciseness, we confine discussion to the attenuation of travelling flexural waves in beams.

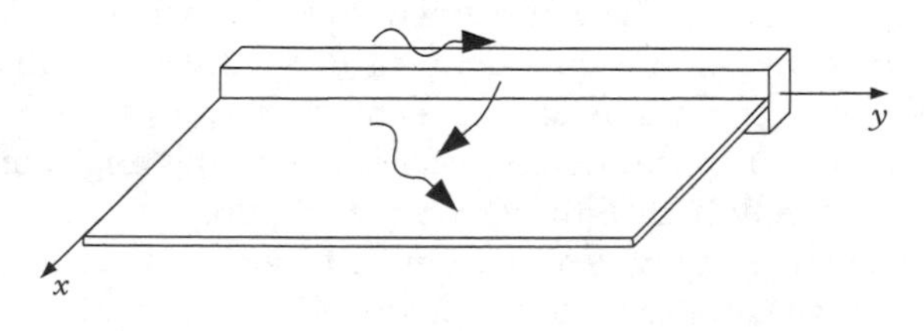

Figure 9.10 *Beam line coupled to a plate.*

Let us consider a plate coupled to a beam along one of its edges and assume that the beam cross-section is symmetrical with respect to the middle surface of the plate. In particular the centroidal axis of the beam coincides with that of the centre of twist applied by the plate. The y-axis is taken along the beam and the x-axis is normal to the beam on the inside of the plate. Let $w_1(y,t)$ be the beam deflection, $\theta_1(y,t)$ the torsion angle of the beam, and $w_2(x,y,t)$ the plate deflection. The governing equations of the beam are (Ungar, 1961)

$$m_1 \frac{\partial^2 w_1}{\partial t^2} + E_1 I_1 \frac{\partial^4 w_1}{\partial y^4} = -D_2 \left(\frac{\partial^3 w_2}{\partial x^3} + (2-\nu) \frac{\partial^3 w_2}{\partial x \partial y^2} \right) \tag{9.127}$$

$$\rho_1 \mathcal{J}_1 \frac{\partial^2 \theta_1}{\partial t^2} - G_1 \mathcal{J}_1 \frac{\partial^2 \theta_1}{\partial y^2} = D_2 \left(\frac{\partial^2 w_2}{\partial x^2} + \nu \frac{\partial^2 w_2}{\partial y^2} \right) \tag{9.128}$$

where ρ_1 is the density of the beam material, m_1 the mass per unit length of the beam, E_1 the Young modulus, G_1 the shear modulus, I_1 the moment of inertia of the beam cross-section with respect to the x-axis, $\mathcal{J}_1$ the polar moment of inertia, $E_1 I_1$ the bending rigidity, $G_1 \mathcal{J}_1$ the torsional stiffness, D_2 the plate bending rigidity, and ν the Poisson ratio of the plate material. In the right-hand side of the first equation appears the effective shear force $V = Q - \partial_y L$ applied to the beam by the plate in accordance with eqn (4.81). The right-hand side of the second equation is the y-component of the bending moment M applied by the plate. In addition to these equations, continuity of deflection along the contact line $x = 0$ and rotation of sections about the y-axis read

$$w_1(y,t) = w_2(0,y,t) \tag{9.129}$$

$$\theta_1(y,t) = -\frac{\partial w_2}{\partial x}(0,y,t) \tag{9.130}$$

The minus sign in the second equation comes from the sign convention for θ_1 and the x-axis direction.

The above set of differential equations and coupling conditions shows that torsional and flexural motions are coupled. So, if a travelling wave exists in the beam, it induces a perturbation of the three fields w_1, w_2, and θ_1. A wave travelling in the beam along the y-axis therefore takes the form

$$w_1(y,t) = a e^{\imath(\omega t + \kappa y)} \tag{9.131}$$

$$\theta_1(y,t) = b e^{\imath(\omega t + \kappa y)} \tag{9.132}$$

$$w_2(x,y,t) = c e^{\imath\left(\omega t + \kappa' x + \kappa y\right)} + d e^{\imath\left(\omega t + \kappa'' x + \kappa y\right)} \tag{9.133}$$

where ω is the angular frequency and κ the in situ wavenumber of the beam that we must yet determine. The y-dependence of all waves is the same by compatibility.

The x-dependence of w_2 is determined by the condition that each exponential must satisfy eqn (9.58). Solving the resulting dispersion equation gives

$$\kappa' = \sqrt{\kappa_2^2 - \kappa^2}, \quad \kappa'' = \imath\sqrt{\kappa_2^2 + \kappa^2} \tag{9.134}$$

where $\kappa_2 = (m_2\omega^2/D_2)^{1/4}$ is the plate wavenumber and m_2 the mass per unit area of the plate material. The wavenumber κ is a priori complex and among the two square roots of $\kappa_2^2 - \kappa^2$ and $\kappa_2^2 + \kappa^2$, one must select the one which ensures that the wave is moving away from the boundary. In particular $\mathrm{Re}(\kappa')$, $\mathrm{Re}(\kappa'') > 0$ (the wave is null at infinity).

The dispersion equation is obtained by substituting eqns (9.131)–(9.133) into (9.127)–(9.130) and writing that the determinant of the resulting set of linear equations is zero:

$$\begin{vmatrix} -m_1\omega^2+\kappa^4 E_1 I_1 & 0 & -\imath D_2\kappa'^3-\imath(2-\nu)D_2\kappa^2\kappa' & -\imath D_2\kappa''^3-\imath(2-\nu)D_2\kappa^2\kappa'' \\ 0 & -\rho_2 \mathcal{J}_2\omega^2+G_2\mathcal{J}_2\kappa^2 & D\left(\kappa'^2+\nu\kappa^2\right) & D_2\left(\kappa''^2+\nu\kappa^2\right) \\ 1 & 0 & -1 & -1 \\ 0 & 1 & \imath\kappa' & \imath\kappa'' \end{vmatrix} = 0 \tag{9.135}$$

The wavenumber κ is a solution to this equation.

A flexural wave travelling in the beam has the form $w_1(y,t) = a\mathrm{e}^{\imath(\omega t+\kappa y)}$ where κ is the in situ wavenumber of the beam that we have just determined. The function $\kappa(\omega)$ is a characteristic of the beam embodied in the plate. For a free beam (without plate) the wave w_1 is purely propagative and κ is real-valued. But in the presence of a surrounding plate, the beam wave generates an out-of-plane motion in the plate. The wave radiates energy in the plate and the in situ wavenumber κ becomes complex-valued $\kappa = \kappa_r + \imath\kappa_i$. In that case,

$$w_1(y,t) = a\mathrm{e}^{-\kappa_i y} \times \mathrm{e}^{\imath(\omega t+\kappa_r y)} \tag{9.136}$$

The presence of a positive imaginary part $\kappa_i > 0$ of the wavenumber implies a decreasing magnitude of the wave representative of a loss of energy during propagation. Equation (9.136) gives the time-average vibrational energy density $1/2 \times m_1 \,|\, \dot{w}_1 \,|^2$ assessed as twice the kinetic energy density,

$$W(x) = \frac{1}{2} m_1\omega^2 \,|\, a \,|^2 \mathrm{e}^{-2\kappa_i y} \tag{9.137}$$

where m_1 is the mass per unit length of the beam.

But for any one-dimensional propagative medium, the decreasing of energy is of the form

$$W(x) = W_0 \mathrm{e}^{-\frac{\eta\omega}{c_g} y} \tag{9.138}$$

where the group speed is

$$c_g = \frac{d\omega}{d\kappa_r} \tag{9.139}$$

So, if one attributes the damping η to the coupling loss factor η_{01}, we have by comparing eqn (9.138) and (9.137) (Bremner and Burton, 1999),

$$\eta_{12} = 2\frac{d\omega}{d\kappa_r}\frac{\kappa_i}{\omega} \tag{9.140}$$

The problem therefore reduces to the calculation of $\kappa(\omega) = \kappa_r + \iota\kappa_i$, which is the solution to the dispersion equation (9.135).

9.8 Coupled Rooms

The results of Section 9.5 concerning two-dimensional systems may be straightforwardly generalized to three dimensions. Let us consider two acoustical cavities separated by a wall and a wave travelling in cavity 1 and impinging on the wall with incidence (θ, φ) where θ is the angle normal to the wall and φ the azimuthal angle (Fig. 9.11). The transmission and reflection efficiencies of the wall are denoted by $T(\theta, \varphi)$ and $R(\theta, \varphi)$.

Let us assume that the sound field in cavity 1 is diffuse. By denoting E_1 the sound energy, the energy density per unit volume and unit solid angle is $E_1/4\pi V_1$ where V_1 is the volume. The radiative intensity vector is therefore constant,

$$I(\theta, \varphi) = \frac{c_1 E_1}{4\pi V_1} \tag{9.141}$$

where c_1 is the sound speed. The power transmitted from cavity 1 to cavity 2 is the flux of $T(\theta, \varphi)I(\theta, \varphi)\cos\theta$ through the wall,

$$P_{1\to 2} = S\frac{c_1 E_1}{4\pi V_1}\int_0^{2\pi}\int_0^{\pi/2} T(\theta, \varphi)\cos\theta\sin\theta\, d\theta d\varphi \tag{9.142}$$

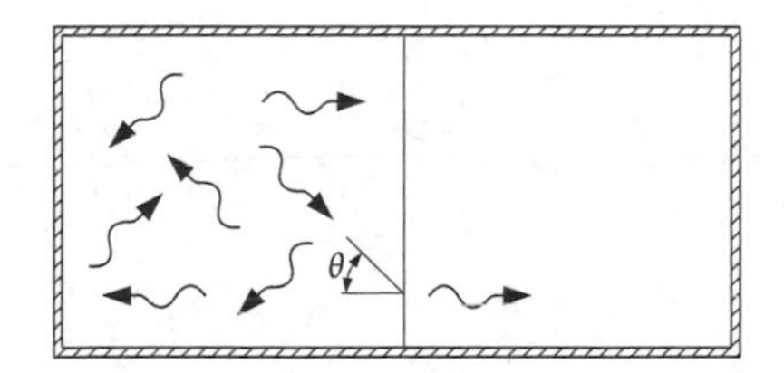

Figure 9.11 *Coupled rooms separated by a wall. The sound field is diffuse.*

where S is the surface of the separating wall and $\sin\theta\, \mathrm{d}\theta \mathrm{d}\varphi$ the infinitesimal solid angle in spherical coordinates. The coupling loss factor follows:

$$\eta_{12} = \frac{Sc_1}{4\pi\omega V_1} \int_0^{2\pi} \int_0^{\pi/2} T(\theta,\varphi) \cos\theta \sin\theta \, \mathrm{d}\theta \mathrm{d}\varphi \tag{9.143}$$

Once again, the determination of the coupling loss factor reduces to that of the transmission efficiency.

Simple wall

In this simple problem, the transmission of sound through a homogeneous single wall is investigated (London, 1949). Two cavities containing the same fluid are separated by a wall of mass per unit area m and bending rigidity D. The fluid has density ρ_0 and sound speed c_0.

Let us consider a harmonic plane wave propagating in the first cavity and impinging on the wall with an incident angle θ (Fig. 9.12). The incident velocity potential is

$$\Phi_0(x,z,t) = \mathrm{e}^{\imath(\omega t + \kappa z \cos\theta - \kappa x \sin\theta)} \tag{9.144}$$

where $\omega^2 = \kappa^2 c_0^2$. The minus sign in the x-term and the plus sign in the z-term indicate that the wave propagates in the positive x-direction and negative z-direction. During the reflection process, a harmonic plane wave is reflected to the first cavity, a second one is transmitted to the other cavity, while a harmonic structural wave propagates in the wall.

Three equations are at our disposal to find the solution. First of all, the continuity conditions are written in terms of normal velocities. According to eqn (4.101) the z-component of the fluid velocity is $\partial_z \Phi_i$. So the equality of fluid velocities and out-of-plane vibration velocity reads

$$\frac{\partial \Phi_1}{\partial z} = \frac{\partial u}{\partial t} = \frac{\partial \Phi_2}{\partial z} \tag{9.145}$$

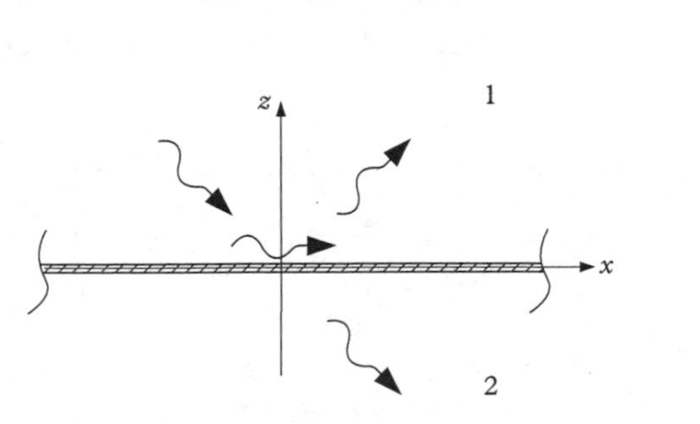

Figure 9.12 *Wave transmission of an acoustical wave through a single wall.*

Secondly, the governing equation of a plate in fluid is

$$m\frac{\partial^2 u}{\partial t^2} + D\frac{\partial^4 u}{\partial x^4} = \rho_0\left(\frac{\partial \Phi_1}{\partial t} - \frac{\partial \Phi_2}{\partial t}\right) \tag{9.146}$$

In the right-hand side of the above equation appears the difference of acoustical pressures $p_2 - p_1$ where $p_i = -\rho_0\dot{\Phi}_i$ from eqn (4.102). Since cavity 2 is located at the bottom and z-axis is oriented upward (Fig. 9.12) a positive p_2 pressure induces a positive acceleration $m\ddot{u}$, but this is the opposite for p_1. To be complete, one must add that Φ_i is the solution to the wave equation while reflected and transmitted waves verify the Sommerfeld condition.

If we introduce the reflection ratio r and transmission ratio t, we expect to find the acoustical potentials as

$$\Phi_1(x, z, t) = \mathrm{e}^{\imath\omega t}\left(\mathrm{e}^{\imath\kappa z\cos\theta} + r\mathrm{e}^{-\imath\kappa z\cos\theta}\right)\mathrm{e}^{-\imath\kappa x\sin\theta} \tag{9.147}$$

$$\Phi_2(x, z, t) = t\mathrm{e}^{\imath(\omega t+\kappa z\cos\theta-\kappa x\sin\theta)} \tag{9.148}$$

In these expressions, we have implicitly admitted that all wavenumbers are equal to κ, and that the reflection angle is the opposite of the incident angle while the transmission angle equals the incident angle. Equality of wavenumbers is a simple consequence of the fact that potentials obey the same wave equation (the fluid is the same on both sides of the wall). Reflection and transmission angles are imposed by the equality $\partial_z\Phi_1 = \partial_z\Phi_2$ which must hold at any time t and any position x. This imposes that the dependence in t and x is done by the same exponentials on both sides of the equality. In particular, ω has the same value in all waves and the x-component $\kappa\sin\theta$ of wavenumbers is the same.

We have also mentioned that a structural wave exists in the wall. So, by denoting s the complex magnitude of this wave, the deflection in the wall is

$$u(x, t) = s\mathrm{e}^{\imath(\omega t-\kappa x\sin\theta)} \tag{9.149}$$

Notice that the coefficients ω and $\kappa\sin\theta$ are imposed by the same argument as above applied to $\partial_z\Phi_1 = \dot{u}$.

To solve in r, t, and s, let us substitute eqns (9.147)–(9.149) into eqns (9.145) and (9.146):

$$\begin{cases} 1 - r = t \\ \kappa\cos\theta \times t = \omega s \\ (-m\omega^2 + D\kappa^4\sin^4\theta)s = \imath\rho_0\omega(1 + r - t) \end{cases} \tag{9.150}$$

This is a set of linear equations whose solution is readily found:

$$r = \frac{(-m\omega^2 + \frac{\omega^4}{c_0^4} D \sin^4\theta)}{(-m\omega^2 + D\frac{\omega^4}{c_0^4}\sin^4\theta) + 2\imath\frac{\rho_0 c_0 \omega}{\cos\theta}} \tag{9.151}$$

$$t = \frac{2\imath\frac{\rho_0 c_0 \omega}{\cos\theta}}{(-m\omega^2 + \frac{\omega^4}{c_0^4} D \sin^4\theta) + 2\imath\frac{\rho_0 c_0 \omega}{\cos\theta}} \tag{9.152}$$

It may be noticed that $|r|^2 + |t|^2 = 1$.

The time-average acoustical intensity is $\mathbf{I}_i = 1/2 \times \mathrm{Re}\,[p_i \overline{\mathbf{v}}_i]$ where $\mathbf{v}_i$ is the fluid velocity. In terms of velocity potential, the z-component of the intensity is

$$I_i = \frac{1}{2}\mathrm{Re}\left[-\rho_0 \frac{\partial \phi_i}{\partial t}\frac{\partial \overline{\phi}_i}{\partial z}\right] \tag{9.153}$$

For the incident harmonic wave given in eqn (9.144),

$$I_0 = \frac{1}{2}\rho_0 \omega \kappa \cos\theta \tag{9.154}$$

and similarly for the transmitted wave given by eqn (9.148),

$$I_2 = \frac{1}{2}\rho_0 \omega \kappa \cos\theta\,|t|^2 \tag{9.155}$$

The transmission efficiency $T(\theta)$ is defined as the ratio of transmitted power to incident power. Since the power is the flux of intensity through the wall, this is also the ratio of normal components of intensity I_2/I_0. It yields

$$T(\theta) = \frac{4\left(\frac{\rho_0 c_0 \omega}{\cos\theta}\right)^2}{(-m\omega^2 + \frac{\omega^4}{c_0^4} D \sin^4\theta)^2 + 4\left(\frac{\rho_0 c_0 \omega}{\cos\theta}\right)^2} \tag{9.156}$$

Remember that in this expression ρ_0 is the fluid density, c_0 the sound speed, m the mass per unit area of the wall, and $D = Eh^3/12(1 - \nu^2)$ the bending rigidity where E is the Young modulus, h the wall thickness, and ν the Poisson coefficient.

The diffuse field transmission efficiency is obtained by integrating over all incident angles as in eqn (9.143). This may generally not be performed easily except in the mass law approximation where the bending rigidity is neglected. The reader may refer to any standard book on acoustics for further details.

Multilayer panel

As a more complicated example, we now solve the general problem of sound transmission through a multilayer panel of infinite extent (Lathuilière and Tarley, 1986). Due to its particular interest in acoustics, the special case of double walls has been studied for a long time (London, 1950). See also Price and Crocker (1970) for a complete statistical model of the problem with both non-resonant and resonant transmission and Sgard et al. (2010) for an application to the acoustical performance of enclosures.

The multilayer panel that we consider here is a succession of fluids separated by plates as shown in Fig. 9.13. The simple wall is of course a special case where two fluids are separated by a single plate. A double-glazing window is made of two glass panels separated by an air space. More generally, we shall consider a stack of fluids of thickness h_i, sound speed c_i, and density ρ_i and plates of mass per unit area m_i and bending stiffness D_i. The fact that two fluids are separated by a plate and that two plates are separated by a fluid is by no means a limitation of the present model. The problem of transmission through the interface of two fluids, air–water for instance, is a special case if we introduce a fictitious plate of bending stiffness $D = 0$ and mass $m = 0$. In the same way, if the panel is made by a superposition of plates of different materials, the present model applies by introducing interstitial fictitious fluids of thickness $h = 0$.

The continuity of normal velocities reads

$$\frac{\partial \Phi_{i-1}}{\partial z} = \frac{\partial u_i}{\partial t} = \frac{\partial \Phi_i}{\partial z}, \qquad i = 1, 2, \ldots, n \tag{9.157}$$

By convention, the fluid wherein the incident wave propagates is numbered 0. The governing equation of plate i is

$$m_i \frac{\partial^2 u_i}{\partial t^2} + D_i \frac{\partial^4 u_i}{\partial x^4} = -\rho_{i-1} \frac{\partial \Phi_{i-1}}{\partial t} + \rho_i \frac{\partial \Phi_i}{\partial t} \tag{9.158}$$

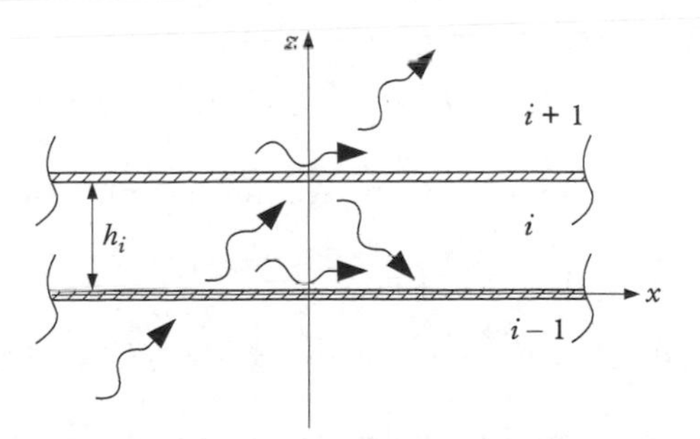

Figure 9.13 *Wave transmission of an acoustical wave through a multilayer panel.*

As is usual, now we consider a harmonic plane wave of frequency ω incident on the panel with angle θ. The incident velocity potential field is

$$\Phi_0(x, z, t) = e^{\iota(\omega t + \kappa_0 z \cos\theta - \kappa_0 x \sin\theta)} \tag{9.159}$$

where the wavenumber is noted $\kappa_0 = \omega/c_0$. We look for the solution as a superposition of harmonic plane waves. In fluid i the velocity potential takes the form

$$\Phi_i(x, z, t) = a_i e^{\iota(\omega t - \lambda_i z - \kappa_0 x \sin\theta)} + b_i e^{\iota(\omega t + \lambda_i z - \kappa_0 x \sin\theta)} \tag{9.160}$$

The magnitude of the incident wave is $a_0 = 1$. By virtue of continuity, the x-component of the wavenumber is the same for all potentials and has the common value $\kappa_0 \sin\theta$. The z-component is imposed by the dispersion equation of the wave equation. We have

$$\lambda_i = \sqrt{\kappa_i^2 - (\kappa_0 \sin\theta)^2} \tag{9.161}$$

where $\kappa_i = \omega/c_i$. The z-component may be real or purely imaginary. In the first case the wave propagates, while in the second case the wave is evanescent.

In the plates, the deflection is

$$u_i(x, t) = c_i e^{\iota(\omega t - \kappa_0 x \sin\theta)} \tag{9.162}$$

Again, the wavenumber is set to $\kappa_0 \sin\theta$ by compatibility.

Let

$$Z_i = \frac{1}{\iota\omega}\left[-m_i\omega^2 + D_i(\kappa_0 \sin\theta)^4\right] \tag{9.163}$$

be the impedance of plate i. By substituting eqns (9.160) and (9.162) into eqns (9.157) and (9.158), it follows

$$\begin{cases} -\iota\lambda_{i-1}(a_{i-1}e_{i-1}^- - b_{i-1}e_{i-1}^+) = \iota\omega c_i \\ -\iota\lambda_i(a_i - b_i) = \iota\omega c_i \\ \iota\omega Z_i c_i = \iota\omega(-\rho_{i-1}a_{i-1}e_{i-1}^- - \rho_{i-1}b_{i-1}e_{i-1}^+ + \rho_i a_i + \rho_i b_i) \end{cases} \tag{9.164}$$

where we have set $e_i^+ = e^{\iota\lambda_i h_i}$ and $e_i^- = e^{-\iota\lambda_i h_i}$. We may simplify by eliminating c_i:

$$\begin{cases} a_{i-1}\lambda_{i-1}e_{i-1}^- - b_{i-1}\lambda_{i-1}e_{i-1}^+ - a_i\lambda_i + b_i\lambda_i = 0 \\ a_{i-1}\rho_{i-1}\omega e_{i-1}^- + b_{i-1}\rho_{i-1}\omega e_{i-1}^+ - a_i\left(Z_i\lambda_i + \rho_i\omega\right) + b_i\left(Z_i\lambda_i - \rho_i\omega\right) = 0 \end{cases} \tag{9.165}$$

To assemble this set of equations in a matrix equation, we form the vector of unknowns $(b_0\ a_1\ b_1\ \ldots\ a_n)^T$. In this vector the coefficients $a_0 = 1$ and $b_n = 0$ do not appear as unknown since their values are imposed. It yields

$$\begin{pmatrix} -\lambda_0 & -\lambda_1 & \lambda_1 & 0 & \dots & 0 & \\ \rho_0\omega & -Z_1\lambda_1-\rho_1\omega & Z_1\lambda_1-\rho_1\omega & & & & \\ & \vdots & & & & \vdots & \\ 0 & \lambda_{i-1}e^-_{i-1} & -\lambda_{i-1}e^+_{i-1} & -\lambda_i & \lambda_i & & \\ & \rho_{i-1}\omega e^-_{i-1} & \rho_{i-1}\omega e^+_{i-1} & -Z_i\lambda_i-\rho_i\omega & Z_i\lambda_i-\rho_i\omega & 0 & \\ \vdots & & & & \vdots & & \\ & & & \lambda_{n-1}e^-_{n-1} & -\lambda_{n-1}e^+_{n-1} & -\lambda_n & \\ 0 & \dots & 0 & \rho_{n-1}\omega e^-_{n-1} & \rho_{n-1}\omega e^+_{n-1} & -Z_n\lambda_n-\rho_n\omega & \end{pmatrix} \begin{pmatrix} b_0 \\ a_1 \\ \vdots \\ a_i \\ b_i \\ \vdots \\ a_n \end{pmatrix} = \begin{pmatrix} -\lambda_0 \\ \rho_0\omega \\ 0 \\ \\ \vdots \\ \\ 0 \end{pmatrix} \tag{9.166}$$

The above matrix has dimension $2n \times 2n$. The $2i - 1$-th and $2i$-th rows contain four components in columns $2i - 2$ to $2i + 1$. The first and last two rows contain only three non-zero entries.

The transmission efficiency is the ratio of transmitted power $1/2 \times \rho_n\omega\lambda_n |a_n|^2$ to incident power $1/2 \times \rho_0\omega\lambda_0$,

$$T(\theta) = \frac{\rho_n\lambda_n}{\rho_0\lambda_0} |a_n|^2 \tag{9.167}$$

This relationship embodies eqn (9.156) as a special case.

Figure 9.14 shows the transmission loss factors $\text{TL} = -10\log T(\theta)$ of respectively a single leaf of 8 mm glass and a double leaf composed of two 4 mm glass panels separated

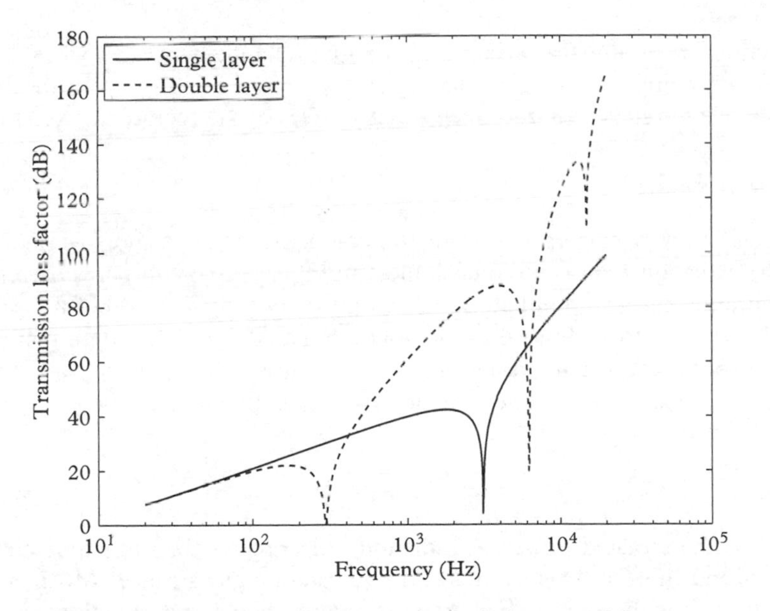

Figure 9.14 *Transmission loss factor of a single-layer panel (8 mm glass) and a double-layer panel (4 mm glass–16 mm air–4 mm glass) at incidence* 45°.

by 16 mm of air at incidence $\theta = 45°$. The curve of a single panel presents two regimes separated by a sharp drop. On the left part transmission is controlled by the mass law with a slope of 6 dB per octave. The coincidence frequency which minimizes the denominator of eqn (9.156) is $\omega_c / \sin^2\theta$ where $\omega_c = c_0^2(m/D)^{1/2}$ is the so-called critical frequency. At coincidence (2990 Hz for the single panel) $T = 1$ and TL = 0 which corresponds to a complete transmission and therefore a loss of efficiency of sound isolation. Beyond coincidence transmission is controlled by the stiffness law with a frequency dependence $T \propto 1/\omega^6$ that is a slope of 18 dB per octave. The double-layer panel has a more rich behaviour. At low frequencies, the mass law gives the same TL as for a single panel (the mass of both panels has been maintained constant) since the movement of both layers is identical and therefore the double layer behaves like an equivalent single layer of the same mass. Coincidence of a layer of 4 mm thickness rather occurs at 5980 Hz and is of course the same for the two identical layers, which waves can cross without loss. This corresponds to the second sharp drop of the double-leaf curve and a total loss of isolation. The first sharp drop is positioned at the mass–air–mass resonance frequency $\omega_0 = \left(2\rho_0 c_0^2/mh\cos^2\theta\right)^{1/2}$ where h is the thickness of the intermediate air layer. At this frequency the air layer acts like a spring and both layers move in opposition phase. The above formula estimates this frequency at 299 Hz. The subsequent sharp drop corresponds to stationary waves in the air layer at frequencies $nc_0/2h\cos\theta$ with $n = 1, 2, \ldots$ the first of which being at 15 158 Hz.

9.9 Radiation of Sound

In this section, we outline the question of energy exchange between a vibrating structure and the surrounding air. Sound radiation is a transfer of energy from the structure to acoustics while the structural response usually refers to the reciprocal exchange.

Radiation ratio

The most important concept in sound radiation problems is the so-called **radiation ratio**. The radiation ratio is a dimensionless number which compares the efficiency of a given vibrating surface to radiate sound with that of a rigid piston of the same area. It may be shown (Norton, 1989) that the power radiated from a rigid and large piston of area S and root-mean square vibrational velocity v_0 is $\rho_0 c_0 S v_0^2$ where c_0 is the speed of sound and ρ_0 the air density. The radiation ratio σ is therefore defined as

$$\sigma = \frac{P_{\text{rad}}}{\rho_0 c_0 v_0^2 S} \tag{9.168}$$

where P_{rad} is the radiated power. A radiation ratio greater than one indicates a surface radiating sound more efficiently than an equivalent rigid surface. Most often the radiation ratio is less than unity. For many standard shapes the radiation ratio is known from exact or approximate solutions of the Helmholtz equation with imposed boundary conditions.

Coupling loss factors

The coupling loss factor associated with sound radiation is noted η_{rad}. From the definition of radiation ratio given in eqn (9.168) the radiated power is

$$P_{\text{rad}} = \rho_0 c_0 \sigma v_0^2 S \tag{9.169}$$

But the vibrational energy in the structure is twice the kinetic energy,

$$E_1 = m v_0^2 S \tag{9.170}$$

where m is the mass per unit area of the structure. The radiation coupling loss factor is obtained from eqn (9.1) by combining the two above equations,

$$\eta_{\text{rad}} = \frac{\rho_0 c_0 \sigma}{m\omega} \tag{9.171}$$

The calculation of a sound radiation coupling loss factor therefore reduces to that of the radiation ratio.

For the reciprocal phenomenon, the structural response, the coupling loss factor is not so easily obtained by a direct wave approach. But the constraint of reciprocity gives

$$\eta_{\text{rad}} n_1 = \eta_{\text{resp}} n_0 \tag{9.172}$$

where n_0 and n_1 are respectively the modal densities of the room and of the structure. We get

$$\eta_{\text{resp}} = \frac{\rho_0 c_0 \sigma n_1}{m\omega n_0} \tag{9.173}$$

Infinite plate

The simplest situation for which the radiation ratio may be evaluated is that of an infinite plate. Let us consider a wave propagating in an infinite plate coupled to a fluid (Fig. 9.15). Since we are looking for the coupling loss factors between the plate and a

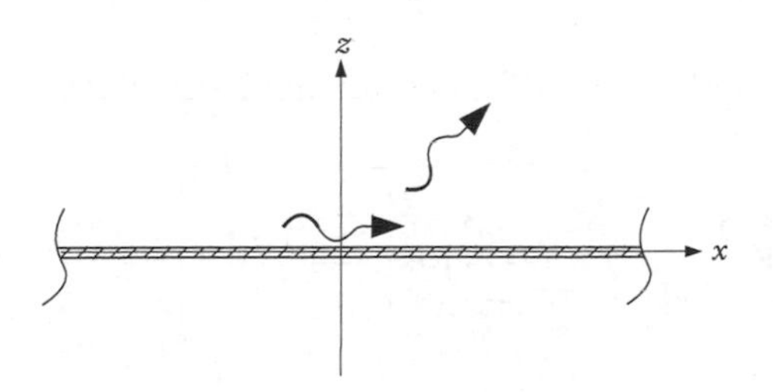

Figure 9.15 *Radiation of a structural wave.*

single acoustical cavity, we consider that the fluid occupies only one side of the plate, say the semi-infinite space $z > 0$. The continuity condition reads

$$\frac{\partial \Phi}{\partial z} = \frac{\partial u}{\partial t} \tag{9.174}$$

while the governing equation of the plate is

$$m\frac{\partial^2 u}{\partial t^2} + D\frac{\partial^4 u}{\partial x^4} = \rho_0 \frac{\partial \Phi}{\partial t} \tag{9.175}$$

where, as usual, u is the deflection of the plate and Φ the velocity potential in fluid.

If a single propagating wave travels in the plate, the deflection has the form

$$u(x,t) = \mathrm{e}^{\imath(\omega t - \kappa x)} \tag{9.176}$$

and Φ may be found as

$$\Phi(x,z,t) = -\frac{\omega}{\lambda}\mathrm{e}^{\imath(\omega t - \lambda z - \kappa x)} \tag{9.177}$$

where the magnitude $-\omega/\lambda$ has been set to match with the continuity condition. Since Φ is the solution to the wave equation, κ, λ, and ω are linked by the acoustic dispersion equation $\kappa^2 + \lambda^2 = \omega^2/c_0^2$. Thus,

$$\lambda = \sqrt{\frac{\omega^2}{c_0^2} - \kappa^2} \tag{9.178}$$

In the determination of the complex square root, the Sommerfeld condition imposes to select the root which verifies $\mathrm{Re}\,(\lambda) > 0$. The z-component of the acoustical intensity is $I = 1/2 \times \mathrm{Re}\left[p\bar{\dot{u}}\right]$ by eqn (5.112). Employing eqns (4.102) and (9.174),

$$I = \frac{1}{2}\mathrm{Re}\left[-\rho_0 \frac{\partial \Phi}{\partial t}\frac{\partial \overline{\Phi}}{\partial z}\right] \tag{9.179}$$

which, after substitution of eqn (9.177) and further simplifications, gives

$$I = \frac{\rho_0 \omega^3}{2}\mathrm{Re}\left[\frac{1}{\sqrt{\frac{\omega^2}{c_0^2} - \kappa^2}}\right] \tag{9.180}$$

The radiated power of a piece of plate of area S is $P_{\mathrm{rad}} = IS$.

On the other hand, the mean square value of the vibrational velocity is $v_0^2 = 1/2 \times \mathrm{Re}\left[\dot{u}\bar{\dot{u}}\right]$. Substitution of eqn (9.176) gives $\rho_0 c_0 v_0^2 S = 1/2 \times \rho_0 c_0 \omega^2 S$. The radiation ratio is therefore

$$\sigma = \mathrm{Re}\left[\frac{1}{\sqrt{1 - \frac{\kappa^2 c_0^2}{\omega^2}}}\right] \tag{9.181}$$

In the above equation both ω and κ appear. But these quantities are linked by the dispersion equation of the loaded plate. This equation is found by substituting eqns (9.176) and (9.177) into eqn (9.175),

$$-m\omega^2 + D\kappa^4 + \imath\frac{\rho_0\omega^2}{\sqrt{\frac{\omega^2}{c_0^2} - \kappa^2}} = 0 \tag{9.182}$$

This equation has several solutions which may be identified to propagating or evanescent waves. But when the fluid is light (air for instance), a good approximation for the radiation ratio is obtained by taking the *in vacuo* dispersion relationship $\kappa^2 = \omega(m/D)^{1/2}$. By introducing the coincidence frequency,

$$\omega_c = c_0^2\sqrt{\frac{m}{D}} \tag{9.183}$$

we get the radiation ratio of the infinite plate in air,

$$\sigma = \begin{cases} \frac{1}{\sqrt{1-\frac{\omega_c}{\omega}}} & \text{if } \omega > \omega_c \\ 0 & \text{otherwise} \end{cases} \tag{9.184}$$

An infinite plate does not radiate sound below the coincidence frequency but has a radiation ratio greater than unity above coincidence.

9.10 Comments

The study of wave transmission at the interface between two media is certainly ancient and may be connected with the problem of refraction in optics or isolation by walls in acoustics. However, similar problems for structural transmission were rarely outlined in the literature before the 1960s. But the advent of statistical energy analysis has given rise to a renewal of interest in calculating the transmission factors.

From the very start of statistical energy analysis, the wave approach is clearly mentioned as an efficient method to calculate the coupling loss factors. The former texts of R.H. Lyon and his co-workers (Lyon and Eichler, 1964; Lyon and Scharton,

1965) introduced the assumption of diffuse field and applied the wave approach to transmission problems in some cases of beam–plate couplings.

In their book, Cremer and Heckl (1988) solved a large number of couplings between plates and, incidentally, between beams, by specializing to normal incidence. They give many expressions of transmission and reflection efficiencies in Chapter V, entitled *Attenuation of Structure-Borne Sound.* Cross-section discontinuities, corners, branches at right-angles, elastic interlayers, and blocking masses are treated in detail amongst other subjects. Even in the time of numerical engineering, their closed-form relationships remain useful for a rapid estimation of transmission or to verify software results. See also Craik et al. (2004) for a benchmark on plate junctions.

Most of these special cases of transmission may be synthesized in a small number of canonical problems. These are beams coupled at their extremities, plate–beam coupling, plates coupled together by a common edge with arbitrary angles, and sound transmission through multilayer walls. Junctions may also have their own characteristics such as mass or stiffness. A significant improvement very useful in practice is to consider that the middle planes of plates may be not coincident at the junction. This question is discussed by Langley and Heron (1990) for the problem of plates coupled to a beam taking into account shear deformation and rotary inertia in the beam.

The choice of beam and plate theories is particularly important when dealing with high frequencies. Basically, beam and plate theories are approximate theories derived from elastodynamics. Their validity is usually limited to a certain frequency bandwidth. The most popular theories for flexural waves are Euler–Bernoulli for beams and Love theory for plates. Both neglect shear forces and rotatory inertia effects. The most correct theories are respectively Timoshenko's theory for beams and Mindlin's (1951) theory for plates. These theories include shear and rotatory effects and give a correct asymptotic behaviour of wave speed when frequency goes to infinity. The importance of shear and rotatory inertia to energy transmission in L-plate configuration was underlined by McCollum and Cushieri (1990). Their use in statistical energy analysis should therefore represent a significant improvement. The most general calculation of energy transmission for an arbitrary number of beams coupled at a common end, taking into account traction-compression, torsion, and flexion in the framework of Timoshenko's theory and without assuming coincidence of shear centre and centroid axis, is performed by De Langhe (1996). The same reference also provides the calculation of energy transmission for in-plane and out-of-plane motion in plates coupled along a common edge with an arbitrary angle using Mindlin's theory and eventually with non-coincidence of the middle plane of plates.

10
Energy Balance

The implementation of statistical energy analysis on actual complex systems provides a simple method to assess vibrational energy and its localization in subsystems. Application of statistical energy analysis consists of three steps:

1. divide the system into subsystems;
2. evaluate the subsystem parameters; and
3. evaluate the modal and total energies in the subsystems.

The statistical nature of the method restricts its range of application to systems having a large number of preferentially disordered modes. The internal damping of subsystems must be light, couplings must be weak, and external excitations must be random, uncorrelated, and with a large spectral bandwidth.

10.1 Substructuring

The first step in developing a statistical model is to divide the system into subsystems. An example of separation into subsystems is shown in Fig. 10.1. In practice, this is certainly the most important and difficult step. The quality of results will greatly depend on the ability to identify the appropriate subsystems. An automatic procedure has even been proposed by Gagliardini et al. (2005), and Totaro and Guyader (2006) have developed an objective criterion, the so-called MIR index, to identify subsystems.

The simplest way to identify subsystems is to separate the structure into physical components, e.g. beams, pipes, plates, shells, acoustical cavities, etc. More complex components can also be used as long as their wave characteristics are known. Most often, the components are studied separately to determine their group velocity and modal density. In this respect, a numerical method, the so-called **wave finite element** may be employed to identify waves and dispersion curves in complex waveguides (Finnveden, 2004; Ichchou et al., 2007) and even two-dimensional structures (Mace and Manconi, 2008). Most software based on statistical energy analysis works using a large database of subsystems and related coupling loss factors (Bremner et al., 2000).

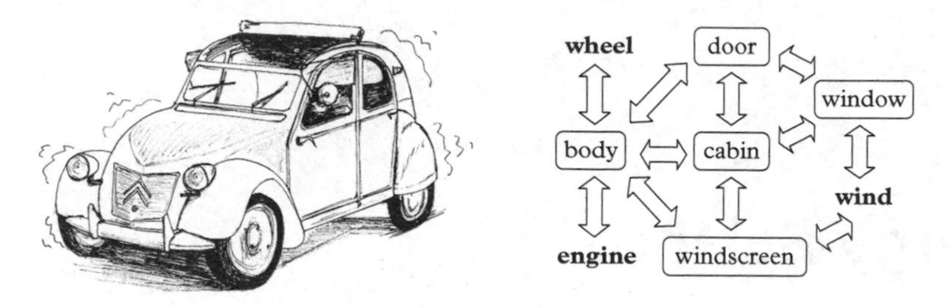

Figure 10.1 *Example of substructuring a truly vibrating system.*

Three rules must be carefully applied during the substructuring procedure.

- All subsystems are excited by rain-on-the-roof forces.
- The subsystems are weakly coupled.
- All subsystems contain a large number of resonant modes.

The weak coupling condition is a strong constraint when substructuring a system. It imposes that junctions between subsystems have mechanical properties (stiffness, mass, or gyroscopic constant) that are small compared with the internal properties of connected subsystems. For instance, two plates assembled by a welding point may clearly be identified to two subsystems. But a single plate cannot be artificially separated into two parts constituting different subsystems since the two half-plates would be strongly coupled. The readers may consult Finnveden (2011) for a general discussion on the notion of weak coupling in structures and a presentation of the so-called modal interaction strength.

The condition of a large number of modes is also a strong constraint in practice. Physical components which are too stiff or small cannot constitute subsystems. The problem is commonly encountered when a structure is made of stiff and compliant constituents. The compliant components have a large modal density, while the stiff ones generally have too low a number of modes to be taken as a subsystem. In the same way, components that are too small cannot constitute a subsystem. This highlights that it is not possible to subdivide a subsystem into an artificially large number of small components, even if this is feasible for instance by the finite element method.

A subsystem should rather be thought of as a group of similar modes within a physical component of a system. When several types of wave propagate in a structural component, the group velocities may differ by several orders of magnitude. This is the case for instance for in-plane waves in plates which travel faster than flexural waves. In such a case, each mode group becomes a single subsystem. A plate is therefore separated into three subsystems; one for bending waves, one for longitudinal waves, and one for transverse waves. They are coupled together by mode conversion occurring at the edges during reflection.

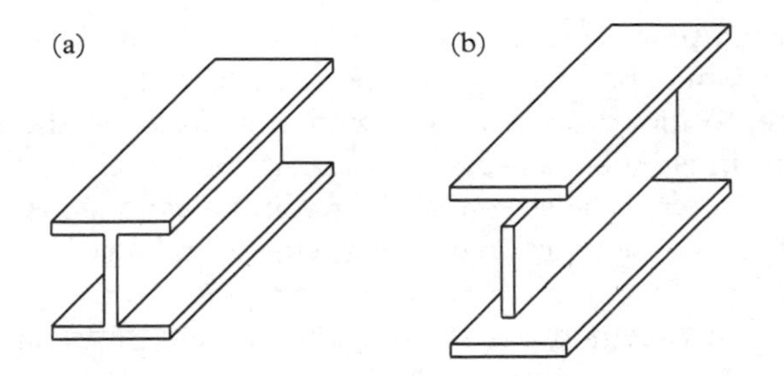

Figure 10.2 *I-beam. (a) a single beam-like subsystem; (b) three plate-like subsystems.*

The definition of subsystems may also depend on the frequency range of the model. An interesting example shown in Fig. 10.2 is discussed in Lyon and DeJong (1995, p.132). At low frequencies, the I-beam has four types of travelling waves, associated with torsion about the axis of the beam, flexion (propagating and evanescent waves) in two orthogonal directions, and longitudinal motion. This model applies as long as the beam cross-section remains undeformed. But at higher frequencies, the flanges and web start to deform and cross-sections of the I-beam are no longer flat. The classical theories of beams (Euler–Bernoulli and Timoshenko) no longer apply. In this frequency range, the appropriate subsystems are rather three plates in flexion instead of a single beam. At still higher frequencies, in-plane motion in the plates starts and the number of subsystems is again increased to take into account longitudinal and transverse waves in the plates.

10.2 Subsystem Parameters

These are the modal densities, coupling loss factors, damping loss factors, and supplied powers.

Modal densities

Most often, the modal densities may be estimated by formula found in Chapter 6, at least for the simplest structural components. But when subsystems are complex, such as curved, ribbed, or honeycomb panels, it may be difficult to assess the wave types and their modal densities.

The empirical approach followed by Bourgine (1973) consists in directly measuring the modal density of subsystems. The method is based on the relationship between modal density and frequency-average point mobility. So, using eqn (8.30),

$$n(\omega) = \frac{2M}{\pi} \times \frac{1}{A\Delta\omega} \int_A \int_{\Delta\omega} \mathrm{Re}[Y(\omega)] \, \mathrm{d}\omega \mathrm{d}A \tag{10.1}$$

To measure the mean point conductance (real part of the mobility), one may employ an impedance head which provides acceleration and force by two piezoelectric sensors placed at the same position between the vibration source and the structure. An analogue or digital time integration is necessary to transform acceleration into velocity. The vibrator is driven by a white noise source filtered in $\Delta\omega$ and a spectrum analyser provides the frequency response function between velocity and force in the frequency band. The mean conductance is simply obtained by averaging the real part of the frequency response function. This method has to be applied several times at different points of the same structure to get accurate results. This is particularly well suited when the modal density is high or when modes overlap so that an individual count of peaks on the frequency response function is not possible.

Resonant transmission

The determination of coupling loss factors between subsystems may usually be done by applying the relationships discussed in Chapter 9. Many other relationships are available in the literature and it is of course beyond the scope of the present book to cover all of them. Most often, the quality of a piece of statistical energy analysis software is determined by its database extent. It may also include direct measurements of connections usual in a particular field of engineering such as the aerospace industry, buildings, and so on.

An alternative approach is to use numerical methods to solve locally the problem of transmission through joints of propagating waves. The method combines the so-called wave finite element for propagation in waveguides and the classical finite element method for joints. This approach allows for calculating the reflection and transmission coefficients of certain discontinuities (Ichchou et al., 2009) or arbitrary joints (Renno and Mace, 2013; Ben Souf et al., 2013).

Non-resonant transmission

It may transpire that subsystems which are not physically connected exhibit a non-zero coupling loss factor. This phenomenon, sometimes called the **tunnelling phenomenon** by analogy with quantum mechanics, stems from the simplifying assumption that only resonant modes are involved in the energy transport process.

Let us consider three subsystems in series as in Fig. 10.3. Subsystem 1 is excited by a wide-band noise confined in $\Delta\omega$. Only resonant modes of subsystem 1 are excited but these modes are coupled with all modes of subsystem 2 including non-resonant modes. These modes, in turn, are coupled with all modes of subsystem 3 and in particular resonant modes. So, a part of this non-resonant energy of subsystem 2 is transmitted to resonant modes of subsystem 3. Such a path of energy is called non-resonant transmission. Due to the approximation made in the determination of the coupling loss factors, this non-typical path is normally not included. This results in a power apparently transmitted from subsystem 1 to subsystem 3 which in fact passes through subsystem 2.

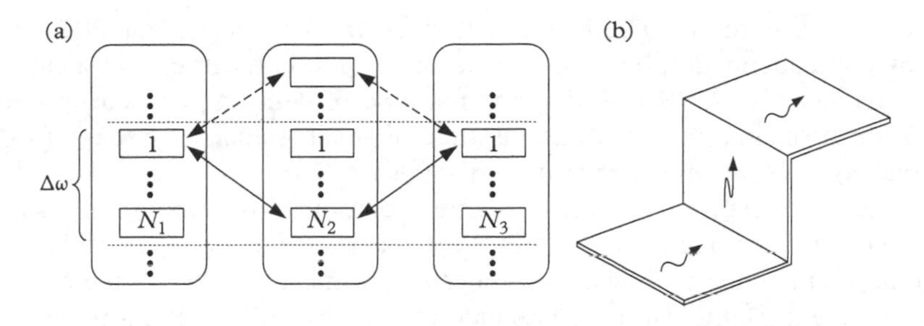

Figure 10.3 *Non-resonant transmission. (a) non-resonant modes of subsystem 2 may receive energy from subsystem 1 and transmit it to subsystem 3 (dashed line), while a resonant path transmits energy only through resonant modes (solid line). (b) example in a three-plate system.*

A simple example of non-resonant transmission, described in Lalor (1999), is realized with three plates in series coupled at right angles as shown in Fig. 10.3. The flexural wave is normally transmitted at the junction by the equality of bending moment and the continuity of rotation angles. This is the resonant transmission path. But, at the same time, the out-of-plane motion at the junction induces an in-plane motion in the intermediate plate which, in turn, induces an out-of-plane motion in the last plate. Since the wavelength of longitudinal waves is usually very large, the intermediate plate remains rigid and mainly acts as a rigid mass.

Indirect coupling

The appearance of non-zero coupling loss factors between disconnected subsystems may have other reasons than non-resonant transmission. In a general manner, we may admit under a few restrictive assumptions that energies and powers are linked by linear equations whose coefficients are called **energy influence coefficients** (Guyader et al., 1982). Mace (2003) distinguishes between the proper SEA and quasi-SEA models. Both are based on an energy balance but the former verifies the consistency relationship (9.2) and the coupling power proportionality (9.1) while the second constraint is relaxed in the latter. Indirect coupling is therefore interpreted as the failure of the coupling power proportionality in which case the energy exchanged between two subsystems may also depend on energies of other subsystems. Strongly non-diffuse fields may therefore induce indirect coupling (Heron, 1994) as well as strong or non-conservative couplings (Fredö, 1997).

Damping loss factors

The damping of vibration in subsystems may have various physical origins which can be difficult to identify. One could cite the elastic hysteresis or relaxation in materials due to internal friction, air-pumping (Maidanik, 1966), or dry friction (Le Bot, 2011) in mechanical joints, and attenuation by atmospheric absorption induced by the shear viscosity of air. Let us also mention an unconventional type of damping caused by the energy transfer from the main structure to a multitude of small attached substructures (Soize, 1993; Strasberg, 1997; Carcaterra and Akay, 2004).

All these phenomena contribute to the value of damping loss factors, but their relative importance is often of no importance. Since the theories of damping are rather complex and difficult to apply due to the large number of parameters involved, the most reliable approach to determine the damping loss factors remains a direct measurement. Several techniques are available based on the determination of the damping ratio ζ_i of individual modes, the half-power bandwidth $\Delta_i = 2\zeta_i\omega_i$, the complex modulus $E(1 + \imath\eta)$, or the reverberation time $T_r = 2.2 \times 2\pi/\eta\omega$.

Identification procedure

A purely experimental procedure has been proposed to identify simultaneously all coupling and damping loss factors in Lyon (1975, p. 217). The method consists in exciting a single subsystem with a known injected power and measuring the vibrational energies in all subsystems. When all subsystems are successively excited, we get n values of energy in n different situations which permits the writing of n^2 balance equations. By considering that the unknowns are the loss factors η_{ij}, $i, j = 1, \ldots, n$, it is possible to solve the set of linear equations and therefore to identify indirectly all loss factors. The method has been tested by Bies and Hamib (1980) and is now referred to as the power injection method.

Supplied powers

The power supplied to subsystems constitutes input data in statistical energy analysis in the same way that external forces are assumed to be known when applying Newton's second law for describing movement. As such, input powers must be known. The equations of Chapter 8 are useful. They require knowing separately the mean mobility and the expectation of the square force. Alternatively, similar relationships with the mean impedance are available if the expectation of the square velocity is known (Goyder and White 1980a, 1980b, 1980c). Otherwise, a direct measurement is sometimes possible, especially when external forces are concentrated. In this case a piezoelectric sensor such as an impedance head provides both force and acceleration at the same point.

10.3 Evaluate the Energies

Let us denote by E_i, $i = 1, \ldots, n$ the vibrational energy of subsystem i, and by $N_i = n_i\Delta\omega$ the number of resonant modes where n_i is the modal density. The modal energy, also called **vibrational temperature**, is defined as the mean energy per mode $T_i = E_i/N_i$. Both variables E_i and T_i can be adopted as primary variables in statistical energy analysis. The total vibrational energy E_i is more important for a practical analysis of the system, whereas the modal energy T_i has a more profound physical meaning.

In steady-state conditions, the energy balance for subsystem i reads

$$P_{\text{diss},i} + \sum_{j \neq i} P_{ij} = P_i. \tag{10.2}$$

where P_i denotes the power being injected into subsystem i by external random forces, $P_{\text{diss},i}$ the power being dissipated, and P_{ij} the power being exchanged between subsystems i and j.

The power being dissipated by internal losses is

$$P_{\text{diss},i} = \omega \eta_i E_i, \tag{10.3}$$

where η_i is the damping loss factor.

The power supplied by the subsystem i to the subsystem j is

$$P_{i \to j} = \omega \eta_{ij} E_i, \tag{10.4}$$

where η_{ij} is the coupling loss factor. The net exchanged power between subsystems i and j is $P_{ij} = P_{i \to j} - P_{j \to i}$ and therefore

$$P_{ij} = \omega \left(\eta_{ij} E_i - \eta_{ji} E_j \right). \tag{10.5}$$

Substituting eqns (10.5) and (10.3) into eqn (10.2) gives a set of linear equations on E_i:

$$\omega \begin{pmatrix} \eta_{11} & -\eta_{21} & \cdots & -\eta_{n1} \\ -\eta_{12} & \eta_{22} & & \\ \vdots & \ddots & \ddots & \vdots \\ & & & -\eta_{n,n-1} \\ -\eta_{1n} & \cdots & -\eta_{n-1,n} & \eta_{nn} \end{pmatrix} \begin{pmatrix} E_1 \\ \\ \vdots \\ \\ E_n \end{pmatrix} = \begin{pmatrix} P_1 \\ \\ \vdots \\ \\ P_n \end{pmatrix} \tag{10.6}$$

where the total loss factor $\eta_{ii} = \eta_i + \sum_{j \neq i} \eta_{ij}$ appears on the diagonal.

Another form of this set of equations is often preferred. To introduce the vibrational temperature as a variable, we first remark that reciprocity,

$$N_i \eta_{ij} = N_j \eta_{ji} \tag{10.7}$$

allows us to make apparent the role of $T_i = E_i / N_i$ in the exchanged power. Substituting eqn (10.7) into eqn (10.5) leads to

$$P_{ij} = \omega \eta_{ij} N_i \left(T_i - T_j \right). \tag{10.8}$$

The net exchanged power is proportional to the difference of modal energies or vibrational temperatures.

The equations concerning the vibrational temperature are simply obtained by substituting eqns (10.3) and (10.8) into eqn (10.2):

$$\omega \begin{pmatrix} N_1\eta_{11} & -N_2\eta_{21} & \dots & -N_n\eta_{n1} \\ -N_1\eta_{12} & N_2\eta_{22} & & \\ \vdots & \ddots & \ddots & \vdots \\ & & & -N_n\eta_{n,n-1} \\ -N_1\eta_{1n} & \dots & -N_{n-1}\eta_{n-1,n} & N_n\eta_{nn} \end{pmatrix} \begin{pmatrix} T_1 \\ \vdots \\ T_n \end{pmatrix} = \begin{pmatrix} P_1 \\ \vdots \\ P_n \end{pmatrix} \tag{10.9}$$

where the ith diagonal entry is N_i times the total loss factor and the extra-diagonal entry of row k and column l is $-N_l\eta_{lk}$ (note reversed indices). The matrix is symmetric. The advantage of this form over eqn (10.6) is that a symmetric matrix can be useful in speeding up the numerical solution.

Whatever the form of energy equations, the total or modal energy is determined in each frequency band by the knowledge of the coupling loss factors, the damping loss factors, and the injected powers. The situation is illustrated in Fig. 10.4. Each subsystem behaves like an energy tank which may receive energy from external sources, dissipate energy by internal damping, and exchange energy with neighbouring subsystems.

An abundant literature is available on applications of the present procedure to various industrial mechanisms. The reader may refer to Culla and Sestieri (2006) for a particularly convincing study on a plate/cavity system and a three plate system. These authors

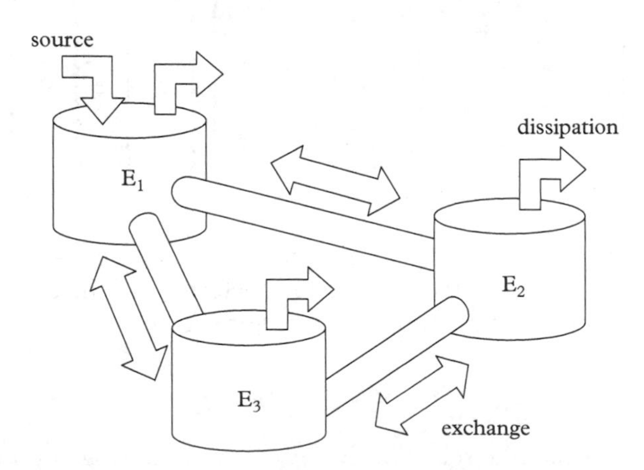

Figure 10.4 *Energy balance. Energy is supplied to subsystems by sources, extracted by dissipation, and exchanged between subsystems.*

show that the numerical solution to the governing equation converges to the solution of eqn (10.9) and discuss the validity conditions.

The determination of uncertainty concerning vibrational energies (or temperatures) is also of interest in practice. Although several causes of fluctuation may be responsible for errors when determining the coupling loss factors, they all result in differences between actual energies observed in a particular system and those predicted by eqn (10.9). See Cotoni et al. (2005) and Culla et al. (2011) on this topic.

Example

Consider two coupled subsystems and assume that $P_2 = 0$, that is no energy is supplied to subsystem 2 except by coupling to subsystem 1. The energy balance of subsystem 2 then reduces to

$$\omega\eta_{21}N_2\,(T_1 - T_2) = \omega\eta_2 N_2 T_2 \tag{10.10}$$

The ratio T_2/T_1 follows,

$$\frac{T_2}{T_1} = \frac{\eta_{21}}{\eta_{21} + \eta_2} \tag{10.11}$$

This relationship highlights that when an indirectly excited subsystem is weakly damped, its vibrational level is controlled by the exchanges with other subsystems rather than by internal dissipation. In the limit case $\eta_2 = 0$, the equilibrium $T_2 = T_1$ is reached. By analogy with thermodynamics, this state could be called thermal equilibrium in the sense that it is characterized by an equality of vibrational temperatures. Around this equilibrium, an increase of the damping loss factor η_2, for instance by adding a damping material, will have no significant effect on the vibrational level of the subsystem. This is important when optimizing the mass of damping material.

An example of equilibrium is provided by transverse and dilatational waves in a solid when a diffuse state is reached for both. Under this condition, $T_1 = T_2$, the ratio of energies, or energy densities since the volume is the same for both, is simply the ratio of modal densities. If we use eqn (6.60) and values of c_1 and c_2 from Table 5.1, the ratio of transversal to dilatational energy is $2\,[(2-2\nu)/(1-2\nu)]^{3/2}$. The extra factor 2 comes from the polarization of shear waves. We get the value 13 for $\nu = 0.3$ (Weaver, 1982).

11

Energy Path

The application of statistical energy analysis also delivers crucial information on the different paths by which energy flows through subsystems. The former definitions on energy path have been given by Luzzato and Ortola (1988), Craik (1990), and Magrans (1993). See Guash and Cortés (2009) and Guash and Aragonès (2011) for recent developments on the subject.

11.1 Single Path

Let us consider first a simple example with only one possible path. The system is sketched in Fig. 11.1. Three sub-structures are numbered from 1 to 3. The power is injected in subsystem 1 and the receiver is subsystem 3. It is further assumed that the links between 1 and 2 and 2 and 3 are one-way, that is the energy can flow from 2 to 3 but not from 3 to 2. Indeed, this situation is not possible in strict statistical energy analysis since the reciprocity relationship imposes that if η_{ij} is zero, then η_{ji} is also zero. But the study of this idealized case before the general case is very instructive.

Thus, we assume $\eta_{12} \neq 0$ and $\eta_{23} \neq 0$, but $\eta_{21} = 0$ and $\eta_{32} = 0$. The only path for the energy to flow is therefore 123. The energy balance equations are

$$\begin{cases} \eta_{11}E_1 = P/\omega \\ \eta_{22}E_2 = \eta_{12}E_1 \\ \eta_{33}E_3 = \eta_{23}E_2 \end{cases} \tag{11.1}$$

where $\eta_{ii} = \eta_i + \sum_{j\neq i} \eta_{ij}$ is the total loss factor of subsystem i, η_i being as usual the damping loss factor. Let us examine this set of equations. The third equation gives E_3 in terms of E_2 but E_2 is known from the second equation in terms of E_1 which, in turn, is known from the first equation. Substituting the first into the second, then the second into the third, and multiplying by η_3 yields

$$\eta_3 \omega E_3 = \frac{\eta_3}{\eta_{33}} \frac{\eta_{23}}{\eta_{22}} \frac{\eta_{12}}{\eta_{11}} P \tag{11.2}$$

Foundation of Statistical Energy Analysis in Vibroacoustics. First Edition. A. Le Bot.

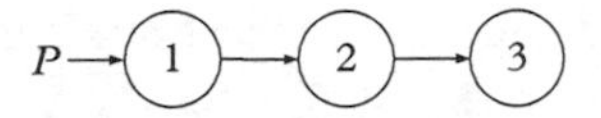

Figure 11.1 *System with a single path of energy.*

The above equation gives the energy dissipated in subsystem 3 resulting from a power P injected into subsystem 1. We may introduce the path efficiency, $\tau_{123} = \eta_{12}\eta_{23}\eta_3/\eta_{11}\eta_{22}\eta_{33}$ as the ratio of energy dissipated in subsystem 3 received through the path 123 to input power P. The energy dissipated in the receiver subsystem is then

$$\eta_3 \omega E_3 = \tau_{123} P \tag{11.3}$$

Since the dissipated energy is less than the input power, the efficiency τ_{123} is less than unit.

The generalization for a chain $123 \ldots n$ of n subsystems is immediate. The efficiency of the path $123 \ldots n$ is

$$\tau_{12\ldots n} = \frac{\eta_{12}\eta_{23} \cdots \eta_{n-1,n}\eta_n}{\eta_{11}\eta_{22} \cdots \eta_{nn}} \tag{11.4}$$

And the energy dissipated in the receiver subsystem is

$$\eta_n \omega E_n = \tau_{123\ldots n} P \tag{11.5}$$

11.2 Multiple Paths

The second example is a little more complicated. The links are always one-way so that the energy is forced to go on. The system shown in Fig. 11.2 has two paths of energy, 124 and 134. The set of equations for such a system is

$$\begin{cases} \eta_{11}E_1 = P/\omega \\ \eta_{22}E_2 = \eta_{12}E_1 \\ \eta_{33}E_3 = \eta_{13}E_1 \\ \eta_{44}E_4 = \eta_{24}E_2 + \eta_{34}E_3 \end{cases} \tag{11.6}$$

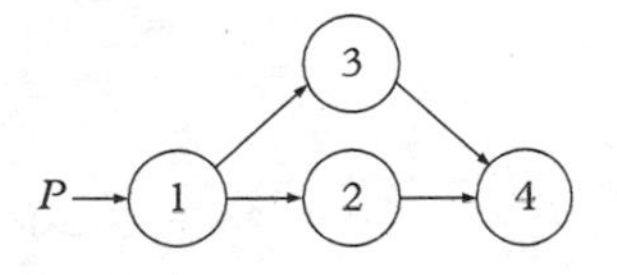

Figure 11.2 *System with two energy paths.*

Substituting the first three equations into the last one gives

$$\eta_4 \omega E_4 = \left(\frac{\eta_4}{\eta_{44}} \frac{\eta_{24}}{\eta_{22}} \frac{\eta_{12}}{\eta_{11}} + \frac{\eta_4}{\eta_{44}} \frac{\eta_{34}}{\eta_{33}} \frac{\eta_{13}}{\eta_{11}} \right) P \tag{11.7}$$

where we can recognize the efficiencies of paths 124 and 134. The first property of path efficiencies is now apparent. The total contribution to the dissipated energy $\eta_4 \omega E_4$ is simply the sum of contributions of all paths. In this case,

$$\eta_4 \omega E_4 = (\tau_{124} + \tau_{134}) P \tag{11.8}$$

11.3 Path with Loops

Let us now consider a system with a loop of energy. The system is shown in Fig. 11.3. The shortest path from subsystem 1 to receiver 4 is 1324. But when the energy reaches subsystem 2, a part also flows toward subsystem 1, giving rise to the loop 1321. Once again, when this energy reaches subsystem 2, a part flows to the receiver 4 with the path 132134 and the rest flows toward subsystem 1 for another loop. The arithmetic of such a scenario is the following. The set of equations is

$$\begin{cases} \eta_{11} E_1 = P/\omega + \eta_{21} E_2 \\ \eta_{22} E_2 = \eta_{32} E_3 \\ \eta_{33} E_3 = \eta_{13} E_1 \\ \eta_{44} E_4 = \eta_{24} E_2 \end{cases} \tag{11.9}$$

Substituting E_2 from the second equation into the last equation, then E_3 from the third equation, and finally E_1 from the first equation gives

$$\eta_4 \omega E_4 = \frac{\eta_4}{\eta_{44}} \frac{\eta_{24}}{\eta_{22}} \frac{\eta_{32}}{\eta_{33}} \frac{\eta_{13}}{\eta_{11}} \left(P + \omega \frac{\eta_{21}}{\eta_{11}} E_2 \right) \tag{11.10}$$

The energy path 1324 is apparent. But a term E_2 remains and the dissipated energy $\eta_4 \omega E_4$ is not obtained in terms of P solely. A further substitution sequence gives

$$\eta_4 \omega E_4 = \left(\frac{\eta_4}{\eta_{44}} \frac{\eta_{24}}{\eta_{22}} \frac{\eta_{32}}{\eta_{33}} \frac{\eta_{13}}{\eta_{11}} + \frac{\eta_4}{\eta_{44}} \frac{\eta_{24}}{\eta_{22}} \frac{\eta_{32}}{\eta_{33}} \frac{\eta_{13}}{\eta_{11}} \frac{\eta_{21}}{\eta_{22}} \frac{\eta_{32}}{\eta_{33}} \frac{\eta_{13}}{\eta_{11}} \right) \left(P + \omega \frac{\eta_{21}}{\eta_{11}} E_2 \right) \tag{11.11}$$

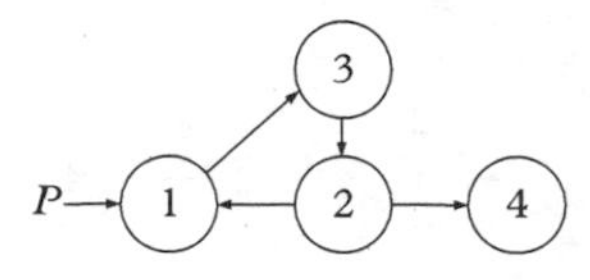

Figure 11.3 *System with an energy loop.*

It is clear that the process can be repeated an infinite number of times. Introducing the path efficiencies, we finally get

$$\eta_4 \omega E_4 = (\tau_{1324} + \tau_{1321324} + \tau_{1321321324} + \ldots) P \tag{11.12}$$

Since the power dissipated $\eta_4 \omega E_4$ in subsystem 4 is less than the injected power P, the sum of efficiencies of all paths from 1 to 4 is less than unit.

11.4 Complex Paths

A complete example is shown in Fig. 11.4. This is basically the same system as in Fig. 11.3 except that all links are now two-way. In a strict statistical energy analysis system, coupling loss factors should in addition verify the reciprocity relationship $\eta_{ij} n_i = \eta_{ji} n_j$, but this not important for the enumeration of energy paths. The set of equations is

$$\begin{cases} \eta_{11} E_1 = P/\omega + \eta_{21} E_2 + \eta_{31} E_3 \\ \eta_{22} E_2 = \eta_{12} E_1 + \eta_{32} E_3 + \eta_{42} E_4 \\ \eta_{33} E_3 = \eta_{13} E_1 + \eta_{23} E_2 \\ \eta_{44} E_4 = \eta_{24} E_2 \end{cases} \tag{11.13}$$

Starting from the fourth equation, we substitute E_2 from the second equation and then E_1, E_3, and E_4 from respectively the first, the third, and the fourth equations. It yields

$$\eta_4 \omega E_4 = \tau_{124} (P + \eta_{21} \omega E_2 + \eta_{31} \omega E_3) + \tau_{324} (\eta_{13} \omega E_1 + \eta_{23} \omega E_2) + \tau_{424} \eta_{24} \omega E_2 \tag{11.14}$$

The first term is the contribution of the path 124 which is the shortest path from 1 to 4. A further substitution gives

$$\begin{aligned} \eta_4 \omega E_4 = \tau_{124} P &+ \tau_{2124} (\eta_{12} \omega E_1 + \eta_{32} \omega E_3 + \eta_{42} \omega E_4) + \tau_{3124} (\eta_{13} \omega E_1 + \eta_{23} \omega E_2) + \\ & \tau_{1324} (P + \eta_{21} \omega E_2 + \eta_{31} \omega E_3) + \tau_{2324} (\eta_{12} \omega E_1 + \eta_{32} \omega E_3 + \eta_{42} \omega E_4) + \\ & \tau_{2424} (\eta_{12} \omega E_1 + \eta_{32} \omega E_3 + \eta_{42} \omega E_4) \end{aligned} \tag{11.15}$$

The second path appears in the first term of the second line. This is 1324 which is the only path from 1 to 4 with four nodes. A further substitution gives

$$\eta_4 \omega E_4 = (\tau_{124} + \tau_{1324} + \tau_{12124} + \tau_{13124} + \tau_{12324} + \tau_{12424}) P + \ldots \tag{11.16}$$

It is found that four paths exist with five nodes. All remaining terms (not written in the previous expression) will give paths with more than five nodes. Table 11.1 shows the paths up to 6 nodes. The subsequent terms in the development (11.16) may be found from such a table. This highlights that $\eta_4 \omega E_4$ may be developed as a series of

Table 11.1 *Energy paths of the system shown in Figure 11.4.*

# nodes	# paths	paths
2	0	
3	1	124
4	1	1324
5	4	12124, 13124, 12324, 12424
6	6	123124, 132124, 132324, 131324, 121324, 132424

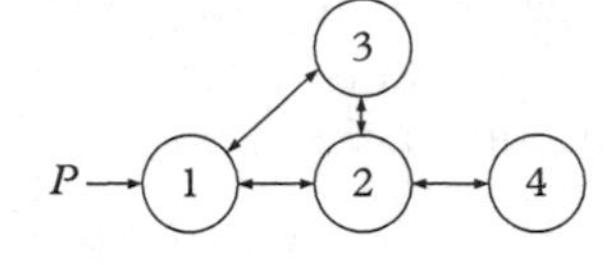

Figure 11.4 *Example of complex system with four subsystems.*

path efficiencies. Of course, the number of paths with a higher number of nodes grows rapidly. Since their efficiencies are multiplied by factors less than unit at each step, their respective contributions rapidly decrease.

11.5 Path Decomposition

We can now turn to the general case. A statistical energy analysis system as obtained in eqn (10.6) has the form

$$\omega \begin{pmatrix} \eta_{11} & & -\eta_{ji} \\ & \ddots & \\ -\eta_{ji} & & \eta_{nn} \end{pmatrix} \begin{pmatrix} E_1 \\ \vdots \\ E_n \end{pmatrix} = \begin{pmatrix} P_1 \\ \vdots \\ P_n \end{pmatrix} \tag{11.17}$$

where i is the index of row and j of column. As usual, $\eta_{ii} = \eta_i + \sum_{j \neq i} \eta_{ij}$ denotes the total loss factor of subsystem i. The first matrix can be separated in two matrices,

$$\omega \left[\begin{pmatrix} \eta_{11} & & 0 \\ & \ddots & \\ 0 & & \eta_{nn} \end{pmatrix} - \begin{pmatrix} 0 & & \eta_{ji} \\ & \ddots & \\ \eta_{ji} & & 0 \end{pmatrix} \right] \begin{pmatrix} E_1 \\ \vdots \\ E_n \end{pmatrix} = \begin{pmatrix} P_1 \\ \vdots \\ P_n \end{pmatrix} \tag{11.18}$$

By factorizing the matrix $\mathbf{diag}(\eta_{11}, \ldots, \eta_{nn})$ on the right,

$$\omega \left[\begin{pmatrix} 1 & & 0 \\ & \ddots & \\ 0 & & 1 \end{pmatrix} - \begin{pmatrix} 0 & & \frac{\eta_{ji}}{\eta_{jj}} \\ & \ddots & \\ \frac{\eta_{ji}}{\eta_{jj}} & & 0 \end{pmatrix} \right] \begin{pmatrix} \eta_{11} & & 0 \\ & \ddots & \\ 0 & & \eta_{nn} \end{pmatrix} \begin{pmatrix} E_1 \\ \vdots \\ E_n \end{pmatrix} = \begin{pmatrix} P_1 \\ \vdots \\ P_n \end{pmatrix} \tag{11.19}$$

If

$$\mathbf{S} = \begin{pmatrix} 0 & & \frac{\eta_{ji}}{\eta_{ii}} \\ & \ddots & \\ \frac{\eta_{ji}}{\eta_{ii}} & & 0 \end{pmatrix}, \quad \mathbf{D} = \begin{pmatrix} \eta_{11} & & 0 \\ & \ddots & \\ 0 & & \eta_{nn} \end{pmatrix}, \quad \text{and} \quad \mathbf{C} = \begin{pmatrix} \eta_1 & & 0 \\ & \ddots & \\ 0 & & \eta_n \end{pmatrix} \tag{11.20}$$

eqn (11.19) reads

$$\omega\,(\mathbf{I} - \mathbf{S})\,\mathbf{DE} = \mathbf{P} \tag{11.21}$$

Now multiplying both hand sides by $\mathbf{CD}^{-1}\,(\mathbf{I} - \mathbf{S})^{-1}$ gives

$$\omega\mathbf{CE} = \mathbf{CD}^{-1}\,(\mathbf{Id} - \mathbf{S})^{-1}\,\mathbf{P} \tag{11.22}$$

A general theorem valid in any Banach algebra states that the matrix $\mathbf{I} - \mathbf{S}$ is invertible if $\|\mathbf{S}\| < 1$. In that case,

$$(\mathbf{I} - \mathbf{S})^{-1} = \sum_{k=0}^{\infty} \mathbf{S}^k \tag{11.23}$$

where the series in the right-hand side is convergent. In the problem at hand, the matrix $\mathbf{S}$ has zero diagonal entries and off-diagonal entries η_{ji}/η_{jj} in the i-th row and j-th column. The norm $\|S\|$ may be defined in several ways but the most appropriate for the present purpose is

$$\|\mathbf{S}\| = \max_j \sum_i \left| \frac{\eta_{ji}}{\eta_{jj}} \right| \tag{11.24}$$

which corresponds to the norm of the space $L(E)$ of linear operators on the vector space $E = \mathbb{R}^n$ with the norm $\|.\|_1$. If all subsystems have a non-zero damping loss factor $\eta_j > 0$ then $\eta_{jj} > \sum_i \eta_{ji}$. It is therefore apparent that $\sum_i \eta_{ji}/\eta_{jj} < 1$ for all j. The norm of the matrix $\mathbf{S}$ is therefore less than 1. The above theorem applies, with the result

$$\omega\mathbf{CE} = \mathbf{CD}^{-1}\left(\mathbf{I} + \mathbf{S} + \mathbf{S}^2 + \ldots \mathbf{S}^k + \ldots\right)\mathbf{P} \tag{11.25}$$

In the right-hand side of this equation, the products $\mathbf{CD}^{-1}\mathbf{S}^k$ can be expanded,

$$\begin{pmatrix} \eta_1 & & 0 \\ & \ddots & \\ 0 & & \eta_n \end{pmatrix} \begin{pmatrix} \eta_{11} & & 0 \\ & \ddots & \\ 0 & & \eta_{nn} \end{pmatrix}^{-1} \begin{pmatrix} 0 & & \frac{\eta_{ji}}{\eta_{jj}} \\ & \ddots & \\ \frac{\eta_{ji}}{\eta_{jj}} & & 0 \end{pmatrix}^k = \begin{pmatrix} \cdots & \sum_{\text{path}} \tau_{j...i} \\ & \vdots \end{pmatrix} \tag{11.26}$$

The entry in row i and column j of the matrix $\mathbf{CD}^{-1}\mathbf{S}^k$ is the sum of all efficiencies of paths of order k starting from j and ending at i.

11.6 Path Efficiency

The general definition of a **path efficiency** is now apparent. The efficiency of a path is the ratio of energy dissipated in the receiver subsystem which has travelled through the path to the power supplied to the source subsystem. Since all subsystems may dissipate energy, the power dissipated in the receiver is less than the total dissipated power which equals the supplied power by energy balance. A path efficiency is therefore less than unit.

The efficiency of the path $i, a, b \dots, z, j$ is

$$\tau_{iab...j} = \frac{\eta_{ia}\eta_{ab} \dots \eta_{zj}}{\eta_{ii}\eta_{aa} \dots \eta_{zz}} \frac{\eta_j}{\eta_{jj}} \tag{11.27}$$

But the total loss factor $\eta_{ii} = \eta_i + \sum_{k \neq i} \eta_{ik}$ is always greater than η_i and greater than η_{ik} for any k. The path efficiency is therefore less than unit. Hence,

$$0 \leq \tau_{iab...j} \leq 1 \tag{11.28}$$

This is of course a necessary condition for an efficiency.

The second property is the behaviour under decomposition. If a path $i \dots j \dots k$ is separated into two paths, $i \dots j$ and $j \dots k$, then eqn (11.27) gives

$$\tau_{i...j...k} = \tau_{i...j} \times \frac{\eta_{jj}}{\eta_j} \times \tau_{j...k} \tag{11.29}$$

This relationship may be interpreted as follows. The first efficiency gives the partial power dissipated $P\tau_{i...j}$. Therefore the partial energy which has reached j is $E_{i...j} = P\tau_{i...j}/\eta_j\omega$. But a subsystem with energy $E_{i...j}$ releases a power $\eta_{jj}\omega E_{i...j}$ to its surrounding subsystems. The partial power dissipated in subsystem k is therefore $\eta_{jj}\omega E_{i...j}\tau_{j...k}$. Equation (11.29) follows.

A further property is obtained on the reciprocal path. By applying the reciprocity relationship $N_a\eta_{ab} = N_b\eta_{ba}$ to each coupling loss factor in the numerator, we immediately get the first property of the path efficiency,

$$N_j\eta_j\tau_{j...bai} = N_i\eta_i\tau_{iab...j} \tag{11.30}$$

This is the reciprocity relationship for path efficiencies.

The product $P\tau_{iab...zj}$ gives only the part of dissipated energy in subsystem j which has travelled through the particular path $i, a, b \ldots, z, j$. Of course the total power dissipated in the receiver is the sum of efficiencies of all paths which start from the source and end at the receiver times the supplied power,

$$\eta_j \omega E_j = P \sum_{a,\ldots,z} \tau_{ia\ldots zj} \tag{11.31}$$

where the sum runs on all possible paths from i to j of any order. But since the global power balance reads $\sum_j \eta_j \omega E_j = P$, we obtain by a further summation,

$$\sum_j \sum_{a,\ldots,z} \tau_{ia\ldots zj} = 1 \tag{11.32}$$

The sum of all path efficiencies from i to any fixed j is unit.

11.7 Comments

The main interest in determining the energy paths and classifying their relative contribution is to deliver crucial information on which subsystem must be treated as a priority to efficiently reduce the vibrational level in the receiver subsystem. In practice, sound and vibration are reduced by adding damping material. The knowledge of the principal path by which energy flows to the receiver allows an optimization of the mass of damping material and permits a most efficient result.

A former criterion for path classification is the noise reduction, introduced by Craik (1977, 1979). This is defined as the ratio of energy in the receiver subsystem to the source subsystem. This leads to

$$\frac{E_j}{E_i} = \frac{\eta_{ia}\eta_{ab} \cdots \eta_{zj}}{\eta_{aa}\eta_{bb} \cdots \eta_{jj}} \tag{11.33}$$

We have preferred to introduce the path efficiency $\tau_{iab\ldots j}$ defined as a ratio of powers instead of energies. This has two advantages. First, a path efficiency is normalized (< 1) and the sum of all efficiencies is unit (eqn (11.32)). Second, the sum of each column of the matrix $\mathbf{S}$ is less than unit. This provides a simple proof of the convergence of the series $\sum_{k=0}^{\infty} \mathbf{S}^k$. Nevertheless, noise reduction and path efficiency differ only by the factor η_j/η_{ii} which does not depend on intermediate nodes of the path from i to j. Both criteria lead therefore to the same path classification.

12

Transient Phenomena

This chapter outlines transient phenomena in room acoustics and statistical energy analysis. The most common problem is the prediction of sound (or vibration) time decay after the sources have been switched off. This problem has been studied by Sabine (1922) in the context of architectural acoustics.

12.1 Sabine's Formula

In architectural acoustics, the most important criterion for the quality of rooms is the so-called **reverberation time**. This is the duration for which the sound decays by 60 dB after its emission. This duration is representative of reverberation in rooms and must be adjusted according to the usage of the room. For instance speech typically requires a shorter reverberation time than music.

To establish a theory of reverberation we must notice first that attenuation of sound is generally not dominated by atmospheric absorption during propagation but rather by energy loss when rays hit walls or other obstacles. The sound decay is therefore driven by the ability of materials to absorb sound during reflection and the mean number of times a ray encounters an obstacle per unit time.

In a diffuse sound field, the energy is uniformly and isotropically distributed (see Fig. 12.1). The radiative intensity I, defined as the power per unit solid angle and unit length normal to the ray, is constant and therefore depends on neither the position nor the direction. Thus, the energy density contained in an infinitesimal solid angle $d\Omega$ is $dW = I\, d\Omega/c$ where c is the speed of sound. The energy density carried out by the rays in all directions is

$$W = \frac{4\pi}{c} I \tag{12.1}$$

Applying homogeneity, this energy density does not depend on position. The total sound energy is obtained by multiplying by the room volume V:

$$E = \frac{4\pi}{c} IV \tag{12.2}$$

Foundation of Statistical Energy Analysis in Vibroacoustics. First Edition. A. Le Bot.

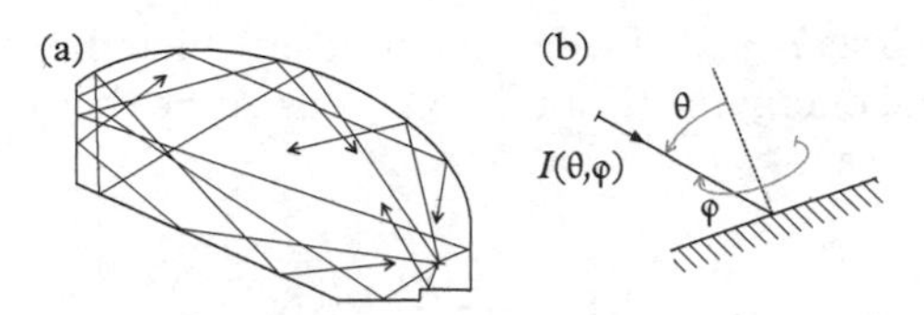

Figure 12.1 *Diffuse sound field in a concert hall. (a) rays are mixed by successive reflections so that radiative intensity $I(\theta, \varphi)$ becomes homogeneous and isotropic; (b) definition of angles θ, φ.*

The energy B received by the walls per unit time and area is the integral of the normal component $I(\varphi, \theta) \cos\theta$ of the radiative intensity over the hemisphere of incident directions. Noting θ the incidence angle and φ the azimuthal angle, the infinitesimal solid angle is $\sin\theta \, d\theta d\varphi$. By integration,

$$B = \int_0^{2\pi} \int_0^{\frac{\pi}{2}} I(\varphi, \theta) \cos\theta \sin\theta \, d\theta d\varphi = \pi I \tag{12.3}$$

since $I(\varphi, \theta)$ has a constant value. During reflection, rays lose a part of their energy. We introduce the **mean absorption coefficient** α as the ratio of absorbed power to incident power. The total power absorbed by boundaries is

$$P_{\text{diss}} = \alpha \int_S B \, dS = \alpha BS \tag{12.4}$$

Combining eqns (12.2), (12.3), and (12.4) gives

$$P_{\text{diss}} = \alpha c \frac{S}{4V} E \tag{12.5}$$

One recognizes the mean free path $4V/S$ of the room. The factor $cS/4V$ is the number of reflections of rays per unit time.

In the absence of sound sources, the power balance is

$$-\frac{dE}{dt} = P_{\text{diss}} \tag{12.6}$$

By substituting eqn (12.5),

$$-\frac{dE}{dt} = \alpha c \frac{S}{4V} E \tag{12.7}$$

The solution to this first-order differential equation with the initial condition E_0 is

$$E(t) = E_0 e^{-\frac{cS\alpha}{4V} t} \tag{12.8}$$

This equation shows that in a diffuse field sound decays according to an exponential law.

The reverberation time T_r is defined as the time at which the residual energy $E(T_r)$ is a millionth of the initial energy E_0. Equivalently, this is the duration of a decay of 60 dB of the sound pressure level,

$$10 \log \frac{E(T_r)}{E_0} = -60 \tag{12.9}$$

Employing eqn (12.8), one obtains

$$T_r = \frac{24 \log 10}{c} \frac{V}{\alpha S} \tag{12.10}$$

With the usual value $c = 343$ m/s for the speed of sound,

$$T_r = 0.161 \frac{V}{\alpha S} \tag{12.11}$$

in the SI unit system. This is the famous Sabine formula which links reverberation with wall absorption.

This relation tacitly assumes that many modes are involved in the dynamics. In particular, the whole sound decay is not dominated by the behaviour of any particular mode. The frequency limit for applying the reverberation theory is defined in terms of modal overlap. When wall absorption is low, resonance peaks are acute and modes are well separated. They can therefore be individually excited. But if the mean spacing of resonance frequencies is smaller than the half-power bandwidth, then the frequency response function of the room is flat and the resonance peaks are no longer distinguishable. A limit commonly admitted in room acoustics is a modal overlap of 3. The modal overlap is $M = \eta\omega n(\omega)$ where $n(\omega) = V\omega^2/2\pi^2c^3$ is the modal density by eqn (6.60) and η the damping loss factor. The damping law (12.5) compared with eqn (10.3) immediately shows that $\eta\omega = \alpha cS/4V$. Combined with eqn (12.10) this also gives $\eta\omega = 6 \log 10/T_r$. Thus,

$$M = \frac{6 \log 10}{T_r} \frac{V\omega^2}{2\pi^2c^3} \tag{12.12}$$

Solving the equation $M = 3$ gives the so-called Schröder frequency (Schröder, 1954):

$$\omega_s = \sqrt{\frac{\pi^2c^3}{\log 10}} \times \sqrt{\frac{T_r}{V}} \tag{12.13}$$

With $c = 343$ m/s,

$$f_s = 2100\sqrt{\frac{T_r}{V}} \tag{12.14}$$

where $f_s = \omega_s/2\pi$ is in Hz, T_r in s, and V in m^3. This frequency gives the lower limit of applicability of the reverberation theory.

Example

Consider a classroom of 8 m length, 5 m width, and 4 m height (about 30 seats) for a volume $V = 160$ m^3. The walls and the ceiling are highly reverberant (hard plaster or window) with an absorption factor $\alpha_1 = 0.03$ at 1000 Hz, and the floor is a wood parquet upon which are tables whose equivalent absorption factor is $\alpha_2 = 0.1$. The absorption area is $\alpha S = \sum_i \alpha_i S_i = 5.52$ m^2. The resulting reverberation time is $T_r = 0.16V/\alpha S = 3.1$ s. This value is too large for use in schools to ensure a correct intelligibility and a recommended value is $T_r = 0.7$ s. The classical solution is to install a false ceiling at a height of 3.3 m which is highly absorptive ($\alpha_3 = 0.6$). The volume is reduced to $V' = 132$ m^3 and the new absorption area is $S' = 30.6$. The final reverberation time $T_r = 0.7$ s is correct. It may be checked that in both cases the Schröder frequency is below 300 Hz.

12.2 Reverberation in Structures

Reverberation in structures also gives rise to an exponential decay law. The principal difference with room acoustics concerns dissipation mechanisms. Structural rays are most often attenuated during their propagation by various phenomena that are all included in the damping loss factor η. We assume that a diffuse vibrational field is generated by some random forces whose spectrum is centred on the frequency ω. At time $t = 0$, the vibrational energy is E_0 and the excitation is removed. The vibrational energy starts to decay.

The damping law is

$$P_{diss} = \omega\eta E \tag{12.15}$$

Since no energy is supplied to the structure, the power balance of the whole system is

$$-\frac{dE}{dt} = \omega\eta E \tag{12.16}$$

The solution to this differential equation is

$$E(t) = E(0)e^{-\eta\omega t} \tag{12.17}$$

We obtain again an exponential decay for which a reverberation time may be defined as in architectural acoustics,

$$T_r = \frac{6\log 10}{\eta\omega} = \frac{13.8}{\eta\omega} \tag{12.18}$$

This equation provides a general method to measure damping loss factors.

12.3 Coupled Subsystems

We now consider the most general case of several subsystems connected by conservative couplings. We assume that the initial state of vibration is such that vibrational fields are in diffuse state in *all* subsystems. At time $t = 0$, all sources are switched off. Absorption of vibration by internal damping then induces a decreasing of energy in all subsystems.

Transient energy balance

At any time t, the power balance of subsystem i reads

$$-\frac{\mathrm{d}E_i}{\mathrm{d}t} = P_{\mathrm{diss},i} + \sum_{j \neq i} P_{ij} \tag{12.19}$$

In this equation, the left-hand side is the transient term while the right-hand side is the sum of all losses either by internal damping or by exchange with other subsystems.

The power being dissipated by internal damping is

$$P_{\mathrm{diss},i} = \eta_i \omega E_i \tag{12.20}$$

where η_i is the damping loss factor.

The exchanged energy between subsystems i and j is given by the coupling power proportionality,

$$P_{ij} = \omega \left(\eta_{ij} E_i - \eta_{ji} E_j \right), \tag{12.21}$$

where η_{ij} is the coupling loss factor. Reciprocity of coupling loss factors imposes

$$N_i \eta_{ij} = N_j \eta_{ji}. \tag{12.22}$$

Substituting eqns (12.21) and (12.20) into eqn (12.19) gives a set of linear differential equations on E_i:

$$-\omega \begin{pmatrix} \eta_{11} & -\eta_{21} & \cdots & -\eta_{n1} \\ -\eta_{12} & \eta_{22} & & \\ \vdots & \ddots & \ddots & \vdots \\ & & & -\eta_{n,n-1} \\ -\eta_{1n} & \cdots & -\eta_{n-1,n} & \eta_{nn} \end{pmatrix} \begin{pmatrix} E_1 \\ \\ \vdots \\ \\ E_n \end{pmatrix} = \frac{\mathrm{d}}{\mathrm{d}t} \begin{pmatrix} E_1 \\ \\ \vdots \\ \\ E_n \end{pmatrix} \tag{12.23}$$

where the diagonal entries are the total loss factor $\eta_{ii} = \eta_i + \sum_{j \neq i} \eta_{ij}$.

Transient response

By denoting $\mathbf{A}$ the above matrix of coupling loss factors, the differential system becomes $\dot{\mathbf{E}} = -\omega \mathbf{A}\mathbf{E}$, where $\mathbf{E} = (E_1, \ldots, E_n)^T$ is the column vector of energies. This is a first-order linear differential system with constant coefficients. The solution to this system is well known (Appendix A):

$$\mathbf{E}(t) = \mathrm{e}^{-\omega t \mathbf{A}} \mathbf{E}(0) \tag{12.24}$$

Thus, the knowledge of the initial state and the coupling loss factors determines the state of the system at any later time t.

To explicitly calculate the exponential $\mathrm{e}^{-\omega t \mathbf{A}}$, one must diagonalize $\mathbf{A}$.

Let us prove that $\mathbf{A}$ is diagonalizable. Let $\mathbf{\Delta} = \mathbf{diag}(1/N_1, \ldots, 1/N_n)$ be the diagonal matrix whose entries are $1/N_i$. This matrix is symmetric positive-definite. Let $\langle \mathbf{X}, \mathbf{Y} \rangle = \mathbf{X}^T \mathbf{\Delta} \mathbf{Y}$ be the scalar product whose matrix is $\mathbf{\Delta}$. Then $\langle \mathbf{A}\mathbf{X}, \mathbf{Y} \rangle = \mathbf{X}^T \mathbf{A}^T \mathbf{\Delta} \mathbf{Y}$. The entries of $\mathbf{A}^T \mathbf{\Delta}$ are

$$[\mathbf{A}^T \mathbf{\Delta}]_{ij} = \begin{cases} -\eta_{ij}/N_j & \text{if } j \neq i \\ \eta_{ii}/N_i & \text{otherwise} \end{cases} \tag{12.25}$$

Similarly $\langle \mathbf{X}, \mathbf{A}\mathbf{Y} \rangle = \mathbf{X}^T \mathbf{\Delta} \mathbf{A} \mathbf{Y}$ and the entries of $\mathbf{\Delta}\mathbf{A}$ are

$$[\mathbf{\Delta}\mathbf{A}]_{ij} = \begin{cases} -\eta_{ji}/N_i & \text{if } j \neq i \\ \eta_{ii}/N_i & \text{otherwise} \end{cases} \tag{12.26}$$

By the reciprocity relationship, both matrices are equal and $\langle \mathbf{A}\mathbf{X}, \mathbf{Y} \rangle = \langle \mathbf{X}, \mathbf{A}\mathbf{Y} \rangle$. The matrix $\mathbf{A}$ is symmetric with respect to the scalar product $\langle ., . \rangle$ and is therefore diagonalizable by the spectral theorem (Lang, 2002).

Since $\mathbf{A}$ is diagonalizable, there exists an invertible matrix $\mathbf{P}$ and a diagonal matrix $\Lambda = \mathbf{diag}(\lambda_1, \ldots, \lambda_n)^T$ such that $\mathbf{A} = \mathbf{P}\Lambda\mathbf{P}^{-1}$. The solution (12.24) becomes

$$\mathbf{E}(t) = \mathbf{P}\mathrm{e}^{-\omega t \Lambda} \mathbf{P}^{-1} \mathbf{E}(0) \tag{12.27}$$

Therefore, the time decay of energy in subsystem i has the form

$$E_i(t) = \sum_j a_j \mathrm{e}^{-\omega \lambda_j t} \tag{12.28}$$

where the coefficients a_j are determined by the matrix $\mathbf{P}$ and the initial value $E(0)$. This formula is a direct generalization of eqn (12.17) established in the special case of a single subsystem. The time decay of energies is therefore a linear combination of exponentials whose characteristic times are $1/\omega\lambda_j$ where λ_j are the eigenvalues of the matrix $\mathbf{A}$. Note that since $\mathbf{A}$ is diagonalizable, solutions of the type $t^k \mathrm{e}^{-\omega \lambda_j t}$ where $k > 1$ are not allowed.

Example

Three rooms have the same volume $V = 100\ \text{m}^3$ but different reverberation times $T_1 = 0.5$ s, $T_2 = 2.5$ s, and $T_3 = 12.5$ s at 1 kHz. They are aligned and coupled by small apertures of area $S = 0.125\ \text{m}^2$.

This system is obviously divided into three subsystems whose internal damping loss factors are $\eta_i = 13.8/T_i\omega$ where $\omega = 6280$ rad/s. We obtain respectively $\eta_1 = 4.4 \times 10^{-3}$, $\eta_2 = 0.87 \times 10^{-3}$, and $\eta_3 = 0.17 \times 10^{-3}$. The coupling loss factors are given by eqn (9.143) where $T(\theta, \varphi) = 1$. One obtains $\eta_{ij} = cS/4V\omega = 0.017 \times 10^{-3}$. Observe that $\eta_{ij} \ll \eta_i$ so that the coupling between rooms is weak. Furthermore the Schröder frequencies $2000\sqrt{T_i/V}$ are respectively 148 Hz, 332 Hz, and 742 Hz. The statistical analysis therefore applies at 1 kHz.

The matrix $\mathbf{A}$ is found:

$$\mathbf{A} = \begin{pmatrix} 4.41 & -0.016 & 0 \\ -0.16 & 0.91 & -0.016 \\ 0 & -0.016 & 0.19 \end{pmatrix} \times 10^{-3} \tag{12.29}$$

and the energy decay in each room if calculated by diagonalizing $\mathbf{A}$ and computing $e^{-\omega t \mathbf{A}} E(0)$. The result with $E(0) = (1,\ 0,\ 0)^T$ is shown in Fig. 12.2.

The energy in room 1 decays rapidly during the first half-second because the walls of room 1 are highly absorptive and the influence of the other rooms is negligible. During this first phase, rooms 2 and 3 receive sound through the apertures and their energies

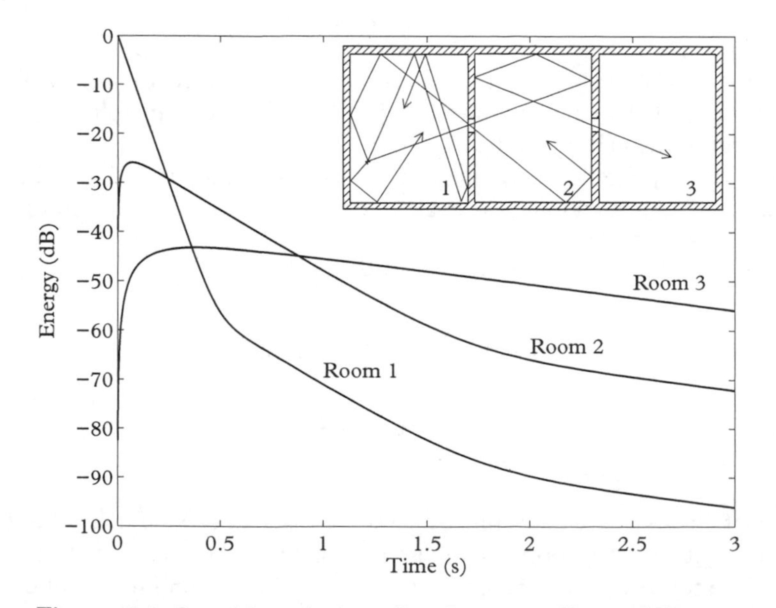

Figure 12.2 *Sound decay in three aligned rooms weakly coupled by small apertures.*

rapidly increase. An equilibrium is reached at $t = 0.5$ s after which the sound decay is imposed by the second room whose energy then dominates. This second phase continues until $t = 1.7$ s. By a similar process, the energy then dominates in room 3 which imposes the decay during the third phase $t > 1.7$ s. The three decay rates are clearly visible in Fig. 12.2, especially on curve 1.

12.4 Comments

The use of the coupling power proportionality (12.21) in transient conditions was proposed by Manning and Lee (1968). Several authors with different points of view have investigated the validity of such an approach.

Mercer et al. (1971) determined the transient energy flow between two linearly coupled oscillators, one of which was being excited by an impulse force. They used the perturbation method as in Newland's analysis. They found that the energy transfer is a second-order effect and is 'inversely proportional to the difference between the blocked natural frequencies', as in the stationary case.

Lai and Soom (1990a, 1990b) also investigated the time-varying energy flow between oscillators and concluded that the coupling power proportionality, if applied, requires time-varying coupling loss factors.

Pinnington and Lednik (1996a) studied the transient response of a two-degree-of-freedom system and compared the exact solution to the transient statistical energy response. They concluded that the solution (12.24) correctly predicts the peak value of energy and the long-range time decay, but not the rising time. These observations are confirmed for a two-beam system (Pinnington and Lednik, 1996b).

Ichchou et al. (2001) adopted a local wave approach, assuming incoherence. They showed that generally the power flow is not proportional to the energy density gradient and therefore that a local version of the coupling power proportionality does not apply.

Carcaterra (2005) investigated the transient régime of a statistical population of resonators. By performing an ensemble average having assumed random natural frequencies, he obtained relationships between the mean exchanged energy and the difference of modal energies. He concluded that the coupling power proportionality is applicable under certain conditions, but that in general, a term proportional to the time derivative of modal energy difference must be added.

All these studies raise the question of the validity of an exponential law for the time decay. As quoted by Weaver in Wright and Weaver (2010, Chapter 8), 'there are numerous sources for deviations from simple exponentials'.

13

Thermodynamics of Sound and Vibration

We have already remarked that modal energy plays the role of a vibrational temperature and that coupling power proportionality fixes the flow of energy from hot to cold subsystems. In this chapter we develop the thermodynamical analogy—in particular we give a strict definition of vibrational temperature—and we introduce the concept of entropy.

13.1 The Clausius Principle

We have demonstrated that vibrational energy flows from subsystems with high modal energy to subsystems with low modal energy. This is what we have named the coupling power proportionality,

$$P_{ij} = \omega\eta_{ij}N_i\left(\frac{E_i}{N_i} - \frac{E_j}{N_j}\right) \tag{13.1}$$

If we further admit that the modal energy E_i/N_i plays the role of a vibrational temperature, this result shows that the energy flows from hot to cold subsystems. In these terms, the coupling power proportionality is nothing other than the Clausius principle in thermodynamics which states that *it is impossible for heat to transfer spontaneously from a colder to a hotter body without causing other changes.*

The Clausius principle, or another equivalent form such as the Thompson, Kelvin, or Caratheodory principles, therefore applies in statistical vibroacoustics in the same terms as in classical thermodynamics. And since all classical thermodynamics is founded on this principle it is now apparent that a thermodynamics of sound and vibration exists.

13.2 Heat and Work

In thermodynamics, energy takes two forms: heat and work. Work is the energy supplied by external mechanical forces. This is the case for instance when a pressure p is applied to the boundary and induces a volume variation dV. The work supplied to the system is $\delta W = -p\,dV$. Conversely when a system of mass m falls from a height dh, the work released by the system is $\delta W = mg\,dh$.

Fundamentally, heat and work are not two distinct forms of energy. In the classical kinetic theory of gases, internal energy is interpreted as molecular agitation and is therefore kinetic energy at the scale of molecules. Heat is a transfer of molecular agitation energy at the scale of individual collisions (Fig. 13.1(a)). This is achieved by putting the gas into contact with a hotter system. The movement of molecules being disordered, collisions are independent events. The length of correlation of this process is small. Furthermore, the time between two successive collisions is very short. The correlation duration is therefore also very short. On the contrary, the gas releases mechanical energy when the collision of molecules into a wall induces a macroscopic movement of this surface (Fig. 13.1(b)). At the microscopic scale, the transfer of momentum through collisions relates only to the normal component. All molecules act in the same direction normal to the surface and the correlation length is therefore very large. Similarly, the movement of the wall is slow compared with the molecular speed and the correlation duration is very large. In solids, heat is stored as the vibration of atoms. Of course, the frequency of vibration of atoms is very high and their mean distance very small so that numerous orders of magnitude in both space and time separate heat and work.

We have seen that the coupling power proportionality (13.1) is formally equivalent to the Clausius principle. The modal energy E_i/N_i plays the role of temperature and the exchanged energy P_{ij} is analogous to heat. Therefore we define **vibrational heat** as the type of vibrational energy which is transferred by fully respecting the coupling power proportionality. We know the assumptions necessary for the coupling power proportionality. These are random, uncorrelated, and wide-band forces; light coupling, light damping, a diffuse field, and a large number of modes. So, vibrational heat is a vibrational energy at high frequency and is randomly distributed among a large number of modes.

We may also raise the question of the existence of vibrational work. Such an energy must correspond to mechanical energy (at low frequency) and must be supplied by

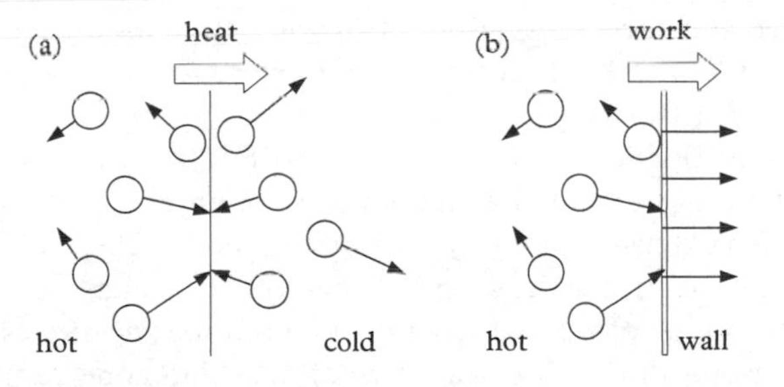

Figure 13.1 *(a) heat transfer is a direct exchange of kinetic energy at the molecular scale. (b) work production is a transformation of kinetic molecular energy into a slow and large-scale movement of the wall.*

deterministic sources. An example of conversion of mechanical energy into vibrational heat, that is vibrational energy in the audio frequency band, is realized by friction noise. Rubbing a solid with a rough surface on another rough surface gives rise to a typical wide-band noise. The energy is supplied by the steady-state movement (kinetic energy of the solid) and transformed into vibration. The vibration is induced by local impacts between antagonist asperities of surfaces. Due to the random nature of rough surfaces, the local forces are uncorrelated in space and time. This is a realization of a rain-on-the-roof excitation. The produced vibrational energy is therefore wide-band and disordered and in that sense may be qualified as vibrational heat. Interested readers may consult Le Bot and Bou Chakra (2010); Ben Abdelounis et al. (2010); Ben Abdelounis et al. (2011); Le Bot et al. (2011); and Dang et al. (2013).

13.3 Thermodynamic Entropy

'Vibrational entropy' can now be introduced in the same way Clausius did a century and a half ago in thermodynamics. Entropy is a state function which depends on the extensive variables of the subsystem, that is the vibrational energy E and the number of modes N.

For an isolated subsystem, if sources of power P supply an infinitesimal vibrational heat $\mathrm{d}E = P\,\mathrm{d}t$ during $\mathrm{d}t$ then, by Clausius' definition, the variation of entropy is

$$\mathrm{d}S = \frac{\mathrm{d}E}{T} \tag{13.2}$$

where $T = E/N$ is the vibrational temperature. The variation of entropy between two states of energy E_2 and E_1 is therefore

$$\Delta S = N \log\left(\frac{E_2}{E_1}\right) \tag{13.3}$$

This relationship gives the variation of **vibrational entropy** of a subsystem. It has been established for reversible processes. But entropy is a function of state and therefore the same relationship must also hold for irreversible processes.

It could be inferred from eqn (13.3) that the absolute entropy $S(E, N)$ is $N \log E$. This expression is compatible with the variation of entropy derived here, but it is inconsistent with the requirement that the entropy is an extensive quantity in classical thermodynamics. An extensive quantity must verify the equality $S(2E, 2N) = 2S(E, N)$. This is obviously not the case with $N \log E$. Equation (13.3) has been derived by considering a small variation $\mathrm{d}E$ of internal energy. The number of modes N has been tacitly assumed to be constant during this transformation. Thus, one cannot obtain the variation of entropy for different numbers of modes by Clausius' approach, or, in particular, fully derive the function $S(E, N)$. This is what is done in the following section, which introduces statistical entropy.

13.4 Statistical Entropy

To define the entropy concept as adapted to statistical vibroacoustics, we must first specify the dynamical system. A subsystem is a packet of N modes whose natural frequencies are ω_α and modal mass m_α where $\alpha = 1, \ldots, N$. The Hamiltonian is (Carcaterra, 2002)

$$H(X_1, \ldots, X_N, P_1, \ldots, P_N) = \sum_{\alpha=1}^{N} \frac{m_\alpha \omega_\alpha^2}{2} X_\alpha^2 + \frac{1}{2m_\alpha} P_\alpha^2 \tag{13.4}$$

where X_α is the modal deflection and $P_\alpha = m_\alpha \dot{X}_\alpha$ the momentum. In the phase space, the trajectory of the subsystem is determined by the time-functions $X_\alpha(t), P_\alpha(t)$, $\alpha = 1, \ldots, N$, solutions to the equations of motion. The phase space Γ has therefore dimension $2N$. A **microstate** is defined as the exact position of the subsystem in the phase space specified by the values X_α, P_α at time t. A full knowledge of microstates is reached by solving exactly the N governing equations. But in statistical vibroacoustics, forces are random and it is impossible to predict the microstate. The position $X_1, \ldots, X_N, P_1, \ldots, P_N$ becomes a random vector and we have abandoned the idea of determining it exactly. The exact repartition of vibrational energy over modes is therefore not known. What is claimed in statistical energy analysis is that *we are not interested in the exact repartition of vibrational energy over modes but only in the knowledge of the number of modes N and the total vibrational energy E of a subsystem* as shown in Fig. 13.2. A **macrostate** is therefore given by the values of the total vibrational energy E and the number of modes N, and nothing else.

Another consequence of the random nature of excitations is that the energy level of individual modes is not constant. Energy is shared between modes in a random manner, so, if at a given time a particular mode can receive much energy to the detriment of others, at a following time it will probably receive less energy. This perpetual modification of the repartition of vibrational energy is not due to a direct exchange of energy between modes as is the case in the kinetic theory of gases where collisions of molecules are responsible for energy redistribution among molecules. But the important fact is that

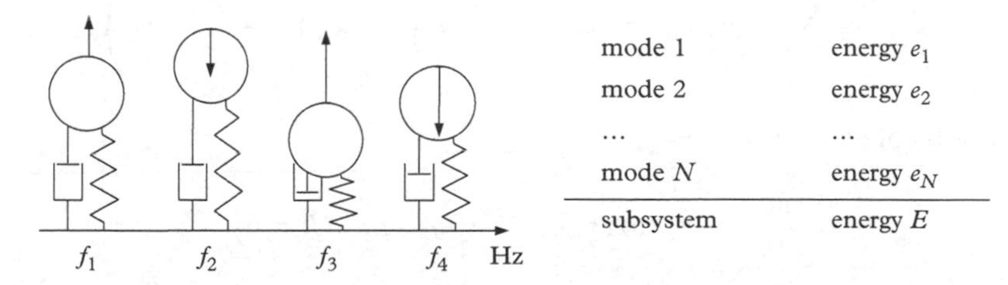

Figure 13.2 *Subsystems in statistical energy analysis. A microstate is determined by the repartition of vibrational energy e_i, $i = 1, \ldots, n$ among the modes, while a macrostate is the total vibrational energy E and the number of modes N.*

the trajectory of a subsystem in the phase space explores the entire surface of constant energy $H(X_1,\ldots,X_N,P_1,\ldots,P_N) = E$ and its neighbourhood.

Thus, the situation described above comes down to a somewhat classical problem in statistical mechanics. The phase space is divided into small regions of size h^N where h is Planck's constant. The fact that Planck's constant appears in a classical context is not important in what follows. The state of the subsystem is only known by its probability of presence p_i in region i of phase space. The probabilities p_i must of course verify the normalization condition

$$\sum_i p_i = 1 \tag{13.5}$$

where the sum runs over all cells of phase space. The mean vibrational energy E is

$$E = \sum_i p_i H_i \tag{13.6}$$

where H_i is the actual value of energy (Hamiltonian) at position $X_1,\ldots,X_N, P_1,\ldots,P_N$ in cell i. The entropy attached to a distribution p_i is defined by

$$S = -k_B \sum_i p_i \log p_i \tag{13.7}$$

where k_B is Boltzmann's constant.

The determination of the probabilities p_i stems from the maximum entropy principle (Jaynes, 1957), a modern version of Laplace's principle of insufficient reason. Application of this principle in physics is somewhat controversial but we shall not discuss its philosophical justification. The debate on this question has been extensively discussed by Uffink (1997).

The problem is to maximize S of eqn (13.7) under the constraints (13.5) and (13.6). We may introduce two Lagrange multipliers α and β for the constraints and the function to maximize becomes

$$\mathcal{S} = -k_B \sum_i p_i \log p_i + \alpha k_B \left(1 - \sum_i p_i\right) + \beta k_B \left(E - \sum_i p_i H_i\right) \tag{13.8}$$

The conditions of extremum are

$$\frac{\partial \mathcal{S}}{\partial p_i} = -k_B(\log p_i + 1) - \alpha k_B - \beta k_B H_i = 0 \tag{13.9}$$

Thus,

$$p_i = \mathrm{e}^{-1-\alpha-\beta H_i} \tag{13.10}$$

Define $Z = e^{1+\alpha}$ for convenience. The above relationship becomes

$$p_i = \frac{e^{-\beta H_i}}{Z} \tag{13.11}$$

This is the Boltzmann distribution giving the probability of presence of the dynamical system in cell i of the phase-space. The constants Z and β are determined by the constraints (13.5)

$$\sum_i \frac{e^{-\beta H_i}}{Z} = 1 \tag{13.12}$$

and (13.6)

$$\sum_i \frac{e^{-\beta H_i}}{Z} H_i = E \tag{13.13}$$

We are now going to calculate Z and β successively.

From eqn (13.12), we may introduce the **partition function** $Z(\beta) = \sum_i e^{-\beta H_i}$. Since the cells have a small size h^N, the sum approximates an integral over the full phase space Γ,

$$Z(\beta) = \frac{1}{h^N} \int_\Gamma e^{-\beta H}\, dV \tag{13.14}$$

where $dV = dX_1 \ldots dX_N dP_1 \ldots dP_N$. When evaluating the above integral, we may take advantage of the fact that the phase function $H(X_1, \ldots, X_N, P_1, \ldots, P_N)$ is constant over the hyper-surface of constant energy $H(X_1, \ldots, X_N, P_1, \ldots, P_N) = e$. If we introduce the **structure function** $\Omega(e)$,

$$\Omega(e) = \frac{dV}{de} \tag{13.15}$$

where

$$V(e) = \int_{H \leq e} dV \tag{13.16}$$

is the volume enclosed by the hyper-surface $H = e$, then the partition function becomes

$$Z(\beta) = \frac{1}{h^N} \int_0^\infty e^{-\beta e} \Omega(e)\, de \tag{13.17}$$

The partition function is, apart from the proportionality constant $1/h^N$, the Laplace transform of the structure function.

From eqn (13.4), the hyper-surface of constant energy $H(X_1, \cdots, X_N, P_1, \cdots, P_N) = e$ has the equation

$$\sum_{\alpha=1}^{N} \frac{X_\alpha^2}{2e/m_\alpha \omega_\alpha^2} + \frac{P_\alpha^2}{2m_\alpha e} = 1 \tag{13.18}$$

This is an ellipsoid of dimension $2N-1$ with semi-axes $(2m_\alpha e)^{1/2}$ and $(2e/m_\alpha \omega_i^2)^{1/2}$. The volume $V(e)$ enclosed by this ellipsoid is

$$V(e) = \frac{(2\pi)^N}{N! \prod_{i=1}^{N} \omega_i} e^N \tag{13.19}$$

In statistical energy analysis, the set of modes is considered to be a random population. The frequencies ω_α have a uniform probability density over the band $[\omega_{\min}, \omega_{\max}]$ of span $\Delta\omega$ and centre frequency ω. The product $\prod_i \omega_i$ can therefore be substituted by its expected value ω^N. The volume becomes

$$V(e) = \left(\frac{2\pi}{\omega}\right)^N \frac{e^N}{N!} \tag{13.20}$$

and the structure function $\Omega(e) = \mathrm{d}V/\mathrm{d}e$ is

$$\Omega(e) = \left(\frac{2\pi}{\omega}\right)^N \frac{e^{N-1}}{N-1!} \tag{13.21}$$

But the Laplace transform of $e \mapsto e^{N-1}/N-1!$ is $\beta \mapsto 1/\beta^N$. The partition function is therefore

$$Z(\beta) = \left(\frac{2\pi}{h\omega\beta}\right)^N \tag{13.22}$$

This is the partition function of a subsystem.

To calculate β, we make use of the second constraint (13.13). By deriving $Z(\beta) = \sum_i \mathrm{e}^{-\beta H_i}$ with respect to β, we may transform eqn (13.13) into

$$E = -\frac{\mathrm{d}}{\mathrm{d}\beta} \log Z(\beta) \tag{13.23}$$

By using eqn (13.22), we immediately get

$$E = \frac{N}{\beta} \tag{13.24}$$

The entropy is obtained by introducing Bolzmann's distribution (13.11) in eqn (13.7) and by simplifying with conditions (13.5) and (13.6):

$$S = k_B\left[\beta E + \log Z(\beta)\right] \tag{13.25}$$

Now, by substituting the expressions of Z and β, it yields

$$S(E,N) = k_B N\left[1 + \log\left(\frac{2\pi E}{h\omega N}\right)\right] \tag{13.26}$$

This is the expression of the **vibrational entropy** of a subsystem having vibrational energy E and N modes (Le Bot, 2011). The quantities E and N are extensive and we can verify that $S(2E, 2N) = 2S(E, N)$ is also an extensive quantity.

The temperature is obtained as for any thermodynamic system with

$$\frac{1}{T} = \left(\frac{\partial S}{\partial E}\right)_N \tag{13.27}$$

This gives

$$T = \frac{E}{k_B N} \tag{13.28}$$

which is the expected result. The **vibrational temperature** is the modal energy divided by Boltzmann's constant k_B.

The heat capacity is

$$C_V = \left(\frac{\partial E}{\partial T}\right)_N = k_B N \tag{13.29}$$

For a frequency band $\Delta\omega$, the number of modes is $N = n\Delta\omega$ where $n(\omega)$ is the modal density. Therefore the heat capacity $C_V = k_B n(\omega)\Delta\omega$ is proportional to the modal density. As we remarked in Chapter 8 after eqn (8.30), the modal density represents the capacity of a subsystem to absorb energy. Furthermore, the modal density is proportional to the length, surface, or volume of the subsystem. Since C_V has been calculated with N constant, C_V must be understood as the volumetric heat capacity.

The differential form dS for an energy variation dE is calculated from eqn (13.26),

$$\mathrm{d}S = k_B N\frac{\mathrm{d}E}{E} = \frac{\mathrm{d}E}{T}, \tag{13.30}$$

valid for d$N = 0$. This is the variation of vibrational entropy dS when an infinitesimal vibrational energy dE is supplied to the system. Of course, this expression matches well with Clausius' definition of entropy.

13.5 Second Law of Thermodynamics

The second law of thermodynamics states that the entropy of an isolated system cannot decrease. This property is also true in the framework of statistical vibroacoustics.

Let us consider two isolated subsystems having respectively energy E_1 and E_2, and N_1 and N_2 modes. We assume that thermal equilibrium is realized within the subsystems, that is, the vibrational field is diffuse. Since the two subsystems are disconnected, their entropies are respectively $S(E_1, N_1)$ and $S(E_2, N_2)$, given by eqn (13.26).

Now, imagine that the two subsystems are connected. For instance, the two subsystems may be two rooms filled with sound of different levels that are connected by opening a window separating them. The subsystems then start to exchange energy until a new equilibrium is reached. The total system has energy $E = E_1 + E_2$ and number of modes $N = N_1 + N_2$. When the equilibrium is reached, the entropy is $S(E, N)$, again given by eqn (13.26).

The difference between the final entropy and the sum of initial entropy is therefore the entropy created during the mixing process:

$$\Delta S = S(E, N) - S(E_1, N_1) - S(E_2, N_2) \tag{13.31}$$

By substituting eqn (13.26), it yields

$$\Delta S = k_B(N_1 + N_2) \log \frac{E_1 + E_2}{N_1 + N_2} - k_B N_1 \log \frac{E_1}{N_2} - k_B N_2 \log \frac{E_2}{N_2} \tag{13.32}$$

This is the entropy created by mixing the energy of the two subsystems, and the entropy production is non-negative $\Delta S \geq 0$. This result stems from the convexity of the function $f(x, y) = -y \log(x/y)$.

We then arrive at the expression of the second law of thermodynamics. The entropy of an isolated system increases and reaches a maximum at equilibrium.

13.6 Entropy Balance

So far, we have only considered isolated subsystems in equilibrium. Of course, this situation is an ideal case which has only been introduced to obtain the explicit expression of the entropy of a single subsystem. But in statistical vibroacoustics, a subsystem can never be considered as isolated. The theory developed in this book makes sense only in the presence of the dissipation and exchange of energy. The subsystems are therefore not isolated in general.

The subsystems are assumed to be in steady-state conditions. All macroscopic quantities such as input power, dissipated power, vibrational energy, vibrational temperature, and entropy are constant in time. The subsystems are not strictly speaking in thermodynamic equilibrium since a flux of energy can flow through them. But if the flux is not too important, the subsystems are in local equilibrium. This means that the relaxation processes inside subsystems, such as the mixing of rays, are more rapid than the breakdown of equilibrium imposed by exchanges with the exterior. This assumption of local equilibrium allows us to define a unique vibrational temperature per subsystem. The problem therefore belongs to linear non-equilibrium thermodynamics.

A subsystem can gain or lose vibrational heat by three processes: source, internal loss, or exchange with neighbours. Each of these processes leads to a variation of vibrational entropy.

Energy is provided to subsystem i by a source of power P_i. During $\mathrm{d}t$, the infinitesimal energy $\mathrm{d}E_i = P_i\,\mathrm{d}t$ is supplied. Clausius' relation (13.30) then gives the variation of vibrational entropy,

$$\mathrm{d}S_{\mathrm{inj},i} = \frac{P_i\,\mathrm{d}t}{T_i} = k_B N_i \frac{P_i\,\mathrm{d}t}{E_i} \tag{13.33}$$

The variation $\mathrm{d}S_i$ is positive. The sources provide heat to the subsystem, warm it, and therefore increase its vibrational entropy.

The energy lost by internal dissipation in subsystem i is $\eta_i \omega E_i\,\mathrm{d}t$. The variation of energy is negative and again, by Clausius' relation, the variation of entropy is

$$\mathrm{d}S_{\mathrm{diss},i} = \frac{\eta_i \omega E_i\,\mathrm{d}t}{T_i} = -k_B N_i \eta_i \omega\,\mathrm{d}t \tag{13.34}$$

The variation $\mathrm{d}S_{\mathrm{diss},i}$ is negative. Internal dissipation extracts heat, cools down the subsystem, and therefore decreases the vibrational entropy. The status of dissipation processes of vibration in statistical vibroacoustics is therefore different from those in thermodynamics. In classical thermodynamics, dissipation processes transform work into heat and are the cause of irreversibility which increases the thermodynamic entropy. Exactly the converse is true in statistical vibroacoustics.

At the interface between two subsystems i and j, the exchange of energy $P_{ij}\,\mathrm{d}t$ during $\mathrm{d}t$ is realized from the hot subsystem, say i, to the cold subsystem, say j. For i, this is a loss of energy which induces a decrease of entropy $-P_{ij}\,\mathrm{d}t/T_i$ where $T_i = E_i/k_B N_i$ is the vibrational temperature of subsystem i. But for subsystem j, this is a gain of vibrational heat and the entropy is increased by $P_{ij}\,\mathrm{d}t/T_j$. The net variation of entropy is therefore

$$\mathrm{d}S_{\mathrm{irr},ij} = P_{ij}\,\mathrm{d}t\left(\frac{1}{T_j} - \frac{1}{T_i}\right) \tag{13.35}$$

Since $P_{ij} > 0$ when $T_i > T_j$, $\mathrm{d}S_{\mathrm{irr},ij} > 0$. Conversely, $P_{ij} < 0$ when $T_i < T_j$ and again $\mathrm{d}S_{\mathrm{irr},ij} > 0$. The net variation of entropy at interfaces is always positive. The exchange of energy at the interface between two subsystems having different vibrational temperatures is therefore an irreversible process.

The coupling power proportionality allows this irreversible entropy to be expressed in another form. Following eqn (13.1), we have

$$\mathrm{d}S_{\mathrm{irr},ij} = k_B \eta_{12} \omega N_i \left(\frac{E_i}{N_i} - \frac{E_j}{N_j} \right) \left(\frac{N_j}{E_j} - \frac{N_i}{E_i} \right) \mathrm{d}t \tag{13.36}$$

Now, consider the entire system. Since entropy is an extensive variable, the entropy of the whole system is the sum of the entropy exchanged with the exterior by subsystems and the entropy created at interfaces by irreversible processes:

$$\mathrm{d}S = \sum_{i=1}^{N} \left(\mathrm{d}S^i_{\mathrm{inj},i} + \mathrm{d}S_{\mathrm{diss},i} \right) + \sum_{i>j} \mathrm{d}S_{\mathrm{irr},ij} \tag{13.37}$$

The last sum runs for $i > j$ since each interface must be counted only once. By substituting eqns (13.33–13.35) into eqn (13.37), it yields

$$\frac{\mathrm{d}S}{\mathrm{d}t} = \sum_{i=1}^{N} \left(\frac{P_i}{T_i} - \frac{\eta_i \omega E_i}{T_i} \right) + \sum_{i>j} P_{ij} \left(\frac{1}{T_j} - \frac{1}{T_i} \right) \tag{13.38}$$

By splitting the last sum and noting that $P_{ij} = -P_{ji}$,

$$\frac{\mathrm{d}S}{\mathrm{d}t} = \sum_{i=1}^{N} \frac{1}{T_i} \left(P_i - \eta_i \omega E_i + \sum_{j \neq i} P_{ij} \right) \tag{13.39}$$

But the energy balance of any subsystem reads

$$P_i - \eta_i \omega E_i + \sum_{j \neq i} P_{ij} = 0 \tag{13.40}$$

Finally,

$$\frac{\mathrm{d}S}{\mathrm{d}t} = 0 \tag{13.41}$$

This is the entropy balance for the entire system (Le Bot, 2009, 2011). Globally, there is no production of entropy. The exchange of entropy with the exterior either by sources or dissipation exactly balances the production of entropy by irreversible processes at interfaces. The situation is illustrated in Fig. 13.3.

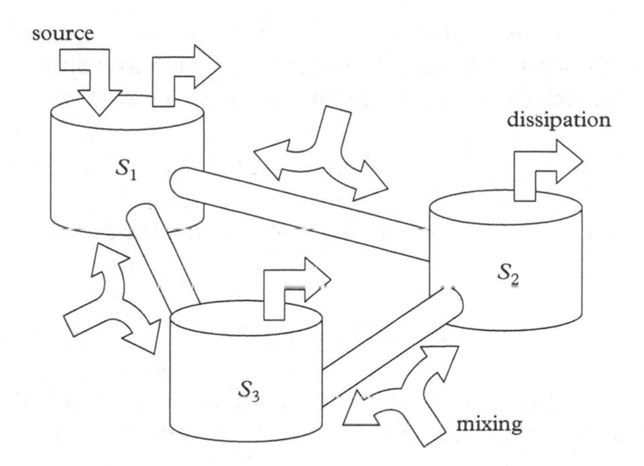

Figure 13.3 *Entropy balance. Entropy is supplied to subsystems by sources, extracted by dissipation, and created by mixing.*

13.7 Linear Non-Equilibrium Thermodynamics

We have shown in eqn (13.36) that coupling power proportionality implies that the net production of entropy at the interface of two subsystems with different temperatures is always positive. But the converse is also true. If we admit the second principle of thermodynamics and in particular that $dS \geq 0$ at interfaces, then the exchanged energy is proportional to the temperature difference (Carcaterra, 2002).

Let us consider a simple situation in which two subsystems of energy E_1, E_2 and entropy S_1, S_2 are in weak interaction. Both energy and entropy are extensive quantities. Therefore by additivity, the entire system has energy $E = E_1 + E_2$ and entropy $S = S_1 + S_2$. Let us remark that additivity and extensiveness come from the weak coupling assumption which states that energy of coupling is negligible.

The conjugate of an extensive quantity is an intensive quantity. In the case of energy, we obtain the temperature by

$$\frac{1}{T_i} = \frac{\partial S_i}{\partial E_i} \tag{13.42}$$

In thermodynamics, a **flux** is defined by the rate of variation of an extensive quantity. The exchanged power,

$$P_{12} = -\frac{dE_1}{dt} \tag{13.43}$$

is therefore the flux associated with energy.

During a process, the two subsystems exchange energy. The variations of energy and entropy follow the first and second principle of thermodynamics. First, the energy conservation principle imposes that the global energy variation is zero,

$$\mathrm{d}E = \mathrm{d}E_1 + \mathrm{d}E_2 = 0 \tag{13.44}$$

And the second principle of thermodynamics imposes that the global variation of entropy is non-negative,

$$\mathrm{d}S = \mathrm{d}S_1 + \mathrm{d}S_2 \geq 0 \tag{13.45}$$

Let us first examine the equilibrium state. Equilibrium is reached when entropy is maximum. In particular, the variation $\mathrm{d}S$ is zero. By dividing eqn (13.45) by $\mathrm{d}E_1 = -\mathrm{d}E_2$, it yields

$$\frac{\mathrm{d}S_1}{\mathrm{d}E_1} - \frac{\mathrm{d}S_2}{\mathrm{d}E_2} = 0 \tag{13.46}$$

or

$$\frac{1}{T_1} - \frac{1}{T_2} = 0 \tag{13.47}$$

At equilibrium, temperatures are equal $T_1 = T_2$. Furthermore, it is clear that the flux P_{12} is zero at equilibrium. Equilibrium is therefore characterized by $P_{12} = 0$ and $T_1 - T_2 = 0$ simultaneously.

In non-equilibrium condition $\mathrm{d}S > 0$. By combining eqns (13.44) and (13.45),

$$\frac{\mathrm{d}S}{\mathrm{d}t} = \left(\frac{\mathrm{d}S_1}{\mathrm{d}E_1} - \frac{\mathrm{d}S_2}{\mathrm{d}E_2}\right)\frac{\mathrm{d}E_1}{\mathrm{d}t} \geq 0 \tag{13.48}$$

In a general thermodynamical situation, the rate of variation of entropy is the product of thermodynamic forces and fluxes. This is verified in the present case provided that

$$\mathcal{F} = \left(\frac{\mathrm{d}S_2}{\mathrm{d}E_2} - \frac{\mathrm{d}S_1}{\mathrm{d}E_1}\right) = \left(\frac{1}{T_2} - \frac{1}{T_1}\right) \tag{13.49}$$

is defined as the force associated with the flux $P_{12} = -\mathrm{d}E_1/\mathrm{d}t$. For a continuous system, the thermodynamical force $\mathcal{F}$ is rather defined as the gradient of the intensive quantity. The rate of production of entropy density is again the product of fluxes and forces.

So, at equilibrium $\mathcal{F} = P_{12} = 0$ while at non-equilibrium $\mathcal{F} \neq 0$ and $P_{12} \neq 0$. It is therefore natural to raise the question of the existence of a functional dependence of fluxes versus forces,

$$P_{ij} = f(\mathcal{F}) \tag{13.50}$$

If such a relationship exists, a first-order polynomial expansion leads to

$$P_{ij} = a\mathcal{F} + o(\mathcal{F}) \tag{13.51}$$

where a is an appropriate coefficient. The domain of validity of such an expression is the linear non-equilibrium thermodynamics.

All transport phenomena admit a phenomenological law of this type. For example Fourier's law in thermal conduction,

$$Q_x = -k\frac{\partial T}{\partial x} \tag{13.52}$$

states that the heat flux is proportional to the temperature gradient. Similarly Fick's law of diffusion,

$$I_x = -D\frac{\partial \phi}{\partial x} \tag{13.53}$$

relates the diffusive flux I_x to the concentration gradient (amount of substance per unit volume). The flux always goes from regions of high concentration to regions of low concentration. We observe that the coupling power proportionality,

$$P_{12} = \omega\eta_{12}N_1\,(T_1 - T_2) \propto \left(\frac{1}{T_2} - \frac{1}{T_1}\right) + o(T_2 - T_1) \tag{13.54}$$

is a relationship of this type. For this reason we can conclude that statistical energy analysis is the theory of linear non-equilibrium thermodynamics applied to audio frequency vibrations (Le Bot et al., 2010).

Appendix A
Stability

The study of system stability may be found in textbooks on differential equations (Gourmelin and Wadi, 2009) or dynamical systems (Strogatz, 1994; Jean, 2011). For general aspects on differential calculus see Cartan (1985).

Differential equation

Let us consider a differential equation,

$$\dot{\mathbf{Y}} = f(\mathbf{Y}) \tag{A.1}$$

where $f : U \subset \mathbb{R}^N \to \mathbb{R}^N$ is a continuously differentiable function defined on an open set U. A solution $\mathbf{Y} : I \to U$ is a function differentiable on an interval $I \subset \mathbb{R}$ which verifies eqn (A.1) on I. A maximal solution is a solution defined on the largest possible interval. The Cauchy–Lipschitz theorem states the existence and uniqueness of a maximal solution for any initial condition $\mathbf{Y}(0) = \mathbf{Y}_0 \in U$. A point $\mathbf{Y}^* \in U$ is an equilibrium if $f(\mathbf{Y}^*) = 0$.

The differential equation is said to be linear if f is a matrix $\mathbf{A}$,

$$\dot{\mathbf{Y}} = \mathbf{AY} \tag{A.2}$$

For all linear systems, the origin is an equilibrium (but not necessarily the only one). For any $\mathbf{Y}_0 \in \mathbb{R}^N$, the unique solution of eqn (A.2) verifying $\mathbf{Y}(0) = \mathbf{Y}_0$ is

$$\mathbf{Y}(t) = e^{t\mathbf{A}}\mathbf{Y}_0 \tag{A.3}$$

and is defined on $I = \mathbb{R}$. Let us recall that the exponential of matrix $\mathbf{A}$ is defined by the series $\sum_{n=0}^{\infty} \mathbf{A}^n/n!$.

Example A.1

In Chapter 2, we introduced the following N-degrees-of-freedom system,

$$\mathbf{M}\ddot{\mathbf{X}} + (\mathbf{C} + \mathbf{G})\dot{\mathbf{X}} + \mathbf{KX} = 0 \tag{A.4}$$

continued

Example A.1 *continued*

where **M**, **K**, **C**, and **G** are respectively the mass, stiffness, damping, and gyroscopic matrices. They are all real-valued and **M**, **K** are symmetric and positive-definite, i.e. $\mathbf{M}^T = \mathbf{M}$, $\mathbf{K}^T = \mathbf{K}$, $\mathbf{M} > 0$, and $\mathbf{K} > 0$. The gyroscopic matrix is antisymmetric $\mathbf{G}^T = -\mathbf{G}$ and $\mathbf{C} = \mathbf{diag}(c_1, \ldots, c_N)$ is diagonal with non-negative entries $c_i \geq 0$ hence $\mathbf{C}^T = \mathbf{C}$. If all $c_i > 0$ then **C** is positive-definite but only semipositive when at least one $c_i = 0$.

The matrices **M** and **K** are invertible since they are diagonalizable (real-valued and symmetric) and all their eigenvalues are non-zero.

By introducing the state vector,

$$\mathbf{Y} = \begin{pmatrix} \mathbf{X} \\ \dot{\mathbf{X}} \end{pmatrix} \tag{A.5}$$

which has values in $\mathbb{R}^{2N}$, the governing equation (A.4) takes the form (A.2) with

$$\mathbf{A} = \begin{pmatrix} 0 & \mathbf{I} \\ -\mathbf{M}^{-1}\mathbf{K} & -\mathbf{M}^{-1}(\mathbf{C} + \mathbf{G}) \end{pmatrix} \tag{A.6}$$

All properties of system (A.4) are embodied in the matrix **A**.

Stability

The stability of an equilibrium is defined in the sense of Lyapunov by the following.

Definition A.1 *The equilibrium* $\mathbf{Y}^*$ *is stable if for any* $\epsilon > 0$ *there exists* $\delta > 0$ *such that*

$$\text{if} \quad \left\| \mathbf{Y}(0) - \mathbf{Y}^* \right\| < \delta \quad \text{then} \quad \forall t \geq 0 \quad \left\| \mathbf{Y}(t) - \mathbf{Y}^* \right\| < \epsilon \tag{A.7}$$

Stability means that any solution starting in a neighbourhood of equilibrium remains indefinitely in a neighbourhood of equilibrium (Fig. A.1). A stronger definition is the asymptotic stability.

Definition A.2 *The equilibrium* $\mathbf{Y}^*$ *is said to be attracting if there exists* $\delta > 0$ *such that*

$$\text{if} \quad \left\| \mathbf{Y}(0) - \mathbf{Y}^* \right\| < \delta \quad \text{then} \quad \lim_{t \to \infty} \mathbf{Y}(t) = \mathbf{Y}^* \tag{A.8}$$

An equilibrium $\mathbf{Y}^*$ *is asymptotically stable if it is both stable and attracting.*

Asymptotic stability means that any solution starting in a neighbourhood of equilibrium tends to equilibrium by remaining arbitrarily close to it for all time (Fig. A.1). Asymptotic stability trivially implies stability but the converse is not true. An equilibrium may be attracting but not stable. An example of such a pathological case is given by the differential equation $\dot{\theta} = 1 - \cos\theta$ and the equilibrium $\theta^* = 0$. For all initial conditions $\theta_0 \neq 0$, the solution is $\theta(t) = 2 \arctan\left[1/\left(1/\tan(\theta_0/2) - t\right)\right]$. The point 0 is attracting since $\lim \theta(t) = 0$ when $t \to \infty$ for any θ_0. But 0 is not stable since $\lim \theta(t) = \pi$ when $t \to 1/\arctan(\theta_0/2)$. The solution cannot be confined to the neighbourhood of 0.

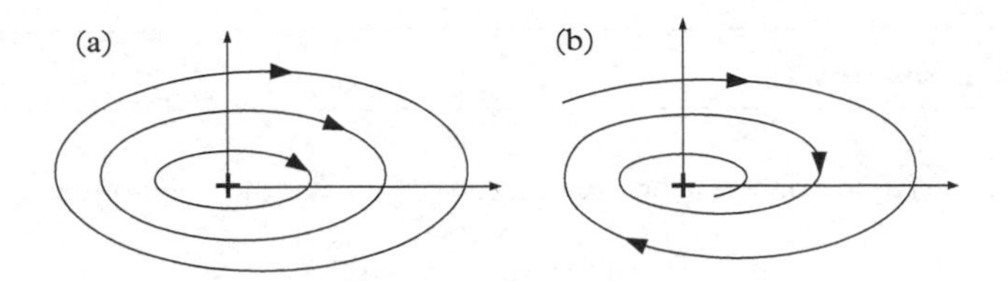

Figure A.1 *Example of (a) stable equilibrium; (b) asymptotically stable equilibrium.*

In the linear case, stability depends on the position of eigenvalues of **A** in the complex plane. The following theorem is established by discussing the solution (A.3).

Theorem A.1 *Let us consider a matrix* **A** *and the linear differential equation (A.2). The origin is an equilibrium. We note* λ_i, $i = 1, \ldots, N$ *the complex eigenvalues of* **A** *and* E_i *the related eigenspaces. Then,*

- *the origin is stable if and only if* $\mathrm{Re}(\lambda_i) \leq 0$ *for all* i *and when* $\mathrm{Re}(\lambda_i) = 0$ *the dimension of* E_i *is equal to the multiplicity of* λ_i *in the characteristic polynomial; and*
- *the origin is asymptotically stable if and only if* $\mathrm{Re}(\lambda_i) < 0$ *for all* i.

Although this criterion only applies to linear systems, it may be also useful for non-linear systems by linearizing the differential equation about equilibrium. But the result is weakened. The following theorem is known as the first Lyapunov method.

Theorem A.2 *Let* **D***f be the Jacobian matrix of a continuously differentiable function* $f : U \rightarrow \mathbb{R}^N$ *at equilibrium* $\mathbf{Y}^*$ *and let* λ_i, $i = 1, \ldots, N$ *be its complex eigenvalues. Then*

- *if* $\mathrm{Re}(\lambda_i) < 0$ *for all* i *then* $\mathbf{Y}^*$ *is asymptotically stable; and*
- *if* $\mathrm{Re}(\lambda_i) > 0$ *for at least one* i *then* $\mathbf{Y}^*$ *is not stable.*

When $\mathrm{Re}(\lambda_i) = 0$ for one or more eigenvalues, one cannot conclude. For instance, the differential equation $\dot{y} = -y^3$ has $Df = 0$ at $y^* = 0$. Theorem 2 does not apply. However, the origin is asymptotically stable since the solution is $y(t) = \mathrm{sgn}(y_0)/\sqrt{2t + 1/y_0^2}$ for $t \geq 0$ and goes to zero when $t \rightarrow \infty$. Conversely, the differential equation $\dot{\theta} = 1 - \cos\theta$ has also $Df = 0$ at $\theta^* = 0$ but we have previously seen that 0 is not stable.

Example A.2

Let us discuss the stability of eqn (A.4) with respect to $Y^* = 0$. We have seen that **A** is given by eqn (A.6). Stability is therefore related to the eigenvalues of **A**, that is the roots of the characteristic polynomial,

$$\begin{aligned}\det(\mathbf{A} - \lambda\mathbf{I}) &= \det(-\lambda\mathbf{I})\det\left(-\mathbf{M}^{-1}(\mathbf{C} + \mathbf{G}) - \lambda\mathbf{I}\right) - \det\left(-\mathbf{M}^{-1}\mathbf{K}\right)\det(\mathbf{I}) \\ &= (\det\mathbf{M})^{-1}\det\left(\lambda^2\mathbf{M} + \lambda(\mathbf{C} + \mathbf{G}) + \mathbf{K}\right)\end{aligned}$$

continued

Example A.2 *continued*

Thus, the system stability depends on the position of zeros of the polynomial,

$$\Delta(\lambda) = \det\left(\lambda^2\mathbf{M} + \lambda(\mathbf{C} + \mathbf{G}) + \mathbf{K}\right) \tag{A.9}$$

with respect to the axis $\mathrm{Re}(z) = 0$ in the complex plane.

Lyapunov functions

To study stability in the general case, we introduce the concept of Lyapunov functions.

Definition A.3 *A continuously differentiable function V defined in an open set U such that $Y^* \in U \subset \mathbb{R}^n$ is said to be a Lyapunov function if*

1. $V(\mathbf{Y}^*) = 0$
2. $V(\mathbf{Y}) > 0$ *for all* $\mathbf{Y} \neq \mathbf{Y}^*$
3. $\dot{\mathbf{V}} = f.\nabla V(\mathbf{Y}) \leq 0$

If, furthermore, $\dot{V}(\mathbf{Y}) < 0$ *for all* $\mathbf{Y} \neq \mathbf{Y}^*$ *then V is a strict Lyapunov function.*

In the above definition, ∇V is the gradient of V defined in U. This is a vector oriented in the direction of increasing values of V. The velocity $\dot{\mathbf{Y}} = f(\mathbf{Y})$ is a vector tangential to the trajectory at point $\mathbf{Y}$. The scalar product $f.\nabla V$ is therefore the time derivative of V along trajectories. The third condition imposes that trajectories are oriented toward lower levels of V (Fig. A.2).

We may now enunciate Lyapunov's second theorem on stability.

Theorem A.3 *Let us consider the differential equation (A.1) and an equilibrium point* $\mathbf{Y}^*$. *If there exists a Lyapunov function in a neighbourhood of* $\mathbf{Y}^*$ *then* $\mathbf{Y}^*$ *is stable. If there exists a strict Lyapunov function in a neighbourhood of* $\mathbf{Y}^*$ *then* $\mathbf{Y}^*$ *is asymptotically stable.*

A common application of Lyapunov functions is concerning the energy. The theorem then states that if the energy is non-increasing (no energy is injected), the equilibrium is stable. Furthermore, if the system dissipates *at all times*, the energy is a strict Lyapunov function and the equilibrium is asymptotically stable.

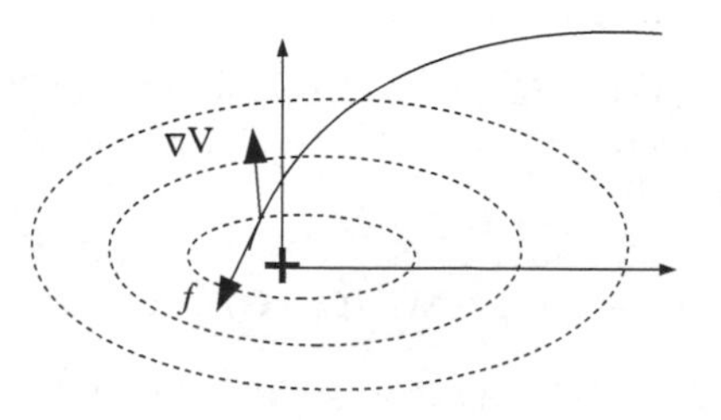

Figure A.2 *Lyapunov function. Velocity along trajectories is oriented towards decreasing values of V.*

Example A.3

To establish the stability of system (A.4), we consider the Lyapunov function $V(\mathbf{Y}) = 1/2 \times \mathbf{Y}^T\mathbf{P}\mathbf{Y}$ where

$$\mathbf{P} = \begin{pmatrix} \mathbf{K} & 0 \\ 0 & \mathbf{M} \end{pmatrix} \tag{A.10}$$

Since **M** and **K** are symmetric and positive-definite, **P** is also symmetric and positive-definite. The first two conditions of a Lyapunov function are therefore fulfilled. By developing the product $\mathbf{Y}^T\mathbf{P}\mathbf{Y}$, we obtain

$$V(\mathbf{Y}) = \frac{1}{2}\mathbf{X}^T\mathbf{K}\mathbf{X} + \frac{1}{2}\dot{\mathbf{X}}^T\mathbf{M}\dot{\mathbf{X}} \tag{A.11}$$

so that $V(\mathbf{Y})$ is simply the energy of system.

To check the third condition, we calculate $\dot{V}(\mathbf{Y})$:

$$\dot{V}(\mathbf{Y}) = \frac{1}{2}\dot{\mathbf{Y}}^T\mathbf{P}\mathbf{Y} + \frac{1}{2}\mathbf{Y}^T\mathbf{P}\dot{\mathbf{Y}} = \frac{1}{2}\mathbf{Y}^T\left(\mathbf{A}^T\mathbf{P} + \mathbf{P}\mathbf{A}\right)\mathbf{Y} \tag{A.12}$$

By denoting $\mathbf{Q} = -1/2 \times \left(\mathbf{A}^T\mathbf{P} + \mathbf{P}\mathbf{A}\right)$, a mere calculation gives

$$\mathbf{Q} = \begin{pmatrix} 0 & 0 \\ 0 & \mathbf{C} \end{pmatrix} \tag{A.13}$$

and $\dot{V}(\mathbf{Y}) = -\dot{\mathbf{X}}^T\mathbf{C}\dot{\mathbf{X}}$. This equality is nothing other than the energy balance. Since **C** is semipositive, we obtain $\dot{V}(\mathbf{Y}) \leq 0$ and the third condition is also fulfilled. By applying the Lyapunov theorem, we conclude that system (A.4) is stable about the origin, that is all zeros of the polynomial $\Delta(\lambda)$ are in the half-plane $\mathrm{Re}(z) \leq 0$. But, unfortunately, **Q** is not positive definite and V is not a strict Lyapunov function. Therefore, we cannot conclude asymptotic stability. To do that, we need a stronger result.

Invariance principle

We now introduce an important extension of Lyapunov's second method. Standard references are Krasovskii (1963), and LaSalle (1960).

Definition A.4 *Let us consider a differential equation* $\dot{\mathbf{Y}} = f(\mathbf{Y})$ *with an equilibrium* $\mathbf{Y}^* \in U$. *A set* $\mathcal{M}$ *is termed* **invariant** *if any solution starting from* $\mathcal{M}$ *is entirely in* $\mathcal{M}$, *i.e.* $\mathbf{Y}(0) \in \mathcal{M} \implies \mathbf{Y}(t) \in \mathcal{M}$ *for all* t, *and termed* **positively invariant** *if* $\mathbf{Y}(0) \in \mathcal{M} \implies \mathbf{Y}(t) \in \mathcal{M}$ *for all* $t \geq 0$.

Theorem A.4 *Let* V *be a continuously differentiable function on* U *such that* $\dot{V}(\mathbf{Y}) \leq 0$. *Let* S *be the set of all points in* U *such that* $\dot{V}(\mathbf{Y}) = 0$. *If* $\mathcal{M}$ *is the largest invariant set in* S *then every solution* $\mathbf{Y}(t)$ *bounded for* $t > 0$ *approaches* $\mathcal{M}$ *as* $t \to \infty$, *i.e.* $\lim_{t\to\infty} d(\mathbf{Y}(t), \mathcal{M}) = 0$.

Let us choose V as the energy of a dissipating system. If no energy is provided to the system, the function V is non-increasing and any trajectory tends to a state which does not dissipate.

Theorem A.5 *Let V be a Lyapunov function on U and Ω a compact neighbourhood of $\mathbf{Y}^*$. If Ω is positively invariant and if the only solution $\mathbf{Y}(t)$ contained in the set $S = \{\mathbf{Y} \in U, \dot{V}(\mathbf{Y}) = 0\}$ is the trivial solution $\mathbf{Y}(t) = \mathbf{Y}^*$, then $\mathbf{Y}^*$ is asymptotically stable. Furthermore all points in Ω tend to $\mathbf{Y}^*$.*

What is interesting about this theorem is that asymptotic stability is obtained with a non-strict Lyapunov function.

Finally, let us mention this last result.

Theorem A.6 *Let V be a Lyapunov function on $\mathbb{R}^N$ such that $\lim_{\|\mathbf{Y}\|\to\infty} V(\mathbf{Y}) = \infty$. If the only solution $\mathbf{Y}(t)$ contained in the set $S = \left\{\mathbf{Y} \in \mathbb{R}^N, \dot{V}(\mathbf{Y}) = 0\right\}$ is the trivial solution $\mathbf{Y}(t) = \mathbf{Y}^*$, then $\mathbf{Y}^*$ is globally asymptotically stable.*

Globally asymptotic stability means that all solutions tend to $\mathbf{Y}^*$, not only those starting in a neighbourhood of $\mathbf{Y}^*$.

Example A.4

Let us consider the special case of system (A.4) when $\mathbf{C}$ is positive-definite (all resonators are damped). The condition $\dot{V}(\mathbf{Y}) = -\dot{\mathbf{X}}^T\mathbf{C}\dot{\mathbf{X}} = 0$ implies that $\dot{\mathbf{X}} = 0$. Then, the set S reduces to $S = \{\mathbf{Y} = (\mathbf{X}, 0)\}$. A trajectory contained in S must verify $\dot{\mathbf{X}} \equiv 0$ at any time. Therefore $\ddot{\mathbf{X}} \equiv 0$ and by substitution into eqn (A.4), we get $\mathbf{X} \equiv 0$ and finally $\mathbf{Y} \equiv 0$. Applying the invariance principle, we conclude that system (A.4) is asymptotically stable. In particular, the zeros of Δ are all in the half-plane $\mathrm{Re}(z) < 0$.

Example A.5

Let us examine an example with a semipositive damping matrix $\mathbf{C}$. The system is shown in Fig. A.3. The set of governing equations is

$$\begin{cases} m_1\ddot{X}_1 + c_1\dot{X}_1 + (k_1 + K)X_1 - KX_2 = 0 \\ m_2\ddot{X}_2 + (k_2 + K)X_2 - KX_1 = 0 \end{cases} \tag{A.14}$$

The mass, stiffness, and damping matrices are

$$\mathbf{M} = \begin{pmatrix} m_1 & 0 \\ 0 & m_2 \end{pmatrix} \quad \mathbf{K} = \begin{pmatrix} k_1 + K & -K \\ -K & k_2 + K \end{pmatrix} \quad \mathbf{C} = \begin{pmatrix} c_1 & 0 \\ 0 & 0 \end{pmatrix} \tag{A.15}$$

Matrix **C** is only semipositive because resonator 2 is undamped. The system energy is

$$V = \frac{1}{2}m_1\dot{X}_1^2 + \frac{1}{2}m_2\dot{X}_2^2 + \frac{1}{2}(k_1 + K)X_1^2 - KX_1X_2 + \frac{1}{2}(k_2 + K)X_2^2 \tag{A.16}$$

is positive-definite. The time derivative of the energy is

$$\dot{V} = -c_1\dot{X}_1^2 \tag{A.17}$$

and is seminegative. Hence $S = \left\{Y = (X_1, X_2, \dot{X}_1, \dot{X}_2) \in \mathbb{R}^4 \,/\, \dot{X}_1 = 0\right\}$. If a solution $X_1(t)$, $X_2(t)$ stays in S, then $\dot{X}_1 \equiv 0$ and therefore $\ddot{X}_1 \equiv 0$. The first governing equation gives $(k_1+K)X_1 - KX_2 = 0$. Furthermore, X_1 being constant in time, X_2 is also constant and therefore $\ddot{X}_2 = 0$. The second governing equation gives $-KX_1 + (k_2 + K)X_2 = 0$. In a matrix form this reads $\mathbf{K}X = 0$ and since $\mathbf{K}$ is invertible, $X = 0$. Applying the invariance principle, we conclude that the system is asymptotically stable and therefore all zeros of $\Delta(\lambda)$ are in the half-plane $\mathrm{Re}(z) < 0$.

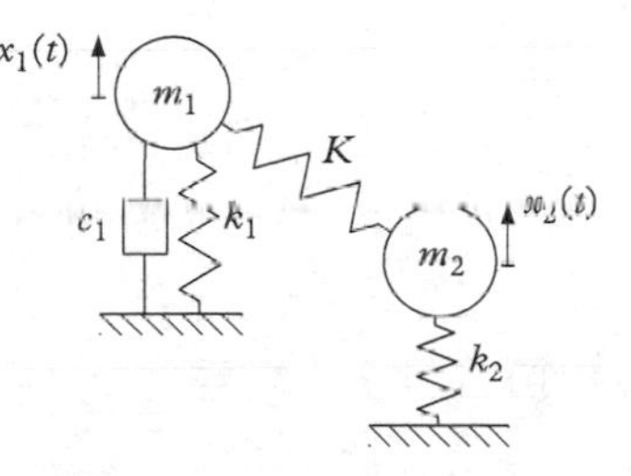

Figure A.3 *Two coupled resonators, one of which is undamped.*

A direct method to check this last result would be to expand Δ,

$$\Delta(\lambda) = \left[m_1\lambda^2 + c_1\lambda + (k_1 + K)\right]\left[m_2\lambda^2 + (k_2 + K)\right] - K^2 \tag{A.18}$$

and to calculate the four complex roots. But this way is evidently more tedious.

In most cases, the stability of a system reduces to localizing the roots of the characteristic polynomial $\Delta(\lambda)$ of **A**. A real polynomial $\Delta(\lambda) \in \mathbb{R}[X]$ is said to be Hurwitz stable if all its roots are located in the left half-plane $\mathrm{Re}(z) < 0$. A first simple property of Hurwitz stable polynomials is the following.

Theorem A.7 *Let $\Delta(\lambda) \in \mathbb{R}[X]$. If $\Delta(\lambda)$ is Hurwitz stable then all its coefficients have the same sign.*

This property is easy to prove by remarking that since $\Delta(\lambda) \in \mathbb{R}[X]$ its complex roots appear in pairs λ_j, $\bar{\lambda}_j$. The factorized form of $\Delta(\lambda)$,

$$\alpha \prod_i (\lambda - \lambda_i) \prod_j \left(\lambda^2 - 2\mathrm{Re}(\lambda_j) + \lambda_j\bar{\lambda}_j\right)$$

where α is the leading coefficient of $\Delta(\lambda)$, shows a product of polynomials whose coefficients are all positive. Developing can only give non-negative coefficients.

The converse is generally false. For instance, the polynomial $\lambda^3+\lambda^2+\lambda+6=(\lambda+2)(\lambda^2-\lambda+3)$ has positive coefficients but two complex roots in the half-plane $\mathrm{Re}(z)>0$.

The Routh–Hurwitz algorithm provides a criterion to recognize a Hurwitz stable polynomial.

Theorem A.8 *Given $a_n\lambda^n+a_{n-1}\lambda^{n-1}+\cdots+a_0$ where a_i are real coefficients, we construct the Routh array,*

$$\begin{array}{llll} a_n & a_{n-2} & a_{n-4} & \cdots \\ a_{n-1} & a_{n-3} & a_{n-5} & \cdots \\ l_{n-2,1} & l_{n-2,2} & l_{n-2,3} & \cdots \\ \vdots & & & \\ l_{0,1} & & & \end{array}$$

completed by zeros on the right. The third and subsequent rows are calculated by the recursive rule,

$$l_{k,i}=\left[l_{k+1,i}l_{k+2,i+1}-l_{k+2,1}l_{k+1,i+1}\right]/l_{k+1,1}$$

If all coefficients in the left column are positive (respectively negative), then $\Delta(\lambda)$ is Hurwitz stable.

We can now revisit the last example.

Example A.6

The characteristic polynomial (A.18), $\Delta(\lambda)=m_1m_2\lambda^4+m_2c_1\lambda^3+[m_1(k_2+K)+m_2(k_1+K)]\lambda^2+c_1(k_2+K)\lambda+k_1k_2+K(k_1+k_2)$ leads to the Routh array,

$$\begin{array}{llll} m_1m_2 & m_1(k_2+K)+m_2(k_1+K) & k_1k_2+K(k_1+k_2) & 0 \\ m_2c_1 & c_1(k_2+K) & 0 & 0 \\ m_2(k_1+K) & k_1k_2+K(k_1+k_2) & 0 & 0 \\ c_1K^2/(k_1+K) & 0 & 0 & 0 \\ k_1k_2+K(k_1+k_2) & 0 & 0 & 0 \end{array}$$

All coefficients of the left column are positive. The polynomial is Routh stable and the system is asymptotically stable.

Appendix B
Residue Theorem

A standard textbook in complex analysis is Lang (1993).

Let $z \to f(z)$ be a complex function on an open set U of the complex plane. We say that f is differentiable or **holomorphic** if the limit

$$\lim_{h \to 0} \frac{f(z+h) - f(z)}{h} \tag{B.1}$$

exists for all $z \subset U$. The limit is noted f'.

Laurent expansion

Let U be an open set and $z_0 \in U$. Let f be a holomorphic function on $U - \{z_0\}$. Then a unique sequence a_n, $n \in \mathbb{Z}$ exists such that

$$f(z) = \sum_{n=-\infty}^{\infty} a_n (z - z_0)^n \tag{B.2}$$

in a non-empty open ball $\mathrm{B}(z_0, R) \subset U$. The series converges absolutely and uniformly

We call a_{-1} the residue of f at z_0 and write

$$a_{-1} = \mathrm{Res}(f, z_0) \tag{B.3}$$

The function f is said to have a pole of order m if and only if $f(z)(z - z_0)^m$ is holomorphic and has no zero at z_0. This is equivalent to saying that the Laurent series has only a finite number m of negative terms.

If f has a pole of order one, then its residue at z_0 is

$$\mathrm{Res}(f, z_0) = \lim_{z \to z_0} f(z)(z - z_0) \tag{B.4}$$

Winding number

Let U be a simply connected open set, z_0 a point in U, and γ a closed path in $U - \{z_0\}$. Then we call the winding number of γ with respect to z_0 the number

$$\mathrm{Wind}(\gamma, z_0) = \frac{1}{2\imath\pi} \int_{\gamma} \frac{\mathrm{d}z}{z - z_0} \tag{B.5}$$

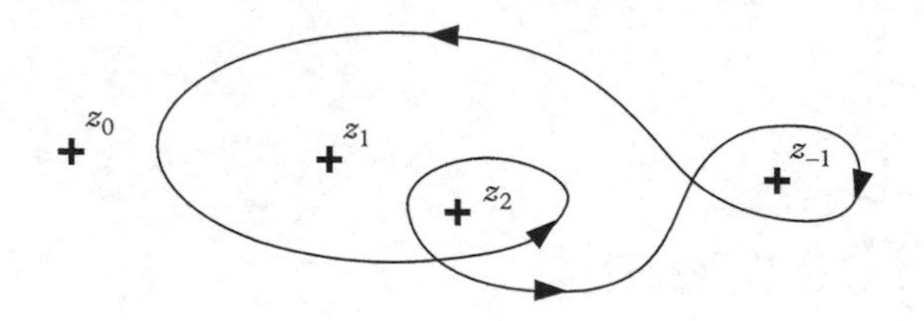

Figure B.1 *Winding number of z_i is i.*

The winding number is a positive or negative integer. Intuitively, the winding number measures the number of times a line starting from z_0 and going to infinity crosses γ with the convention that the count is positive when γ turns counterclockwise and negative otherwise (Fig. B.1).

Residue theorem

Let U be a simply connected set and f a holomorphic on U except at a finite number of points $z_1, \ldots, z_n$. Let γ be a closed path in $U - \{z_1, \ldots, z_n\}$. Then,

$$\int_\gamma f(z)\, dz = 2\imath\pi \sum_{i=1}^{n} \text{Wind}(\gamma, z_i)\text{Res}(z_i, f) \tag{B.6}$$

Jordan's lemma

Let f be a rational function with no real pole. If $\lim_{|z|\to\infty} zf(z) = 0$, then

$$\int_{-\infty}^{\infty} f(x)\, dx = 2\imath\pi \sum_i \text{Res}(z_i, f) \tag{B.7}$$

where the sum runs over all poles z_i of f such that $\text{Im}(z_i) > 0$. The lemma also applies with the factor $-2\imath\pi$ and a sum running over all poles in the half-plane $\text{Im}(z_i) < 0$.

Appendix C
Useful Integrals

The frequency response function of a simple resonator is

$$H_i(\omega) = \frac{1}{m_i\omega_i^2\left[1 + 2\imath\zeta_i\frac{\omega}{\omega_i} - \frac{\omega^2}{\omega_i^2}\right]} \tag{C.1}$$

where $\zeta_i > 0$ is the damping ratio, $\omega_i > 0$ the natural frequency, and $m_i > 0$ the modal mass. This is a complex-valued function defined on $\mathbb{R}$.

The following integral is given in Newland (1975, Appendix 1) and Gradshteyn et al. (2000, Section 3.112, eqn (5)).

$$\int_{-\infty}^{\infty}\left|\frac{b_0 + \imath\omega b_1 - \omega^2 b_2 - \imath\omega^3 b_3}{a_0 + \imath\omega a_1 - \omega^2 a_2 - \imath\omega^3 a_3 + \omega^4 a_4}\right|^2 \mathrm{d}\omega$$
$$= \pi\frac{a_0 b_3^2(a_1a_2 - a_0a_3) + a_0a_1a_4(b_2^2 - 2b_1b_3) + a_0a_3a_4(b_1^2 - 2b_0b_2) + a_4b_0^2(a_2a_3 - a_1a_4)}{a_0a_4(a_1a_2a_3 - a_0a_3^2 - a_1^2a_4)} \tag{C.2}$$

The above integral holds if all roots of the denominator lie in the upper half-plane $\mathrm{Im}(z) > 0$ (stable system).

Calculation of $\int_{-\infty}^{\infty} \omega^p H_i^n H_j^m \,\mathrm{d}\omega$

Equation (C.1) shows that H_i is a rational function with two poles, $\omega_i\alpha_1$ and $\omega_i\alpha_2$,

$$\begin{cases} \alpha_k = \imath\zeta_i \pm \sqrt{1-\zeta_i^2} & \text{when } 0 \le \zeta_i < 1 \\ \alpha_k = \imath\left(\zeta_i \pm \sqrt{\zeta_i^2 - 1}\right) & \text{when } 1 \le \zeta_i \end{cases} \tag{C.3}$$

where $k = 1, 2$ depending on the plus or minus sign. It is clear that $\mathrm{Im}(\alpha_{1,2}) > 0$ and therefore the two poles of H_i are located in the half-plane $\mathrm{Im}(z) > 0$.

Let p, n, and m be three non-negative integers. The function $\omega \mapsto \omega^p H_i^n(\omega) H_j^m(\omega)$ is rational and therefore is holomorphic on $\mathbb{C} - \{z_1, z_2, z_3, z_4\}$ where z_i, $i = 1, \ldots, 4$ are the four poles of H_i and H_j. They are all located in the half-plane $\mathrm{Im}(z) > 0$.

Since H_i has degree −2, the rational function $\omega^p H_i^n H_j^m$ has degree $p - 2(n+m)$. The condition $\lim_{|\omega|\to\infty} \omega^{p+1} H_i^n(\omega) H_j^m(\omega) = 0$ is fulfilled under the assumption $2(n+m) - p \geq 2$. Using Jordan's lemma,

$$\int_{-\infty}^{\infty} \omega^p H_i^n(\omega) H_j^m(\omega)\, \mathrm{d}\omega = 0 \tag{C.4}$$

Calculation of $I(\zeta) = \int_{-\infty}^{\infty} \frac{dz}{(1-z^2)^2 + 4\zeta^2 z^2}$

Let $Q(z)$ be the integrand of I. This is a rational function $Q(z) = 1/P(z^2)$ where P is the polynomial,

$$P(z) = (1-z)^2 + 4\zeta^2 z = z^2 - 2(1-2\zeta^2)z + 1 \tag{C.5}$$

The poles of Q are found by solving the biquadratic equation $P(z^2) = 0$. The discriminant of P is

$$\Delta = 4\zeta^2(\zeta^2 - 1) \tag{C.6}$$

We consider successively the three cases $\zeta < 1$, $\zeta > 1$, and $\zeta = 1$.

When $\zeta < 1$, P has two roots,

$$z_i = (1-2\zeta^2) \pm 2\imath\zeta\sqrt{1-\zeta^2} = (\sqrt{1-\zeta^2} \pm \imath\zeta)^2 \tag{C.7}$$

where $i = 1, 2$. The poles of Q are the square roots of z_i. These are α_0, $\overline{\alpha}_0$, $-\alpha_0$, and $-\overline{\alpha}_0$ where

$$\alpha_0 = \sqrt{1-\zeta^2} + \imath\zeta \tag{C.8}$$

Factorizing Q gives

$$Q(z) = \frac{1}{(z-\alpha_0)(z+\alpha_0)(z-\overline{\alpha}_0)(z+\overline{\alpha}_0)} \tag{C.9}$$

In the complex plane, the two poles located in the half-plane $\mathrm{Im}(z) > 0$ are α_0 and $-\overline{\alpha}_0$. The closed path γ is chosen as shown in Fig. C.1. Then,

$$I = 2\imath\pi\,[\mathrm{Res}(Q, \alpha_0) + \mathrm{Res}(Q, -\overline{\alpha}_0)] \tag{C.10}$$

Since all poles are simple, their residues are calculated by

$$\mathrm{Res}(Q, \alpha_i) = \lim_{z\to\alpha_i} (z-\alpha_i) Q(z) \tag{C.11}$$

where $\alpha_i = \alpha_0, -\overline{\alpha}_0$. One obtains

$$\mathrm{Res}(Q, \alpha_0) = \frac{1}{2\alpha_0 \times 2\imath\mathrm{Im}(\alpha_0) \times 2\mathrm{Re}(\alpha_0)} = \frac{1}{8\imath\zeta\sqrt{1-\zeta^2}\left[\sqrt{1-\zeta^2} + \imath\zeta\right]}$$

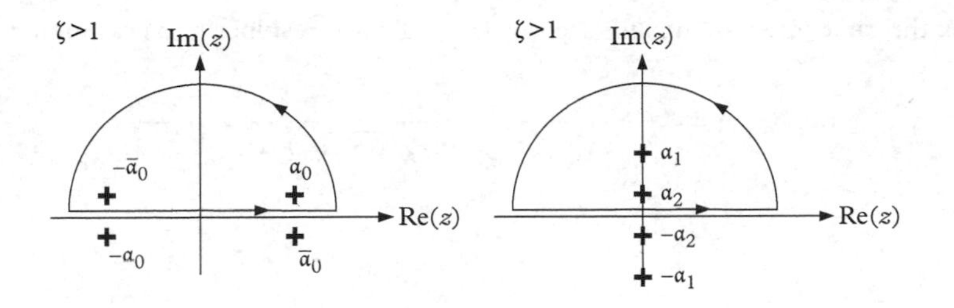

Figure C.1 *Closed path of integration for I.*

and

$$\mathrm{Res}(Q, -\bar{\alpha}_0) = \frac{1}{2\mathrm{Re}(\alpha_0) \times 2\imath\mathrm{Im}(\alpha_0) \times 2\bar{\alpha}_0} = \frac{1}{8\imath\zeta\sqrt{1-\zeta^2}\left[\sqrt{1-\zeta^2} - \imath\zeta\right]}$$

By summing the two above expressions,

$$I = \frac{2\imath\pi}{8\imath\zeta\sqrt{1-\zeta^2}}\left[\frac{1}{\sqrt{1-\zeta^2}+\imath\zeta} + \frac{1}{\sqrt{1-\zeta^2}-\imath\zeta}\right] \tag{C.12}$$

And finally,

$$I = \frac{\pi}{2\zeta} \tag{C.13}$$

When $\zeta > 1$, $\Delta > 0$ and the two roots of P are

$$z_i = (1-2\zeta^2) \pm 2\zeta\sqrt{\zeta^2-1} = -\left(\zeta \pm \sqrt{\zeta^2-1}\right)^2 \tag{C.14}$$

with $i = 1, 2$. The four poles of Q are then $\alpha_1, -\alpha_1, \alpha_2$, and $-\alpha_2$ where

$$\alpha_1 = \imath\left(\zeta + \sqrt{\zeta^2-1}\right) \tag{C.15}$$

$$\alpha_2 = \imath\left(\zeta - \sqrt{\zeta^2-1}\right) \tag{C.16}$$

and Q becomes

$$Q(z) = \frac{1}{(z-\alpha_1)(z+\alpha_1)(z-\alpha_2)(z+\alpha_2)} \tag{C.17}$$

By choosing the same integration path (see Fig. C.1), the two residues to be computed are

$$\mathrm{Res}(Q,\alpha_1) = \frac{1}{2\alpha_1(\alpha_1^2-\alpha_2^2)} = \frac{-1}{8\imath\zeta\sqrt{1-\zeta^2}\left[\zeta+\sqrt{\zeta^2-1}\right]}$$

and

$$\mathrm{Res}(Q,\alpha_2) = \frac{1}{2\alpha_2(\alpha_2^2-\alpha_1^2)} = \frac{1}{8\imath\zeta\sqrt{1-\zeta^2}\left[\zeta-\sqrt{\zeta^2-1}\right]}$$

Hence,

$$I = \frac{2\imath\pi}{8\imath\zeta\sqrt{1-\zeta^2}}\left[\frac{1}{\zeta-\sqrt{\zeta^2-1}} - \frac{1}{\zeta+\sqrt{\zeta^2-1}}\right] \tag{C.18}$$

And finally,

$$I = \frac{\pi}{2\zeta} \tag{C.19}$$

The last case $\zeta = 1$ is straightforward. Since $Q(z) = 1/(1+z^2)^2$, the change of variable $z = \tan u$, $\mathrm{d}z = (1 + \tan^2 u)\,\mathrm{d}u$ gives

$$I = \int_{-\pi/2}^{\pi/2} \frac{(1+\tan^2 u)}{(1+\tan^2 u)^2}\,\mathrm{d}u = \int_{-\pi/2}^{\pi/2} \cos^2 u\,\mathrm{d}u = \frac{\pi}{2} \tag{C.20}$$

The same result is therefore obtained for all cases,

$$\int_{-\infty}^{\infty} \frac{\mathrm{d}z}{(1-z^2)^2 + 4\zeta^2 z^2} = \frac{\pi}{2\zeta} \tag{C.21}$$

for all $\zeta > 0$. This integral is also given in Gradshteyn et al. (2000, Section 3.112, eqn (3)).

Application to $\int |H_i|^2\,\mathrm{d}\omega$

Substituting the expression of $H_i(\omega)$ given by eqn (C.1),

$$\int_{-\infty}^{\infty} |H_i|^2(\omega)\,\mathrm{d}\omega = \frac{1}{m_i^2\omega_i^4}\int_{-\infty}^{\infty} \frac{\mathrm{d}\omega}{\left|1 + 2\imath\zeta_i\frac{\omega}{\omega_i} - \frac{\omega^2}{\omega_i^2}\right|^2} \tag{C.22}$$

By the change of variable $z = \omega/\omega_i$,

$$\int_{-\infty}^{\infty} |H_i|^2(\omega)\,\mathrm{d}\omega = \frac{1}{m_i^2\omega_i^3}\int_{-\infty}^{\infty} \frac{\mathrm{d}z}{(1-z^2)^2 + 4\zeta^2 z^2} = \frac{I(\zeta)}{m^2\omega_i^3} \tag{C.23}$$

And finally,

$$\int_{-\infty}^{\infty} |H_i|^2(\omega)\, \mathrm{d}\omega = \frac{\pi}{2\zeta_i m_i^2 \omega_i^3} \tag{C.24}$$

Application to $\int \omega_i^2 |H_i|^2 \,\mathrm{d}\omega$

Again with eqn (C.1) and the change of variable $z = \omega/\omega_i$, one obtains

$$\int_{-\infty}^{\infty} \omega^2 |H|^2(\omega)\, \mathrm{d}\omega = \frac{2}{m^2 \omega_0} \int_0^{\infty} \frac{z^2\, \mathrm{d}z}{(1-z^2)^2 + 4\zeta^2 z^2} \tag{C.25}$$

The factor 2 stems from the fact that the integrand is odd. Factorizing z^4 and performing the change of variable $u = 1/z$ gives

$$\int_{-\infty}^{\infty} \omega^2 |H_i|^2(\omega)\, \mathrm{d}\omega = \frac{2}{m^2 \omega_i} \int_0^{\infty} \frac{\mathrm{d}z/z^2}{(1/z^2 - 1)^2 + 4\zeta^2/z^2} \tag{C.26}$$

$$= \frac{2}{m_i^2 \omega_i} \int_0^{\infty} \frac{\mathrm{d}u}{(u^2 - 1)^2 + 4\zeta^2 u^2} \tag{C.27}$$

$$= \frac{1}{m_i^2 \omega_i} I(\zeta) \tag{C.28}$$

And finally,

$$\int_{-\infty}^{\infty} \omega^2 |H_i|^2(\omega)\, \mathrm{d}\omega = \frac{\pi}{2\zeta_i m_i^2 \omega_i} \tag{C.29}$$

Application to $\int \imath\omega H_i \,\mathrm{d}\omega$

Multiplying the numerator and denominator of $\imath\omega H_i(\omega)$ by H_i^* gives

$$\int \imath\omega H_i(\omega)\, \mathrm{d}\omega = \int \imath\omega m_i \omega_i^2 \left(1 - 2\imath\zeta_i \frac{\omega}{\omega_i} - \frac{\omega^2}{\omega_i^2}\right) |H_i|^2(\omega)\, \mathrm{d}\omega \tag{C.30}$$

where the limits of integration are arbitrary and finite. By the change of variable $z = \omega/\omega_i$,

$$\int \imath\omega H_i(\omega)\, \mathrm{d}\omega = \frac{2\zeta_i}{m_i} \int \frac{z^2}{\left(1-z^2\right)^2 + 4\zeta^2 z^2}\, \mathrm{d}z + \frac{\imath}{m_i} \int \frac{z(1-z^2)}{\left(1-z^2\right)^2 + 4\zeta^2 z^2}\, \mathrm{d}z \tag{C.31}$$

The first integral of the right-hand side is convergent when the limits become infinite and its value is $I(\zeta)$. Therefore,

$$\int_{-\infty}^{\infty} \mathrm{Re}[\imath\omega H_i(\omega)]\, \mathrm{d}\omega = \frac{\pi}{m_i} \tag{C.32}$$

But the second integral is clearly divergent at infinite. However, the integrand is an odd function, so

$$\int_{-X}^{X} \mathrm{Im}[\imath\omega H_i(\omega)]\,\mathrm{d}\omega = 0 \qquad \text{for all } X \tag{C.33}$$

We shall write

$$\int_{-\infty}^{\infty} \imath\omega H_i(\omega)\,\mathrm{d}\omega = \frac{\pi}{m_i} \tag{C.34}$$

which must be understood in the sense of principal values.

Calculation of $J_n(r, \zeta_1, \zeta_2) = \int_{-\infty}^{\infty} \frac{\imath z^{2n+1}\,dz}{(1+2\imath\zeta_2 rz - r^2z^2)((1-z^2)^2+4\zeta_1^2 z^2)}$

The integer n has values 0 or 1. Let $Q(z)$ be the integrand of $\mathcal{J}_n$. The function $z \mapsto Q(z)$ is rational and may be written as

$$Q(z) = \frac{\imath z^{2n+1}}{P_{\zeta_2}(rz)P_{\zeta_1}(z)P_{-\zeta_1}(z)} \tag{C.35}$$

where

$$P_{\zeta_i}(z) = 1 + 2\imath\zeta_i z - z^2 \tag{C.36}$$

Since P_{ζ_i} may be factorized as

$$P_{\zeta_i}(z) = -(z-\alpha_i)(z-\beta_i) \tag{C.37}$$

where

$$\begin{cases} \alpha_i = \imath\zeta_i + \sqrt{1-\zeta_i^2}, & \beta_i = \imath\zeta_i - \sqrt{1-\zeta_i^2} \quad \text{if } |\zeta_i| < 1 \\ \alpha_i = \imath(\zeta_i + \sqrt{\zeta_i^2 - 1}), & \beta_i = \imath(\zeta_i - \sqrt{\zeta_i^2 - 1}) \quad \text{otherwise} \end{cases} \tag{C.38}$$

we get

$$Q(z) = \frac{-\imath z^{2n+1}}{(rz-\alpha_2)(rz-\beta_2)(z-\alpha_1)(z-\beta_1)(z-\overline{\alpha}_1)(z-\overline{\beta}_1)} \tag{C.39}$$

The six poles of Q are therefore α_1, β_1, α_2, β_2, $\overline{\alpha}_1$, and $\overline{\beta}_1$. The poles are plotted in Fig. C.2. In all cases, only two poles are in the half-plane $\mathrm{Im}(z) < 0$, while the four others are in $\mathrm{Im}(z) > 0$. The better choice of closed path to minimize the number of residues to be calculated is to turn clockwise in the lower half-plane as shown in Fig. C.2. Application of Jordan's lemma then leads to

$$\mathcal{J}_n = -2\imath\pi\left[\mathrm{Res}(Q, \overline{\alpha}_1) + \mathrm{Res}(Q, \overline{\beta}_1)\right] \tag{C.40}$$

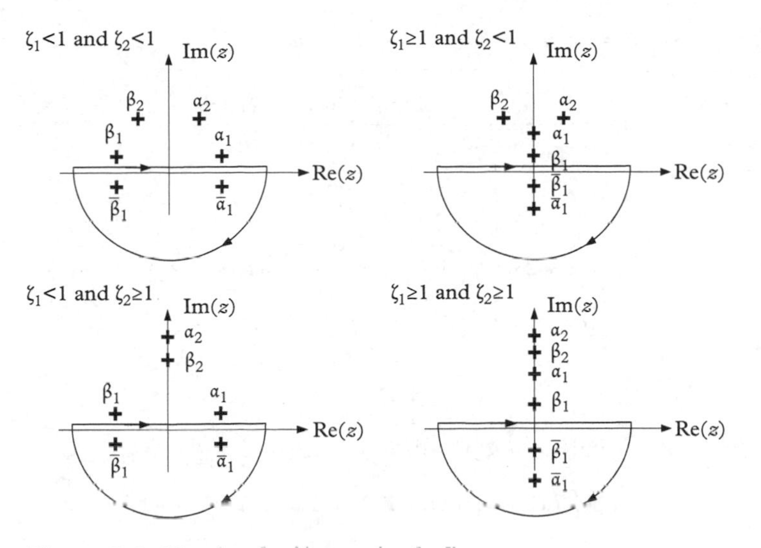

Figure C.2 *Closed path of integration for $\mathcal{J}_n$.*

where

$$\mathrm{Res}(Q,\overline{\alpha}_1) = \frac{-\imath\overline{\alpha}_1^{2n+1}}{(r\overline{\alpha}_1-\alpha_2)(r\overline{\alpha}_1-\beta_2)(\overline{\alpha}_1-\alpha_1)(\overline{\alpha}_1-\beta_1)(\overline{\alpha}_1-\overline{\beta}_1)} \quad \text{(C.41)}$$

and

$$\mathrm{Res}(Q,\overline{\beta}_1) = \frac{-\imath\overline{\beta}_1^{2n+1}}{(r\overline{\beta}_1-\alpha_2)(r\overline{\beta}_1-\beta_2)(\overline{\beta}_1-\alpha_1)(\overline{\beta}_1-\beta_1)(\overline{\beta}_1-\overline{\alpha}_1)} \quad \text{(C.42)}$$

From now on, we consider the case $0 < \zeta_1 < 1$ and $0 \leq \zeta_2 < 1$. Then, from eqn (C.38), $\beta_1 = -\overline{\alpha}_1$ and $\beta_2 = -\overline{\alpha}_2$. And

$$\mathrm{Res}(Q,\overline{\alpha}_1) = \frac{-\imath\overline{\alpha}_1^{2n+1}}{(r\overline{\alpha}_1-\alpha_2)(r\overline{\alpha}_1+\overline{\alpha}_2)(\overline{\alpha}_1-\alpha_1)(\overline{\alpha}_1+\overline{\alpha}_1)(\overline{\alpha}_1+\alpha_1)} \quad \text{(C.43)}$$

and

$$\mathrm{Res}(Q,-\alpha_1) = \frac{\imath\alpha_1^{2n+1}}{(-r\alpha_1-\alpha_2)(-r\alpha_1+\overline{\alpha}_2)(-\alpha_1-\alpha_1)(-\alpha_1+\overline{\alpha}_1)(-\alpha_1-\overline{\alpha}_1)} \quad \text{(C.44)}$$

By adding $\mathrm{Res}(Q,\overline{\alpha}_1)$ and $\mathrm{Res}(Q,-\alpha_1)$ and applying (C.40),

$$\mathcal{J}_n = \left[\frac{\imath\overline{\alpha}_1^{2n}}{(r\overline{\alpha}_1-\alpha_2)(r\overline{\alpha}_1+\overline{\alpha}_2)} + \frac{-\imath\alpha_1^{2n}}{(r\alpha_1+\alpha_2)(r\alpha_1-\overline{\alpha}_2)}\right] \times \frac{-2\imath\pi}{8\imath\mathrm{Im}(\alpha_1)\mathrm{Re}(\alpha_1)} \quad \text{(C.45)}$$

As, $\mathrm{Re}(\alpha_1) = \sqrt{1-\zeta_1^2}$ and $\mathrm{Im}(\alpha_1) = \zeta_1$

$$\mathcal{J}_n = \frac{\pi}{4\zeta_1\sqrt{1-\zeta_1^2}} \times \frac{2\mathrm{Im}[\bar{\alpha}_1^{2n}(r\alpha_1+\alpha_2)(r\alpha_1-\bar{\alpha}_2)]}{|r\bar{\alpha}_1-\alpha_2|^2\,|r\alpha_1+\alpha_2|^2} \tag{C.46}$$

Since

$$r\alpha_1+\alpha_2 = r\sqrt{1-\zeta_1^2}+\sqrt{1-\zeta_2^2}+\imath(r\zeta_1+\zeta_2) \tag{C.47}$$

$$r\alpha_1-\bar{\alpha}_2 = r\sqrt{1-\zeta_1^2}-\sqrt{1-\zeta_2^2}+\imath(r\zeta_1+\zeta_2) \tag{C.48}$$

it yields

$$\mathrm{Im}[(r\alpha_1+\alpha_2)(r\alpha_1-\bar{\alpha}_2)] = 2r(r\zeta_1+\zeta_2)\sqrt{1-\zeta_1^2} \tag{C.49}$$

$$\mathrm{Im}[\bar{\alpha}_1^2(r\alpha_1+\alpha_2)(r\alpha_1-\bar{\alpha}_2)] = 2(\zeta_1+r\zeta_2)\sqrt{1-\zeta_1^2} \tag{C.50}$$

Finally,

$$\mathcal{J}_0 = \frac{\pi}{\zeta_1} \times \frac{r(r\zeta_1+\zeta_2)}{\left[(r\sqrt{1-\zeta_1^2}+\sqrt{1-\zeta_2^2})^2+(r\zeta_1+\zeta_2)^2\right]\left[(r\sqrt{1-\zeta_1^2}-\sqrt{1-\zeta_2^2})^2+(r\zeta_1+\zeta_2)^2\right]} \tag{C.51}$$

and

$$\mathcal{J}_1 = \frac{\pi}{\zeta_1} \times \frac{\zeta_1+r\zeta_2}{\left[(r\sqrt{1-\zeta_1^2}+\sqrt{1-\zeta_2^2})^2+(r\zeta_1+\zeta_2)^2\right]\left[(r\sqrt{1-\zeta_1^2}-\sqrt{1-\zeta_2^2})^2+(r\zeta_1+\zeta_2)^2\right]} \tag{C.52}$$

Application to $\int_{-\infty}^{\infty} \imath\omega \mathrm{H}_j |H_i|^2 \,\mathrm{d}\omega$

$$\int_{-\infty}^{\infty} \imath\omega H_j |H_i|^2 \,\mathrm{d}\omega = \frac{1}{m_j\omega_j^2 m_i^2\omega_i^4}\int_{-\infty}^{\infty} \frac{\imath\omega\,\mathrm{d}\omega}{\left(1+2\imath\zeta_j\frac{\omega}{\omega_j}-\frac{\omega^2}{\omega_j^2}\right)\left|1+2\imath\zeta_i\frac{\omega}{\omega_i}-\frac{\omega^2}{\omega_i^2}\right|^2} \tag{C.53}$$

By introducing the ratio $r = \omega_i/\omega_j$ and performing the change of variable $z = \omega/\omega_i$, it yields

$$\int_{-\infty}^{\infty} \imath\omega H_j |H_i|^2 \,\mathrm{d}\omega = \frac{1}{m_j\omega_j^2 m_i^2\omega_i^2}\int_{-\infty}^{\infty} \frac{\imath z\,\mathrm{d}z}{(1+2\imath\zeta_j rz-r^2z^2)\,|1+2\imath\zeta_i z-z^2|^2} \tag{C.54}$$

We recognize the expression of $\mathcal{J}_0(r,\zeta_i,\zeta_j)$. Therefore by substituting eqn (C.51), it leads to

$$\int_{-\infty}^{\infty} \imath\omega H_j |H_i|^2 \,\mathrm{d}\omega = \frac{1}{m_j\omega_j^2 m_i^2\omega_i^2} \times \frac{\pi}{\zeta_i} \times \frac{r(r\zeta_i+\zeta_j)}{\left[(r\sqrt{1-\zeta_i^2}+\sqrt{1-\zeta_j^2})^2+(r\zeta_i+\zeta_j)^2\right]\left[(r\sqrt{1-\zeta_i^2}-\sqrt{1-\zeta_j^2})^2+(r\zeta_i+\zeta_j)^2\right]} \tag{C.55}$$

After simplification,

$$\int_{-\infty}^{\infty} \imath\omega H_j(\omega)\,|H_i|^2(\omega)\,\mathrm{d}\omega = \frac{\pi}{m_j m_i^2 \omega_i \zeta_i} \times \frac{\omega_i\zeta_i+\omega_j\zeta_j}{\left[\left(\omega_i\sqrt{1-\zeta_i^2}+\omega_j\sqrt{1-\zeta_j^2}\right)^2+(\omega_i\zeta_i+\omega_j\zeta_j)^2\right]\left[\left(\omega_i\sqrt{1-\zeta_i^2}-\omega_j\sqrt{1-\zeta_j^2}\right)^2+(\omega_i\zeta_i+\omega_j\zeta_j)^2\right]} \tag{C.56}$$

This result is valid when $\zeta_i,\ \zeta_j < 1$.

Application to $\int_{-\infty}^{\infty} \imath\omega^3 H_j\,|H_i|^2\,\mathrm{d}\omega$

The calculation is similar. With $r = \omega_i/\omega_j$, we get

$$\int_{-\infty}^{\infty} \imath\omega^3 H_j\,|H_i|^2\,\mathrm{d}\omega = \frac{1}{m_j\omega_j^2 m_i^2}\int_{-\infty}^{\infty}\frac{\imath z^3\,\mathrm{d}z}{(1+2\imath\zeta_j rz - r^2z^2)\,|1+2\imath\zeta_i z - z^2|^2} \tag{C.57}$$

where the integral of the right-hand side is $\mathcal{I}_1(r,\zeta_i,\zeta_j)$. By substituting eqn (C.52),

$$\int_{-\infty}^{\infty} \imath\omega^3 H_j\,|H_i|^2\,\mathrm{d}\omega = \frac{1}{m_j\omega_j^2 m_i^2}\times\frac{\pi}{\zeta_i} \times \frac{\zeta_i+r\zeta_j}{\left[(r\sqrt{1-\zeta_i^2}+\sqrt{1-\zeta_j^2})^2+(r\zeta_i+\zeta_j)^2\right]\left[(r\sqrt{1-\zeta_i^2}-\sqrt{1-\zeta_j^2})^2+(r\zeta_i+\zeta_j)^2\right]} \tag{C.58}$$

After simplification,

$$\int_{-\infty}^{\infty} \imath\omega^3 H_j(\omega)\,|H_i|^2(\omega)\,\mathrm{d}\omega = \frac{\pi\omega_j}{m_j m_i^2 \zeta_i} \times \frac{\omega_i\zeta_j+\omega_j\zeta_i}{\left[(\omega_i\sqrt{1-\zeta_i^2}+\omega_j\sqrt{1-\zeta_j^2})^2+(\omega_i\zeta_i+\omega_j\zeta_j)^2\right]\left[(\omega_i\sqrt{1-\zeta_i^2}-\omega_j\sqrt{1-\zeta_j^2})^2+(\omega_i\zeta_i+\omega_j\zeta_j)^2\right]} \tag{C.59}$$

This result is valid when $\zeta_i,\ \zeta_j < 1$.

Application to $\int_{-\infty}^{\infty} \imath\omega^3 H_j\,|H_i|^2\,\mathrm{d}\omega$ (other method)

Again, by setting $r = \omega_i/\omega_j$ and performing the change of variable $z = \omega/\omega_i$,

$$\int_{-\infty}^{\infty} \imath\omega^3 H_j\,|H_i|^2\,\mathrm{d}\omega = \frac{1}{m_j\omega_j^2 m_i^2}\int_{-\infty}^{\infty}\frac{\imath z^3\,\mathrm{d}z}{(1+2\imath\zeta_j rz - r^2z^2)\,|1+2\imath\zeta_i z - z^2|^2} \tag{C.60}$$

$$= \frac{2}{m_j\omega_j^2 m_i^2}\,\mathrm{Re}\int_{0}^{\infty}\frac{\imath z^3\,\mathrm{d}z}{(1+2\imath\zeta_j rz - r^2z^2)\,|1+2\imath\zeta_i z - z^2|^2} \tag{C.61}$$

where the last equality stems from $Q(-z) = \overline{Q}(z)$ where Q is the integrand. By the change of variable $u = 1/z$,

$$\int_{-\infty}^{\infty} \imath\omega^3 H_j |H_i|^2 \, d\omega = \frac{2}{m_j\omega_j^2 m_i^2} \text{Re} \int_0^{\infty} \frac{\imath \times 1/u^3 \times du/u^2}{\left(1 + 2\imath\zeta_j \frac{r}{u} - \frac{r^2}{u^2}\right) \left|1 + 2\imath\zeta_i \frac{1}{u} - \frac{1}{u^2}\right|^2} \quad (C.62)$$

$$= \frac{2}{m_j\omega_j^2 m_i^2 r^2} \text{Re} \int_0^{\infty} \frac{\imath u \, du}{\left(\frac{u^2}{r^2} + 2\imath\zeta_j \frac{u}{r} - 1\right) |1 + 2\imath\zeta_i u - u^2|^2} \quad (C.63)$$

$$= \frac{1}{m_j\omega_j^2 m_i^2 r^2} \int_{-\infty}^{\infty} \frac{-\imath u \, du}{\left(1 - 2\imath\zeta_j \frac{u}{r} - \frac{u^2}{r^2}\right) |1 + 2\imath\zeta_i u - u^2|^2} \quad (C.64)$$

The integral $\overline{\mathcal{J}}_0(1/r, \zeta_i, \zeta_j)$ may be recognized,

$$\int_{-\infty}^{\infty} \imath\omega^3 H_j |H_i|^2 \, d\omega = \frac{1}{m_j\omega_i^2 m_i^2} \overline{\mathcal{J}}_0\left(\frac{1}{r}, \zeta_i, \zeta_j\right) \quad (C.65)$$

Expanding $\overline{\mathcal{J}}_0(1/r, \zeta_i, \zeta_j)$ in the above equality reestablishes eqn (C.59).

Calculation of $K_n = \int_{-\infty}^{\infty} \frac{z^{2n} \, dz}{((1-\omega_1^2 z^2)^2 + 4\zeta_1^2\omega_1^2 z^2)((1-\omega_2^2 z^2)^2 + 4\zeta_2^2\omega_2^2 z^2)}$

The integer n has values 0, 1, or 2. Let $Q(z)$ be the integrand. This is a rational function $Q(z) = |z^n/P(z)|^2$ where P is the polynomial,

$$P(z) = (1 + 2\imath\zeta_1\omega_1 z - \omega_1^2 z^2)(1 + 2\imath\zeta_2\omega_2 z - \omega_2^2 z^2) \quad (C.66)$$

Factorizing P gives

$$P(z) = (\omega_1 z - \alpha_1)(\omega_1 z - \beta_1)(\omega_2 z - \alpha_2)(\omega_2 z - \beta_2) \quad (C.67)$$

where

$$\begin{cases} \alpha_i = \imath\zeta_i + \sqrt{1 - \zeta_i^2}, & \beta_i = \imath\zeta_i - \sqrt{1 - \zeta_i^2} \quad \text{if } |\zeta_i| < 1 \\ \alpha_i = \imath(\zeta_i + \sqrt{\zeta_i^2 - 1}), & \beta_i = \imath(\zeta_i - \sqrt{\zeta_i^2 - 1}) \quad \text{otherwise} \end{cases} \quad (C.68)$$

Thus, all zeros of P are localized in the upper complex half-plane $\text{Re}(z) > 0$. By developing the polynomial P, we get

$$P(z) = 1 + 2\imath z(\zeta_1\omega_1 + \zeta_2\omega_2) - z^2(\omega_1^2 + 4\zeta_1\zeta_2\omega_1\omega_2 + \omega_2^2) - 2\imath z^3\omega_1\omega_2(\zeta_1\omega_2 + \zeta_2\omega_1) + z^4\omega_1^4\omega_2^4$$

We can therefore apply eqn (C.2). The coefficients a_i and b_i take the values

$$\begin{aligned}
a_0 &= 1 & b_0 &= 1 & b'_0 &= 0 & b''_0 &= 0 \\
a_1 &= 2(\zeta_1\omega_1 + \zeta_2\omega_2) & b_1 &= 0 & b'_1 &= 1 & b''_1 &= 0 \\
a_2 &= \omega_1^2 + 4\zeta_1\zeta_2\omega_1\omega_2 + \omega_2^2 & b_2 &= 0 & b'_2 &= 0 & b''_2 &= 1 \\
a_3 &= 2\omega_1\omega_2(\zeta_1\omega_2 + \zeta_2\omega_1) & b_3 &= 0 & b'_3 &= 0 & b''_3 &= 0 \\
a_4 &= \omega_1^2\omega_2^2
\end{aligned} \tag{C.69}$$

for where the three columns of b_i apply respectively to $n = 0$, $n = 1$, and $n = 2$. Now, substituting these values successively into eqn (C.2) gives

$$K_0 = \pi \frac{a_2 a_3 - a_1 a_4}{a_0(a_1 a_2 a_3 - a_0 a_3^2 - a_1^2 a_4)} \tag{C.70}$$

$$K_1 = \pi \frac{a_0 a_3}{a_0(a_1 a_2 a_3 - a_0 a_3^2 - a_1^2 a_4)} \tag{C.71}$$

$$K_2 = \pi \frac{a_0 a_1}{a_0(a_1 a_2 a_3 - a_0 a_3^2 - a_1^2 a_4)} \tag{C.72}$$

Now the denominator is

$$a_0(a_1 a_2 a_3 - a_0 a_3^2 - a_1^2 a_4) = 4\omega_1\omega_2 D \tag{C.73}$$

where

$$D = (\zeta_1\omega_1 + \zeta_2\omega_2)(\omega_1^2 + 4\zeta_1\zeta_2\omega_1\omega_2 + \omega_2^2)(\zeta_1\omega_2 + \zeta_2\omega_1) - \omega_1\omega_2\left[(\zeta_1\omega_2 + \zeta_2\omega_1)^2 + (\zeta_1\omega_1 + \zeta_2\omega_2)^2\right] \tag{C.74}$$

By separating $4\zeta_1\zeta_2\omega_1\omega_2(\zeta_1\omega_1 + \zeta_2\omega_2)(\zeta_1\omega_2 + \zeta_2\omega_1)$ from the above and developing and simplifying the other terms, we find $\zeta_1\zeta_2(\omega_1^2 - \omega_2^2)^2$. So,

$$a_0(a_1 a_2 a_3 - a_0 a_3^2 - a_1^2 a_4) = 4\omega_1\omega_2\zeta_1\zeta_2\left[(\omega_1^2 - \omega_2^2)^2 + 4\omega_1\omega_2(\zeta_1\omega_1 + \zeta_2\omega_2)(\zeta_1\omega_2 + \zeta_2\omega_1)\right] \tag{C.75}$$

Furthermore,

$$\begin{aligned}
a_2 a_3 - a_1 a_4 &= 2\left[(\omega_1^2 + 4\zeta_1\zeta_2\omega_1\omega_2 + \omega_2^2)\omega_1\omega_2(\zeta_1\omega_2 + \zeta_2\omega_1) - (\zeta_1\omega_1 + \zeta_2\omega_2)\omega_1^2\omega_2^2\right] \\
&= 2\omega_1\omega_2\left[\zeta_1\omega_2^3 + 4\zeta_1\zeta_2\omega_1\omega_2(\zeta_1\omega_2 + \zeta_2\omega_1) + \zeta_2\omega_1^3\right] && \text{(C.76)} \\
a_0 a_3 &= 2\omega_1\omega_2\left[\zeta_1\omega_2 + \zeta_2\omega_1\right] && \text{(C.77)} \\
a_0 a_1 &= 2\left(\zeta_1\omega_1 + \zeta_2\omega_2\right) && \text{(C.78)}
\end{aligned}$$

Finally,

$$K_0 = \frac{\pi}{2\zeta_1\zeta_2} \times \frac{\zeta_1\omega_2^3 + 4\zeta_1\zeta_2\omega_1\omega_2(\zeta_1\omega_2 + \zeta_2\omega_1) + \zeta_2\omega_1^3}{(\omega_1^2 - \omega_2^2)^2 + 4\omega_1\omega_2(\zeta_1\omega_1 + \zeta_2\omega_2)(\zeta_1\omega_2 + \zeta_2\omega_1)} \tag{C.79}$$

$$K_1 = \frac{\pi}{2\zeta_1\zeta_2} \times \frac{\zeta_1\omega_2 + \zeta_2\omega_1}{(\omega_1^2 - \omega_2^2)^2 + 4\omega_1\omega_2(\zeta_1\omega_1 + \zeta_2\omega_2)(\zeta_1\omega_2 + \zeta_2\omega_1)} \tag{C.80}$$

$$K_2 = \frac{\pi}{2\zeta_1\zeta_2} \times \frac{\zeta_1\omega_1 + \zeta_2\omega_2}{\omega_1\omega_2\left[(\omega_1^2 - \omega_2^2)^2 + 4\omega_1\omega_2(\zeta_1\omega_1 + \zeta_2\omega_2)(\zeta_1\omega_2 + \zeta_2\omega_1)\right]} \tag{C.81}$$

Application to $\int_{-\infty}^{\infty} \omega^{2(n+1)} |H_jH_i|^2 \, d\omega$

Since $\omega \mapsto \omega^{2(n+1)} |H_j(\omega)H_i(\omega)|^2$ is an even function, the integration bounds may be limited to 0 and ∞. The change of variable $z = 1/\omega$ then leads to

$$\int_{-\infty}^{\infty} \omega^{2(n+1)} |H_jH_i|^2 \, d\omega = 2\int_0^{\infty} z^{-2(n+1)} \left| H_i\left(\frac{1}{z}\right) H_j\left(\frac{1}{z}\right)\right|^2 \frac{dz}{z^2} \tag{C.82}$$

After developing,

$$\int_{-\infty}^{\infty} \omega^{2(n+1)} |H_jH_i|^2 \, d\omega = \frac{2}{m_i^2 m_j^2} \int_0^{\infty} \frac{z^{4-2n} \, dz}{((1-\omega_i^2 z^2)^2 + 4\zeta_i^2\omega_i^2 z^2)((1-\omega_j^2 z^2)^2 + 4\zeta_j^2\omega_j^2 z^2)} \tag{C.83}$$

One finally recognizes the integral K_{2-n},

$$\int_{-\infty}^{\infty} \omega^{2(n+1)} |H_jH_i|^2 \, d\omega = \frac{K_{2-n}(\omega_i, \omega_j, \zeta_i, \zeta_j)}{m_i^2 m_j^2} \tag{C.84}$$

Or, after substituting the values of K_{2-n},

$$\int_{-\infty}^{\infty} \omega^2 |H_jH_i|^2 \, d\omega = \frac{\pi}{2\zeta_i\zeta_j m_i^2 m_j^2} \times \frac{\zeta_i\omega_i + \zeta_j\omega_j}{\omega_i\omega_j\left[(\omega_i^2 - \omega_j^2)^2 + 4\omega_i\omega_j(\zeta_i\omega_i + \zeta_j\omega_j)(\zeta_i\omega_j + \zeta_j\omega_i)\right]} \tag{C.85}$$

$$\int_{-\infty}^{\infty} \omega^4 |H_jH_i|^2 \, d\omega = \frac{\pi}{2\zeta_i\zeta_j m_i^2 m_j^2} \times \frac{\zeta_i\omega_j + \zeta_j\omega_i}{(\omega_i^2 - \omega_j^2)^2 + 4\omega_i\omega_j(\zeta_i\omega_i + \zeta_j\omega_j)(\zeta_i\omega_j + \zeta_j\omega_i)} \tag{C.86}$$

$$\int_{-\infty}^{\infty} \omega^6 |H_jH_i|^2 \, d\omega = \frac{\pi}{2\zeta_i\zeta_j m_i^2 m_j^2} \times \frac{\zeta_i\omega_j^3 + 4\zeta_i\zeta_j\omega_i\omega_j(\zeta_i\omega_j + \zeta_j\omega_i) + \zeta_j\omega_i^3}{(\omega_i^2 - \omega_j^2)^2 + 4\omega_i\omega_j(\zeta_i\omega_i + \zeta_j\omega_j)(\zeta_i\omega_j + \zeta_j\omega_i)} \tag{C.87}$$

These integrals are valid for any non-negative value of ζ_i and ζ_j if $\zeta_i^2 + \zeta_j^2 \neq 0$.

Calculation of $L(\alpha, \beta) = \int_{-\infty}^{\infty} e^{-\alpha^2(x+\beta)^2} \, dx$

Let α and β be two complex numbers. The convergence of the integral is ensured by the condition $\mathrm{Re}\left(\alpha^2\right) > 0$.

Let us fix α. The function $\beta \mapsto I(\alpha,\beta)$ is holomorphic on $\mathbb{C}$ and its derivative is

$$\begin{aligned}\frac{\partial}{\partial\beta}L(\alpha,\beta) &= \int_{-\infty}^{\infty}\frac{\partial}{\partial\beta}\left[\mathrm{e}^{-\alpha^2(x+\beta)^2}\right]\mathrm{d}x\\ &= \int_{-\infty}^{\infty}-\alpha^2\,(x+\beta)\,\mathrm{e}^{-\alpha^2(x+\beta)^2}\,\mathrm{d}x\\ &= \left[\mathrm{e}^{-\alpha^2(x+\beta)^2}\right]_{x=-\infty}^{x=+\infty} = 0\end{aligned}$$

Therefore $\beta \mapsto L(\alpha,\beta)$ is constant on $\mathbb{C}$. Thus,

$$L(\alpha,\beta) = L(\alpha,0) \tag{C.88}$$

Let us calculate $L(\alpha,0)$ by squaring it:

$$\begin{aligned}\left(\int_{-\infty}^{\infty}\mathrm{e}^{-\alpha^2x^2}\,\mathrm{d}x\right)^2 &= \int_{-\infty}^{\infty}\int_{-\infty}^{\infty}\mathrm{e}^{-\alpha^2(x^2+y^2)}\,\mathrm{d}x\mathrm{d}y\\ &= \int_0^{2\pi}\mathrm{d}\varphi\int_0^{\infty}\mathrm{e}^{-\alpha^2r^2}\,r\,\mathrm{d}r\\ &= 2\pi\left[-\frac{\mathrm{e}^{-\alpha^2r^2}}{2\alpha^2}\right]_{r=0}^{r=+\infty} = \frac{\pi}{\alpha^2}\end{aligned}$$

The first line is obtained by Fubini's theorem and the second line is a change of variable in polar coordinates. Thus,

$$L(\alpha,0) = \frac{\sqrt{\pi}}{\alpha} \tag{C.89}$$

By combining (C.88) and (C.89), it yields

$$\int_{-\infty}^{+\infty}\mathrm{e}^{-\alpha^2(x+\beta)^2}\,\mathrm{d}x = \frac{\sqrt{\pi}}{\alpha} \tag{C.90}$$

for all $\alpha,\ \beta\in\mathbb{C}$ such that $\mathrm{Re}\left(\alpha^2\right) > 0$.

Appendix D
Frequency Response Functions

In Table D.1 are shown the definitions of different types of frequency response functions of linear mechanical systems excited by a point force. Input and output are complex-valued with a harmonic time dependence $e^{\iota\omega t}$. The vibrational displacement U and velocity V are taken in the direction of the applied force F and at the same point.

Table D.1 *Names of mechanical frequency response functions.*

Name	Input	Output	Definition
Receptance	Force	Displacement	$H(\omega) = U(\omega)/F(\omega)$
Dynamic stiffness	Displacement	Force	$D(\omega) = F(\omega)/U(\omega)$
Mobility	Force	Velocity	$Y(\omega) = V(\omega)/F(\omega)$
Conductance	Force	Velocity	$G(\omega) = \mathrm{Re}\,[Y(\omega)]$
Susceptance	Force	Velocity	$\mathrm{Im}\,[Y(\omega)]$
Mechanical impedance	Velocity	Force	$Z(\omega) = F(\omega)/V(\omega)$
Resistance	Velocity	Force	$R(\omega) = \mathrm{Re}\,[Z(\omega)]$
Reactance	Velocity	Force	$\mathrm{Im}\,[Z(\omega)]$

References

Barton, G. (1989). *Elements of Green's Functions and Propagation.* Clarendon Press, Oxford.

Ben Abdelounis, H., Le Bot, A., Perret-Liaudet, J., and Zahouani, H. (2010). 'An experimental study on roughness noise of dry rough flat surfaces'. *Wear*, **268**, 335–45.

Ben Abdelounis, H., Zahouani, H., Le Bot, A., Perret-Liaudet, J., and Ben Tkaya, M. (2011). Numerical simulation of friction noise. *Wear*, **271**, 621–624.

Ben Souf, M.A., Bareille, O., Ichchou, M.N., Troclet, B., and Haddar, M. (2013). 'Variability of coupling loss factors through a wave finite element technique'. *Journal of Sound and Vibration*, **332**, 2179–90.

Bies, D.A. and Hamid, S. (1980). 'In situ determination of loss and coupling loss factors by the power injection method'. *Journal of Sound and Vibration*, **70**, 187–204.

Billingham, J. and King, A.C. (2000). *Wave Motion.* (Cambridge: Cambridge University Press).

Biot, M.A. (1957). 'General theorems on the equivalence of group velocity and energy transport'. *Physical Review*, **105**, 1129–37.

Bleistein, N. and Handelsman, R.H. (1975). *Asymptotic Expansions of Integrals.* (New York: Dover).

Blevins, R.D. (1979). *Formulas for Natural Frequency and Mode Shape.* (New York: Van Nostrand Reinhold Company).

Blevins, R.D. (2006). 'Modal density of rectangular volumes, areas, and lines'. *Journal of the Acoustical Society of America*, **106**, 788–91.

Bogomolny, E. and Hugues, E. (1998). 'Semiclassical theory of flexural vibrations in plates'. *Physical Review E*, 57, 5404–24.

Bolt, R.H. (1939). 'Frequency distribution of eigentones in a three dimensional continuum'. *Journal of the Acoustical Society of America*, **10**, 228–34.

Bourgine, A. (1973). *Sur une approche statistique de la dynamique vibratoire des structures.* Thesis, Université Paris sud, Orsay, France.

Bouthier, O.M. and Bernhard, R.J. (1995). 'Simple models of the energetics of transversely vibrating plates'. *Journal of Sound and Vibration*, **182**, 149–66.

Boutillon, X. and Ege, K. (2013). 'Vibroacoustics of the piano soundboard: Reduced models, mobility synthesis, and acoustical radiation regime'. *Journal of Sound and Vibration*, **332**, 4261–79.

Bremner, P.G. and Burton, T.E. (1999). 'Statistical energy analysis of beams which are line-connected to plates'. In *IUTAM Symposium on Statistical Energy Analysis*, edited by F.J. Fahy and W.G. Price, (Dordrecht: Kluwer Academic Publishers).

Bremner, P., Nelisse, H., and Shorter, P. (2000). 'An overview of autoSEA2'. *Canadian Acoustics*, **28**, 166–7.

Brillouin, L. (1960). *Wave Propagation and Group Velocity.* (New York and London: Academic Press).

Budrin, S.V. and Nikiforov, S. (1964). 'Wave transmission through assorted plate joints'. *Soviet Physics Acoustics*, **9**, 333–6.

Burroughs, C.B., Fisher, R.W., and Kern, F.R. (1997). 'An introduction to statistical energy analysis'. *Journal of the Acoustical Society of America*, **101**, 1779–89.

Carcaterra, A. (2002). 'An entropy formulation for the analysis of energy flow between mechanical resonators'. *Mechanical Systems and Signal Processing*, **16**, 905–20.

Carcaterra, A. (2005). 'Ensemble energy average and energy flow relationships for nonstationary vibrating systems'. *Journal of Sound and Vibration*, **288**, 751–90.

Carcaterra, A. and Akay, A. (2004). 'Transient energy exchange between a primary structure and a set of oscillators: Return time and apparent damping'. *Journal of the Acoustical Society of America*, **115**, 683–96.

Cartan, H. (1985). *Cours de Calcul Différentiel* (3rd edn). (Paris: Hermann).

Cohen-Tannoudji, C., Diu, B., and Laloë, F. (1977). *Mécanique Quantique*. (Paris: Hermann).

Cotoni, V., Langley, R.S., and Kidner, M.R.F. (2005). 'Numerical and experimental validation of variance prediction in the statistical energy analysis of built-up systems'. *Journal of Sound and Vibration*, **288**, 701–28.

Courant, R. and Hilbert, D. (1953). *Methods of Mathematical Physics*. (New York: John Wiley & Sons, Interscience Publishers Inc.).

Craik, R.J.M. (1977). *The effect of a ventilation duct on the transmission of sound between two rooms*. M. Sc. Thesis, Heriot-Watt University, Edinburgh.

Craik, R.J.M. (1979). 'The noise reduction of the acoustic paths between two rooms interconnected by a ventilation duct'. *Applied Acoustics*, **12**, 161–79.

Craik, R.J.M. (1990). 'Sound transmission paths through a statistical energy analysis model'. *Applied Acoustics*, **30**, 45–55.

Craik, R.J.M. (1996). *Sound Transmission Through Buildings Using Statistical Energy Analysis*. (London: Gower).

Craik, R.J.M. (1999). 'The relationship between transmission coefficient and coupling loss factor'. In *IUTAM Symposium on Statistical Energy Analysis*, edited by F.J. Fahy and W.G. Price, (Dordrecht: Kluwer Academic Publishers).

Craik, R.J.M., Bosmans, I., Cabos, C., Heron, K.H., Sarradj, E. , Steel, J.A., and Vermeir, G. (2004). 'Structural transmission at line junctions: A benchmarking exercise'. *Journal of Sound and Vibration*, **272**, 1086–96.

Crandall, S.H. and Lotz, R. (1971). 'On the coupling loss factor in statistical energy analysis'. *Journal of the Acoustical Society of America*, **49**, 352–6.

Cremer, L. and Heckl, M. (1988). *Structure-Borne Sound* (2nd edn). (Berlin: Springer-Verlag).

Crocker, M.J. and Price, A.J. (1969). 'Sound transmission using statistical energy analysis'. *Journal of Sound and Vibration*, **9**, 469–86.

Culla, A. and Sestieri, A. (2006). 'Is it possible to treat confidentially SEA the wolf in sheep's clothing?' *Mechanical Systems and Signal Processing*, **20**, 1372–99.

Culla, A., D'Ambrogio, W., and Fregolent, A. (2011). 'Parametric approaches for uncertainty propagation in SEA'. *Mechanical Systems and Signal Processing*, **25**, 193–204.

Dang, V.H., Perret-Liaudet, J., Scheibert, J., and Le Bot, A. (2013). 'Direct numerical simulation of the dynamics of sliding rough surfaces'. *Computational Mechanics*, **52**, 1169–83.

De Langhe, K. (1996). *High frequency vibrations: Contributions to experimental and computational SEA parameter identification techniques*. Ph.D. thesis, Katholieke Universiteit Leuven, Belgium.

Fahy, F.J. (1969). 'Vibration of containing structures by sound in the contained fluid'. *Journal of Sound and Vibration*, **10**, 490–512.

Fahy, F.J. (1970). 'Energy flow between oscillators: Special case of point excitation'. *Journal of Sound and Vibration*, **11**, 481–3.

Fahy, F.J. (1994). 'Statistical energy analysis: A critical overview'. In *Statistical Energy Analysis: An Overview with Applications in Structural Dynamics*, edited by A.J. Keane and W.G. Price, (Cambridge: Cambridge University Press).

Fahy, F.J. and Yao de-Yuan (1987). 'Power flow between non-conservatively coupled oscillators'. *Journal of Sound and Vibration*, **114**, 1–11.

Fahy, F. and Gardonio, P. (2007). *Sound and Structural Vibration* (2nd edn). (Amsterdam: Academic Press).

Finnveden, S. (1995). 'Ensemble averaged vibration energy flows in a three-element structure'. *Journal of Sound and Vibration*, **187**, 495–529.

Finnveden, S. (2004). 'Evaluation of modal density and group velocity by a finite element method'. *Journal of Sound and Vibration*, **273**, 51–75.

Finnveden, S. (2011). 'A quantitative criterion validating coupling power proportionality in statistical energy analysis'. *Journal of Sound and Vibration*, **330**, 87–109.

Folland, G.B. (1983). *Lectures On Partial Differential Equations*. (Heidelberg: Springer-Verlag).

Fredö, C.R. (1999). 'A note on conservative and non-conservative coupling'. In *IUTAM Symposium on Statistical Energy Analysis*, edited by F.J. Fahy and W.G. Price, (Dordrecht: Kluwer Academic Publishers).

Gagliardini, L., Houillon, L., Borello, G., and Petrinelli, P. (2005). 'Virtual SEA-FEA-based modeling of mid-frequency structure-borne noise'. *Sound and Vibration*, **39**, 22–8.

Gersch, W. (1969). 'Average power and power exchange in oscillators'. *Journal of the Acoustical Society of America*, **45**, 1180–5.

Gourmelin, S. and Wadi, H. (2009). *Équations Différentielles*. (Paris: Hermann).

Goyder, H.G.D. and White, R.G. (1980a). 'Vibrational power flow from machines into built-up structures, Part I: Introduction and approximate analyses of beam and plate-like foundations'. *Journal of Sound and Vibration*, **68**, 59–75.

Goyder, H.G.D. and White, R.G. (1980b). 'Vibrational power flow from machines into built-up structures, Part II: Wave propagation and power flow in beam-stiffened plates'. *Journal of Sound and Vibration*, **68**, 77–96.

Goyder, H.G.D. and White, R.G. (1980c). 'Vibrational power flow from machines into built-up structures, Part III: Power flow through isolation systems'. *Journal of Sound and Vibration*, **68**, 97–117.

Gradshteyn, I.S. and Ryzhik, I.M. (2000). *Table of Integrals, Series, and Products* (6th edn). (London: Academic Press).

Graff, K.F. (1991). *Wave Motion in Elastic Solids*. (New York: Dover).

Guash, O. and Cortés, L. (2009). 'Graph theory applied to noise and vibration control in statistical energy analysis models'. *Journal of the Acoustical Society of America*, **125**, 3657–72.

Guash, O. and Aragonès, A. (2011). 'Finding the dominant energy transmission paths in statistical energy analysis'. *Journal of Sound and Vibration*, **330**, 2325–38.

Guyader, J.L. (2002). *Vibrations des Milieux Continus*. (Lavoisier Paris: Hermes Science Publications).

Guyader, J.L., Boisson, C., and Lesueur, C. (1982). 'Energy transmission in finite coupled plates, part I: Theory'. *Journal of Sound and Vibration*, **81**, 81–92.

Guyader, J.L., Boisson, C. Lesueur, C., and Millot, P. (1986). 'Sound transmission by coupled structures: Application to flanking transmission in buildings'. *Journal of Sound and Vibration*, **106**, 289–310.

Heckl, M. and Lewit, M. (1994). 'Statistical energy analysis as a tool for quantifying sound and vibration paths'. In *Statistical Energy Analysis: An Overview with Applications in Structural Dynamics*, edited by A.J. Keane and W.G. Price. (Cambridge: Cambridge University Press).

Heron, K.H. (1994). 'Advanced statistical energy analysis'. In *Statistical Energy Analysis: An Overview with Applications in Structural Dynamics*, edited by A.J. Keane and W.G. Price, (Cambridge: Cambridge University Press).

Hopkins, C. (2003). 'Vibration transmission between coupled plates using finite element methods and statistical energy analysis. Part 1: Comparison of measured and predicted data for masonry walls with and without apertures'. *Applied Acoustics*, **64**, 955–73.

Hopkins, C. (2009). 'Experimental statistical energy analysis of coupled plates with wave conversion at the junction'. *Journal of Sound and Vibration*, **322**, 155–66.

Ichchou, M.N., Le Bot, A., and Jezequel, L. (2001). 'A transient local energy approach as an alternative to transient SEA: Wave and telegraph equations'. *Journal of Sound and Vibration*, **246**, 829–40.

Ichchou, M.N., Akrout, S., and Mencik, J.M. (2007). 'Guided waves group and energy velocities via finite elements'. *Journal of Sound and Vibration*, **305**, 931–44.

Ichchou, M.N., Mencik, J.M., and Zhou, W. (2009). 'Wave finite elements for low and mid-frequency description of coupled structures with damage'. *Computer Methods in Applied Mechanics and Engineering*, **198**, 1311–26.

Jaynes, E.T. (1957). 'Information theory and statistical mechanics'. *Physical Review*, **106**, 620–30.

Jean, F. (2011). *Stabilité et Commande des Systèmes Dynamiques*. (Paris: Les presses de l'ENSTA).

Joyce, W.B. (1975). 'Sabine's reverberation times and ergodic auditoriums'. *Journal of the Acoustical Society of America*, **58**, 643–55.

Karnopp, D. (1966). 'Coupled vibratory-system analysis, using the dual formulation'. *Journal of the Acoustical Society of America*, **40**, 380–4.

Keane, A.J. and Price, W.G. (1987). 'Statistical energy analysis of strongly coupled systems'. *Journal of Sound and Vibration*, **117**, 363–86.

Keane, A.J. and Price, W.G. (1991). 'A note on the power flowing between two conservatively coupled multi-modal subsystems'. *Journal of Sound and Vibration*, **144**, 185–96.

Kosten, C.W. (1960). 'The mean free path in room acoustics'. *Acoustica*, **10**, 245–50.

Krasovskii, N.N. (1963). *Stability of Motion*. (Stanford: Stanford University Press) (Russian original 1959).

Kuttruff, H. (2000). *Room Acoustics* (4th edn). (London: Spon Press).

Lafont, T., Totaro, N., and Le Bot, A. (2014). 'Review of statistical energy analysis hypotheses in vibroacoustics'. *Proceedings of the Royal Society A*, **470**, 20130515.

Lai, M.L. and Soom, A. (1990a). 'Prediction of transient vibration envelopes using statistical energy analysis techniques'. *Transaction of ASME*, **112**, 127–37.

Lai, M.L. and Soom, A. (1990b). 'Statistical energy analysis for the time-integrated transient response of vibrating systems'. *Transaction of ASME*, **112**, 206–13.

Lalor, N. (1999). 'The practical implementation of SEA'. In *IUTAM Symposium on Statistical Energy Analysis*, edited by F.J. Fahy and W.G. Price. (Dordrecht: Kluwer Academic Publishers).

Lang, S. (1993). *Complex Analysis* (3rd edn). (New York: Springer-Verlag).

Lang, S. (2002). *Algebra* (3rd edn). (New York: Springer-Verlag).

Langley, R.S. (1989). 'A general derivation of the statistical energy equations for dynamic systems'. *Journal of Sound and Vibration*, **135**, 499–508.

Langley, R.S. (1996). 'The modal density of anisotropic structural components'. *Journal of the Acoustical Society of America*, **99**, 3481–7.

Langley, R.S. and Heron, K.H. (1990). 'Elastic wave transmission through plate/beam junctions'. *Journal of Sound and Vibration*, **143**, 241–53.

Langley, R.S. and Brown, A.W.M. (2004a). 'The ensemble statistics of the energy of a random system subjected to random harmonic excitation'. *Journal of Sound and Vibration*, **275**, 823–46.

Langley, R.S. and Brown, A.W.M. (2004b). 'The ensemble statistics of the band-averaged energy of a random system'. *Journal of Sound and Vibration*, **275**, 847–57.

LaSalle, J.P. (1960). 'Some extensions of Lyapunov's second method'. *Institute of Radio Engineers Trans. Circuit Theory*, **CT-7**, 520–7.

Lathuilière, M. and Tarley, A. (1986). *Transparence acoustique de parois multiplaques multicouches*, Projet de fin d'étude INSA de Lyon, Lyon.

Le Bot, A. (1998). 'A vibroacoustic model for high frequency analysis'. *Journal of Sound and Vibration*, **211**, 537–54.

Le Bot, A. (2002). 'Energy transfer for high frequencies in built-up structures'. *Journal of Sound and Vibration*, **250**, 247–75.

Le Bot, A. (2007). 'Derivation of statistical energy analysis from radiative exchanges'. *Journal of Sound and Vibration*, **300**, 763–79.

Le Bot, A. (2009). 'Entropy in statistical energy analysis'. *Journal of the Acoustical Society of America*, **125**, 1473–78.

Le Bot, A. (2011). 'Statistical energy analysis and the second principle of thermodynamics'. In *IUTAM Symposium on the Vibration Analysis of Structures with Uncertainties*, edited by A.K. Belayev and R.S. Langley, IUTAM bookseries volume 27 (Dordrecht: Springer).

Le Bot, A. and Bocquillet, A. (2000). 'Comparison of an integral equation on energy and the ray-tracing technique in room acoustics'. *Journal of the Acoustical Society of America*, **108**, 1732–40.

Le Bot, A. and Bou Chakra, E. (2010). 'Measurement of friction noise versus contact area of rough surfaces weakly loaded'. *Tribology Letters*, **37**, 273–81.

Le Bot, A. and Cotoni, V. (2010). 'Validity diagrams of statistical energy analysis'. *Journal of Sound and Vibration*, **329**, 221–35.

Le Bot, A., Carcaterra, A., and Mazuyer, D. (2010). 'Statistical vibroacoustics and entropy concept'. *Entropy*, **12**, 2418–35.

Le Bot, A., Bou Chakra, E., and Michon, G. (2011). 'Dissipation of vibration in rough contact'. *Tribology Letters*, **41**, 47–53.

Lesueur, C. (1988). *Rayonnement Acoustique des Structures.* (Paris: Editions Eyrolles).

London, A. (1949). 'Transmission of reverberant sound through single walls'. *Journal of Research of the National Bureau of Standards*, **42**, 605.

London, A. (1950). 'Transmission of reverberant sound through double walls'. *Journal of the Acoustical Society of America*, **22**, 270–9.

Lotz, R. and Crandall, S.H. (1973). 'Prediction and measurement of the proportionality constant in statistical energy analysis'. *Journal of the Acoustical Society of America*, **54**, 516–24.

Luzzato, E. and Ortola, E. (1988). 'The characterization of energy flow paths in the study of dynamic systems using SEA theory'. *Journal of Sound and Vibration*, **123**, 189–97.

Lyon, R.H. (1969). 'Statistical analysis of power injection and response in structures and rooms'. *Journal of the Acoustical Society of America*, **45**, 545–65.

Lyon, R.H. (1973). *Méthodes d'Analyse Statistique de l'Énergie.* Lectures given on 16–18 January 1973 in Ecole centrale de Lyon, Ecully, France.

Lyon, R.H. (1975). *Statistical Energy Analysis of Dynamical Systems: Theory and Applications.* (Cambridge, Massachusetts: MIT Press).

Lyon, R.H. (2003). 'Fluctuation theory and very (early) statistical energy analysis (SEA)'. *Journal of the Acoustical Society of America*, **113**, 2401–3.

Lyon, R.H. and Maidanik, G. (1962). 'Power flow between linearly coupled oscillators'. *Journal of the Acoustical Society of America*, **34**, 623–39.

Lyon, R.H. and Eichler, E. (1964). 'Random vibration of connected structures'. *Journal of the Acoustical Society of America*, **36**, 1344–54.

Lyon, R.H. and Scharton, T. (1965). 'Vibrational-energy transmission in a three-element structure'. *Journal of the Acoustical Society of America*, **38**, 253–61.

Lyon, R.H. and DeJong, R.G. (1995). *Theory and Application of Statistical Energy Analysis*. (Newton, Massachusetts: Butterworth–Heinemann).

Maa, D.Y. (1939). 'Distribution of eigentones in a rectangular chamber at low frequency range'. *Journal of the Acoustical Society of America*, **10**, 235–8.

Mace, B.R. (1992). 'Power flow between two continuous one-dimensional subsystems: A wave solution'. *Journal of Sound and Vibration*, **154**, 289–319.

Mace, B.R. (2003). 'Statistical energy analysis, energy distribution models and system modes'. *Journal of Sound and Vibration*, **264**, 391–409.

Mace, B.R. (2005). 'Statistical energy analysis: Coupling loss factors, indirect coupling and system modes'. *Journal of Sound and Vibration*, **279**, 141–70.

Mace, B.R. and Ji, L. (2007). 'The statistical energy analysis of coupled sets of oscillators'. *Proceedings of the Royal Society A*, **463**, 1359–77.

Mace, B.R. and Manconi, E. (2008). 'Modelling wave propagation in two-dimensional structures using finite element analysis'. *Journal of Sound and Vibration*, **318**, 884–902.

Magionesi, F. and Carcaterra, A. (2009). 'Insights into the energy equipartition principle in undamped engineering structures'. *Journal of Sound and Vibration*, **322**, 851–69.

Magrans, F.X. (1993). 'Definition and calculation of transmission paths within an SEA framework'. *Journal of Sound and Vibration*, **165**, 277–83.

Maidanik, G. (1962). 'Response of ribbed panels to reverberant acoustic fields'. *Journal of the Acoustical Society of America*, **34**, 809–26.

Maidanik, G. (1966). 'Energy dissipation associated with gas-pumping in structural joints'. *Journal of the Acoustical Society of America*, **40**, 1064–72.

Maidanik, G. and Dickey, J. (1989). 'Wave derivation of the energetics of driven coupled one-dimensional dynamic systems'. *Journal of Sound and Vibration*, **139**, 31–42.

Manning, J. (1994). 'Formulation of SEA parameters using mobility functions'. In *Statistical Energy Analysis: An Overview with Applications in Structural Dynamics*, edited by A.J. Keane and W.G. Price, (Cambridge: Cambridge University Press).

Manning, J.E. and Lee, K. (1968). 'Predicting mechanical shock prediction'. *Shock and Vibration Bulletin*, **35**, 65–70.

Maxit, L. and Guyader, J.L. (2003). 'Extension of SEA model to subsystems with non-uniform modal energy distribution'. *Journal of Sound and Vibration*, **265**, 337–58.

Maxit, L., Ege, K., Totaro, N., and Guyader, J.L. (2003). 'Non resonant transmission modelling with statistical modal energy distribution analysis'. *Journal of Sound and Vibration*, **333**, 499–519.

McCollum, M.D. and Cushieri, J.M. (1990). 'Bending and in-plane wave transmission in thick connected plates using statistical energy analysis'. *Journal of the Acoustical Society of America*, **88**, 1480–1485.

Mead, D.J. (1994). 'Waves and modes in finite beams: Application of the phase-closure principle'. *Journal of Sound and Vibration*, **171**, 695–702 .

Meirovitch, L. (1967). *Analytical Methods in Vibrations*. (London: MacMillan).

Mercer, C.A., Rees, P.L., and Fahy, F.J. (1971). 'Energy flow between two weakly coupled oscillators subject to transient excitation'. *Journal of Sound and Vibration*, **15**, 373–9.

Mindlin, R.D. (1951). 'Influence of rotary inertia and shear on flexural motion of isotropic elastic plates'. *Journal of Applied Mechanics*, **18**, 31–8.

Morse, P.M. and Ingard, K.U. (1968). *Theoretical Acoustics*. (Princeton, New Jersey: Princeton University Press).

Newland, D.E. (1966). 'Calculation of power flow between coupled oscillators'. *Journal of Sound and Vibration*, 3, 262–76.

Newland, D.E. (1968). 'Power flow between a class of coupled oscillators'. *Journal of the Acoustical Society of America*, 43, 553–9.

Newland, D.E (1975). *Random Vibrations and Spectral Analysis*. (London: Longman).

Norton, M.P. (1989). *Fundamentals of Noise and Vibration Analysis for Engineers*. (Cambridge: Cambridge University Press).

Papoulis, A. (1965). *Probability, Random Variables, and Stochastic Processes*. (New York: McGraw-Hill).

Park, D.H. , Kil, H.G., and Jeon, J.J. (2001). 'Power flow models and analysis of in-plane waves in finite coupled thin plates'. *Journal of Sound and Vibration*, **244**, 651–68.

Peyton Z. Peebles, J.R. (1987). *Probability, Random Variables, and Random Signal Principles* (2nd edn). (New York: McGraw-Hill).

Pierce, J.R. (2006). *Almost All About Waves*. (New York: Dover).

Pinnington, R.J. and Lednik, D. (1996a). 'Transient statistical energy analysis of an impulsively excited two oscillator system'. *Journal of Sound and Vibration*, **189**, 249–64.

Pinnington, R.J. and Lednik, D. (1996b). 'Transient energy flow between two coupled beams'. *Journal of Sound and Vibration*, **189**, 265–87.

Polack, J.D. (1993). 'Playing billiards in the concert hall: The mathematical foundations of the geometrical room acoustics'. *Applied Acoustics*, 38, 235–44.

Price, A.J. and Crocker, M.J. (1970). 'Sound transmission through double panels using statistical energy analysis'. *Journal of the Acoustical Society of America*, 47, 683–93.

Renji, K. (2004). 'On the number of modes required for statistical energy analysis-based calculations'. *Journal of Sound and Vibration*, **269**, 1128–32.

Renji, K. (2005). 'Author's reply'. *Journal of Sound and Vibration*, **281**, 481.

Renno, J.M. and Mace, B.R. (2013). 'Calculation of reflection and transmission coefficients of joints using a hybrid finite element/wave and finite element approach'. *Journal of Sound and Vibration*, 332, 2149–64.

Roe, G.M. (1941). 'Frequency distribution of normal modes'. *Journal of the Acoustical Society of America*, 13, 1–7.

Rudin, W. (1987). *Real and Complex Analysis* (3rd edn). (New York: McGraw-Hill).

Rudin, W. (1991). *Functional Analysis* (2nd edn). (New York: McGraw-Hill).

Sabine, W.C. (1922). *Collected Papers on Acoustics*. (Cambridge: Harvard University Press).

Sabot, J. (2000). *Cours de Vibration Aléatoire*. Cours École Centrale de Lyon.

Scharton, T.D. and Lyon, R.H. (1968). 'Power flow and energy sharing in random vibration'. *Journal of the Acoustical Society of America*, **43**, 1332–43.

Schröder, M. (1954). 'Die statistishen parameter der frequenzkurven von grossen raumen'. *Acustica*, 4, 594–600.

Sestieri, A. and Carcaterra, A. (2013). 'Vibroacoustic: The challenges of a mission impossible?' *Mechanical Systems and Signal Processing*, **34**, 1–18.

Shorter, P. (2011). 'Modeling noise and vibration transmission in complex systems'. In *IUTAM symposium on the Vibration Analysis of Structures with Uncertainties*, edited by A.K. Belayev and R.S. Langley, IUTAM bookseries volume 27, (New York: Springer).

Sgard, F., Nelisse, H., Atalla, N., Cafui, C., and Oddo, R. (2010). 'Prediction of the acoustical performance of enclosures using a hybrid statistical energy analysis: Image source model'. *Journal of the Acoustical Society of America*, **127**, 784–95.

Skudrzyk, E.J. (1958). 'Vibrations of a system with a finite or an infinite number of resonances'. *Journal of the Acoustical Society of America*, **30**, 1140–52.

Skudrzyk, E.J. (1968). *Simple and Complex Vibratory Systems.* (University Park: Pennsylvania State University Press).

Smith, P.W. (1962). 'Response and radiation of structural modes excited by sound'. *Journal of the Acoustical Society of America*, **34**, 640–7.

Soedel, W. (2004). *Vibrations of Shells and Plates* (3rd edn). (New York: Marcel Dekker Inc.).

Soize, C. (1993). 'A model and numerical method in the medium frequency range for vibroacoustic predictions using the theory of structural fuzzy'. *Journal of the Acoustical Society of America*, **94**, 849–66.

Soize, C. (1995). 'Coupling between an undamped linear acoustic fluid and a damped nonlinear structure—Statistical energy analysis considerations'. *Journal of the Acoustical Society of America*, **98**, 373–85.

Soize, C. (2001). *Dynamique des Structures.* (Paris: Ellipses).

Strasberg, M. (1997). 'Is the dissipation induced by 'fuzzy' substructures real or apparent?' *Journal of the Acoustical Society of America*, **102**, 3130.

Strogatz, S.H. (1994). *Nonlinear Dynamics and Chaos.* (Cambridge: Westview Press, Perseus Books Publishing).

Tanner, G. (2009). 'Dynamical energy analysis—Determining wave energy distributions in vibroacoustical structures in the high-frequency regime'. *Journal of Sound and Vibration*, **320**, 1023–38.

Totaro, N., Guyader, J.L. (2006). 'SEA substructuring using cluster analysis: The MIR index'. *Journal of Sound and Vibration*, **290**, 264–89.

Totaro, N., Guyader, J.L. (2012). 'Extension of the statistical modal energy distribution analysis for estimating energy density in coupled subsystems'. *Journal of Sound and Vibration*, **331**, 3114–29.

Uffink, J. (1997). 'Can the maximum entropy principle be explained as a consistency requirement?' *Studies in History and Philosophy of Modern Physics*, **26B**, 223–61.

Ungar, E.E. (1961). 'Transmission of plate flexural waves through reinforcing beams: Dynamic stress concentrations'. *Journal of the Acoustical Society of America*, **33**, 633–9.

Ungar, E.E. (1966). *Fundamentals of Statistical Energy Analysis of Vibrating Systems.* Technical report AFFDL-TR-66-52. (Cambridge, Massachusetts: Bolt Beranek and Newman Inc.).

Ventsel, E. and Krauthammer, T. (2001). *Thin Plates and Shells: Theory, Analysis, and Applications.* (New York: Marcel Dekker Inc.).

Wang, C. and Lai, J.C.S. (2005). 'Discussions on "On the number of modes required for statistical energy analysis-based calculations"'. *Journal of Sound and Vibration*, **281**, 475–80.

Weaver, R.L. (1982). 'On diffuse waves in solid media'. *Journal of the Acoustical Society of America*, **71**, 1608–9.

Weaver, R.L. (1984). 'Diffuse waves in finite plates'. *Journal of Sound and Vibration*, **94**, 319–35.

Weaver, R.L. and O.I. Lobkis (2000). 'Enhanced backscattering and modal echo of reverberant elastic waves'. *Physical Review Letters*, **84**, 4942–5.

Wijker, J. (2009). *Random Vibrations in Spacecraft Structures Design.* (Dordrecht: Springer).

Wöhle, W., Beckmann, T., and Schreckenbach, H. (1981). 'Coupling loss factors for statistical energy analysis of sound transmission at rectangular structural slab joints, Part I'. *Journal of Sound and Vibration*, 77, 323–34.

Woodhouse, J. (1981). 'An introduction to statistical energy analysis of structural vibration'. *Applied Acoustics*, **14**, 455–69.

Wright, M. and Weaver, R. (2010). *New Directions in Linear Acoustics and Vibration.* (Cambridge: Cambridge University Press).

Xie, G., Thompson, D.J., and Jones, J.C. (2002a). *Investigation of the Mode Count of One-Dimensional Systems*. ISVR Technical Memorandum No 882, University of Southampton.

Xie, G., Thompson, D.J., and Jones, J.C. (2002b). *The Influence of Boundary Conditions on the Mode Count and Modal Density of Two-Dimensional Systems*. ISVR Technical Memorandum No 894, University of Southampton.

Xie, G., Thompson, D.J., and Jones, J.C. (2004). 'Mode count and modal density of structural systems: Relationship with boundary conditions'. *Journal of Sound and Vibration*, 274, 621–51.

Index